AF357342

Calcium Signalling
in Alzheimer's Disease

Calcium Signalling
in Alzheimer's Disease

From Pathophysiological Regulation to Therapeutic Approaches

Editor

Mounia Chami

MDPI • Basel • Beijing • Wuhan • Barcelona • Belgrade • Manchester • Tokyo • Cluj • Tianjin

Editor
Mounia Chami
Laboratory of Excellence (LABEX) DistAlz,
INSERM, CNRS, IPMC, Université Côte d'Azur
France

Editorial Office
MDPI
St. Alban-Anlage 66
4052 Basel, Switzerland

This is a reprint of articles from the Special Issue published online in the open access journal *Cells* (ISSN 2073-4409) (available at: https://www.mdpi.com/journal/cells/special_issues/AD_Calcium).

For citation purposes, cite each article independently as indicated on the article page online and as indicated below:

LastName, A.A.; LastName, B.B.; LastName, C.C. Article Title. *Journal Name* **Year**, *Volume Number*, Page Range.

ISBN 978-3-0365-0498-8 (Hbk)
ISBN 978-3-0365-0499-5 (PDF)

Contents

About the Editor

Mounia Chami is currently a Senior Researcher at the University Côte d'Azur, at the Institute of Molecular and Cellular Pharmacology. She received her doctoral degree in molecular and cellular biology from the University of Paris XI. She worked at the University of Ferrara, Italy, as a post-doctoral fellow, served as a junior researcher at Paris V University, France, and as a team leader at the Italian Institute of Italy. Dr. Chami has authored several publications in the field of calcium signalling deregulation in apoptotic cell death, with a focus on the role of the contact sites between the endoplasmic reticulum (ER) and the mitochondria. Her recent studies demonstrated the impact of ER calcium homeostasis alterations and of mitochondria structure, function, and mitophagy defects in Alzheimer's disease pathogenesis. These studies open the possibility of identifying new molecular targets for the treatment of AD.

Editorial

Calcium Signalling in Alzheimer's Disease: From Pathophysiological Regulation to Therapeutic Approaches

Mounia Chami

Laboratory of Excellence DistALZ, INSERM, CNRS, Institute of Molecular and Cellular Pharmacology, Université Côte d'Azur, Sophia-Antipolis, 06560 Valbonne, France; mchami@ipmc.cnrs.fr; Tel.: +33-4939-53453; Fax: +33-4939-53408

Received: 6 January 2021; Accepted: 8 January 2021; Published: 12 January 2021

Alzheimer's disease (AD) is a neurodegenerative pathology representing a socioeconomic challenge, however, the complex mechanism behind the disease is not yet fully understood. AD is commonly defined as a proteinopathy characterized by the accumulation of intracellular neurofibrillary tangles composed of abnormal hyper-phosphorylated, conformated, and truncated tau, as well as extracellular deposits of β-amyloid (Aβ) species forming amyloid plaques in different brain areas [1]. The "amyloidogenic hypothesis" in AD postulates that the accumulation of Aβ plaques acts as a pathological trigger for a cascade that includes neuritic injury, the formation of neurofibrillary tangles via tau protein leading to neuronal dysfunction, and cell death [2]. This hypothesis is supported by genetic, biochemical, and pathological evidence linking familial autosomal dominant mutations in the amyloid precursor protein (APP) and presenilins (PS1 and PS2) genes, triggering an imbalance between Aβ peptide production and clearance and causing early-onset neurodegeneration [3,4]. The main progress in understanding AD pathophysiology was achieved thanks to the identification of disease-causing mutations [3–5]. Then, the generation of cellular and mouse models expressing disease-causing genes mimicking the development of familial forms of AD (FAD) (https://www.alzforum.org/research-models/alzheimers-disease) enabled the formulation of several interconnected mechanistic theories. Among others, the "calcium hypothesis" emerged as a key AD pathogenic pathway, impacting most, if not all, cellular components of the nervous system comprising neurons and glial cells [6–8]. As a second messenger, calcium is critical for proper neuronal synaptic plasticity, governing learning and memory functions [9,10], and commonly described as among the major features characterizing AD [8]. The complexity of the "calcium hypothesis" relies on the fact that disturbances of calcium homeostasis affect different cellular compartments, such as mitochondria, endoplasmic reticulum (ER), lysosomes, and several microdomains within the plasma membrane, occurring through broad interventions of calcium signalling "tool-kits" (receptors, channels, binding protein, etc.). The significance of the "calcium hypothesis" in AD pathogenesis has been formally approved since calcium dyshomeostasis was reported in presymptomatic FAD study mice and thus seemed to occur prior to the development of histopathological markers or clinical symptoms. Noteworthily, disturbances of calcium signalling, largely reported in FAD study models (in vitro and in vivo) [8,11–16], were also observed in human-derived post-mortem brains [17] and fibroblasts [18–20], as well as recently in human-induced neurons [21,22].

In this Special Issue in *Cells*, six reviews address the newest results and advances in calcium signalling deregulation mechanisms in AD, how they are linked to other molecular players involved in AD pathogenesis, and the potential therapeutic approaches to correct calcium alterations to treat AD [23–28].

In the review by John McDaid et al. [26], the authors describe the role of calcium dysregulation in synaptic network dysfunctions in AD. The review focuses on the mechanisms impacting plasma membrane *N*-methyl-D-aspartate receptor (NMDAR), the L voltage-gated calcium channel (VGCC), and the nicotinic acetylcholine receptor (α7nAchR) function [26]. The discussed data draw upon the complexity of calcium-synaptic dysfunction connections observed in AD mice models. The authors point out the role of extracellular Aβ plaques and toxic soluble Aβ oligomers towards synaptic hyperactivity and highlight studies demonstrating the contribution of intracellular tau to synaptic loss and the impairment of synaptic function. They also provide evidence demonstrating that synaptic plasticity dysfunctions in AD are linked to excessive ER calcium release, mainly through the ryanodine receptor (RyR) by the process of calcium-induced calcium release (CICR). In addition, the review by John McDaid et al. provides key elements demonstrating the role of calcium dyshomeostasis in lysosome-autophagosome-mediated protein degradation in AD [26]. Noteworthily, enhanced lysosomal calcium efflux is seen as an early event in AD pathology, contributing to defective lysosome—autophagy degradative function but also to synaptic transmission deficiency [29].

In addition to synaptic plasticity deficits, calcium dyshomeostasis has a profound effect on the function of cell organelles, including ER and mitochondria, both of which play an important role in maintaining cellular and synaptic function. These specific items were discussed in our review [23] and in that by Noemi Esteras and Andrey Y. Abramov [25], respectively.

In our review [23], we describe the main neuronal calcium signalling "tool-kits" and focus on ER calcium handling molecules alterations in AD and the benefit of targeting the aforementioned to alleviate AD pathogenesis. Our review describes the tight link between the "calcium hypothesis" and the amyloidogenic cascade generating Aβ peptides and other APP-derived toxic fragments [30]. ER calcium mishandling in AD includes alterations of the inositol 1,4,5-trisphosphatereceptors (IP$_3$Rs) and ryanodine receptors (RyRs) expression and function, the dysfunction of the sarco-endoplasmic reticulum calcium ATPase (SERCA) activity, and the upregulation of SERCA1 truncated isoform (S1T), as well as presenilins (PS1, PS2), forming the catalytic core of the γ-secretase enzymatic complex cleaving APP [23]. We summarize the neuronal expression, structure, and physiological function for each ER molecular component. The sum of studies discussed offers an outline of the disease-associated remodelling of ER calcium machinery coupled to specific cellular signalling cascades modulating the activity (i.e., post-translational modifications, interactions with regulatory proteins) and/or the expression of ER calcium channels and pump. The depletion of ER calcium content activates the store-operated calcium entry (SOCE) pathway [31]. We then report studies describing the expression and function alterations of the molecular bridge linking ER calcium depletion and the activation of plasma membrane calcium entry implicating STIM and ORAI proteins [23].

The review by Noemi Esteras and Andrey Y. Abramov specifically describes the mechanisms underlying mitochondrial calcium deregulation linked to Aβ and tau pathologies [25]. They first depict the basis of physiological mitochondrial calcium homeostasis and then describe the cytosolic and mitochondrial calcium homeostasis impairments in AD and in tauopathies (neurodegenerative disorders characterized by the deposition of abnormal tau protein in the brain) [25]. The authors specifically discuss the molecular mechanisms underlying mitochondrial calcium disturbances and expose complementary scenarios linking the deleterious mitochondria calcium overload to neuronal death [25]. These mechanisms include the alteration of the expression of mitochondrial calcium-related proteins and of ER–mitochondria interactions, and also the impairment of mitochondrial calcium efflux, and mitochondrial permeability transition pore opening. These mechanisms appear to act in concert in the process of neurodegeneration in AD and tauopathies [25].

In addition to forming the catalytic core of the γ-secretase enzyme, several studies have demonstrated a role of PS1 and PS2 in subcellular calcium signalling. Our review [23] and that by John McDaid et al. [26] extensively highlight the role of PS1 in controlling several aspects of the subcellular calcium signalling deregulation and in synaptic plasticity. The review by Paola Pizzo et al. [28] specifically focuses on the role of PS2 in the modulation of ER and Golgi apparatus calcium handling, calcium entry through the plasma membrane channels, mitochondrial function, ER–mitochondria communication, and autophagy. The authors overview the alterations of calcium homeostasis observed in several cell lines expressing FAD-PS2 mutants, in human-derived fibroblasts, and in PS2 mice and ex vivo models (primary neurons culture and acute hippocampal slices) [28]. Of most interest, they discuss the impact of familial PS2 mutations in the control of multiple aspects of cell and tissue physiology, including cell metabolism and bioenergetic and brain network excitability [28].

The review by Veronika Prikhodko et al. [27] focuses on the TRPC6 (transient receptor potential channel 6), a non-selective cation plasma membrane channel that is permeable to calcium and activated by the emptying of the ER calcium store in a SOCE-dependent manner [32]. The review describes the role of TRPC6 in AD and brain ischemia [33,34]. The authors argue that although the pathophysiological mechanisms causing AD and cerebral ischemia may differ, cerebral ischemia serves as a risk factor for AD development, and vice versa. They postulate that both pathologies share a common mechanism associated with intracellular calcium dyshomeostasis likely implicating TRPC6. The review describes the contribution of the TRPC6 in neuronal hypo- or hyper-activation in both pathologies, with a particular focus on calcium entry alteration. The authors then discuss the potential drug candidates targeting TRPC6 that have shown some beneficial therapeutic effects in different cellular and animal models [27].

The review by Maria Calvo-Rodriguez et al. [24] describes AD-related calcium disturbances in neurons, astrocytes, and microglia. The authors discuss studies demonstrating that enhanced cytosolic calcium levels linked to Aβ and also to APOE4 (a genetic risk factor for sporadic AD forms) likely contribute to astrogliosis [35]. Importantly, the enhanced frequency of spontaneous calcium waves and calcium hyperactivity in astrocytes were observed in the intact brain of AD mice. The authors also report that calcium homeostasis was impaired in microglia isolated from AD mice, likely contributing to their activation. In a specific section, the authors highlight studies using intravital imaging to directly monitor the cytosolic calcium content in transgenic AD mice brains [24]. The review is composed of different chapters describing the contribution and the potential therapeutic effect of distinct calcium channels of the plasma membrane, endoplasmic reticulum, SOCE, mitochondria, and lysosomes [24]. The authors discuss the available therapeutic strategies targeting Aβ and emphasize the potential benefits in the genetic and immunomodulation of tau, and review the different strategies for targeting calcium deregulation, such as therapeutics in AD including human data and those generated from experimental models.

To conclude, this Special Issue provides recent research insights in the field of calcium signalling involvement in AD, which may open new research hypotheses and stimulate the development of therapeutic strategies.

Funding: This work was supported by Fondation Vaincre Alzheimer.

Conflicts of Interest: The author declares no conflict of interest.

References

1. Bloom, G.S. Amyloid-beta and tau: The trigger and bullet in Alzheimer disease pathogenesis. *JAMA Neurol.* **2014**, *71*, 505–508. [CrossRef]
2. Selkoe, D.J.; Hardy, J. The amyloid hypothesis of Alzheimer's disease at 25 years. *EMBO Mol. Med.* **2016**, *8*, 595–608. [CrossRef]
3. Van Cauwenberghe, C.; Van Broeckhoven, C.; Sleegers, K. The genetic landscape of Alzheimer disease: Clinical implications and perspectives. *Genet. Med.* **2016**, *18*, 421–430. [CrossRef]
4. Tanzi, R.E.; Bertram, L. Twenty years of the Alzheimer's disease amyloid hypothesis: A genetic perspective. *Cell* **2005**, *120*, 545–555. [CrossRef] [PubMed]
5. Hardy, J.; Selkoe, D.J. The amyloid hypothesis of Alzheimer's disease: Progress and problems on the road to therapeutics. *Science* **2002**, *297*, 353–356. [CrossRef] [PubMed]
6. Berridge, M.J. Calcium signalling and Alzheimer's disease. *Neurochem. Res.* **2011**, *36*, 1149–1156. [CrossRef] [PubMed]
7. Chakroborty, S.; Stutzmann, G.E. Calcium channelopathies and Alzheimer's disease: Insight into therapeutic success and failures. *Eur. J. Pharmacol.* **2014**, *739*, 83–95. [CrossRef]

8. Tong, B.C.-T.; Wu, A.J.; Li, M.; Cheung, K.-H. Calcium signaling in Alzheimer's disease & therapies. *Biochim. Biophys. Acta Mol. Cell Res.* **2018**, *1865*, 1745–1760. [CrossRef]

9. Berridge, M.J.; Lipp, P.; Bootman, M.D. The versatility and universality of calcium signalling. *Nat. Rev. Mol. Cell Biol.* **2000**, *1*, 11–21. [CrossRef]

10. Bootman, M.D.; Collins, T.J.; Peppiatt, C.M.; Prothero, L.S.; MacKenzie, L.; Smet, P.D.; Travers, M.; Tovey, S.C.; Seo, J.T.; Berridge, M.J.; et al. Calcium signalling—An overview. *Semin. Cell Dev. Biol.* **2001**, *12*, 3–10. [CrossRef]

11. Chakroborty, S.; Stutzmann, G.E. Early calcium dysregulation in Alzheimer's disease: Setting the stage for synaptic dysfunction. *Sci. China Life Sci.* **2011**, *54*, 752–762. [CrossRef] [PubMed]

12. Mattson, M.P. ER calcium and Alzheimer's disease: In a state of flux. *Sci. Signal.* **2010**, *3*, pe10. [CrossRef] [PubMed]

13. LaFerla, F.M. Calcium dyshomeostasis and intracellular signalling in Alzheimer's disease. *Nat. Rev. Neurosci.* **2002**, *3*, 862–872. [CrossRef] [PubMed]

14. Del Prete, D.; Checler, F.; Chami, M. Ryanodine receptors: Physiological function and deregulation in Alzheimer disease. *Mol. Neurodegener.* **2014**, *9*, 21. [CrossRef]

15. Supnet, C.; Bezprozvanny, I. The dysregulation of intracellular calcium in Alzheimer disease. *Cell Calcium* **2010**, *47*, 183–189. [CrossRef]

16. Chami, M.; Checler, F. Targeting Post-Translational Remodeling of Ryanodine Receptor: A New Track for Alzheimer's Disease Therapy? *Curr. Alzheimer Res.* **2020**, *17*, 313–323. [CrossRef]

17. Gibson, G.E.; Peterson, C. Calcium and the aging nervous system. *Neurobiol. Aging* **1987**, *8*, 329–343. [CrossRef]

18. Gibson, G.E.; Vestling, M.; Zhang, H.; Szolosi, S.; Alkon, D.; Lannfelt, L.; Gandy, S.; Cowburn, R.F. Abnormalities in Alzheimer's disease fibroblasts bearing the APP670/671 mutation. *Neurobiol. Aging* **1997**, *18*, 573–580. [CrossRef]

19. Hirashima, N.; Etcheberrigaray, R.; Bergamaschi, S.; Racchi, M.; Battaini, F.; Binetti, G.; Govoni, S.; Alkon, D.L. Calcium responses in human fibroblasts: A diagnostic molecular profile for Alzheimer's disease. *Neurobiol. Aging* **1996**, *17*, 549–555. [CrossRef]

20. Peterson, C.; Ratan, R.R.; Shelanski, M.L.; Goldman, J.E. Altered response of fibroblasts from aged and Alzheimer donors to drugs that elevate cytosolic free calcium. *Neurobiol. Aging* **1988**, *9*, 261–266. [CrossRef]

21. Ghatak, S.; Dolatabadi, N.; Trudler, D.; Zhang, X.; Wu, Y.; Mohata, M.; Ambasudhan, R.; Talantova, M.; Lipton, S.A. Mechanisms of hyperexcitability in Alzheimer's disease hiPSC-derived neurons and cerebral organoids vs. isogenic controls. *Elife* **2019**, *8*, e50333. [CrossRef] [PubMed]

22. Schrank, S.; McDaid, J.; Briggs, C.A.; Mustaly-Kalimi, S.; Bringgs, D.; Houcek, A.; Singer, A.; Bottero, V.; Marr, R.A.; Stutzmann, G.E. Human-Induced Neurons from Presenilin 1 Mutant Patients Model. Aspects of Alzheimer's Disease Pathology. *Int. J. Mol. Sci.* **2020**, *21*, 1030. [CrossRef] [PubMed]

23. Chami, M.; Checler, F. Alterations of the Endoplasmic Reticulum (ER) Calcium Signaling Molecular Components in Alzheimer's Disease. *Cells* **2020**, *9*, 2577. [CrossRef]

24. Calvo-Rodriguez, M.; Kharitonova, E.K.; Bacskai, B.J. Therapeutic Strategies to Target Calcium Dysregulation in Alzheimer's Disease. *Cells* **2020**, *9*, 2513. [CrossRef] [PubMed]

25. Esteras, N.; Abramov, A.Y. Mitochondrial Calcium Deregulation in the Mechanism of Beta-Amyloid and Tau Pathology. *Cells* **2020**, *9*, 2135. [CrossRef] [PubMed]

26. McDaid, J.; Mustaly-Kalimi, S.; Stutzmann, G.E. Ca^{2+} Dyshomeostasis Disrupts Neuronal and Synaptic Function in Alzheimer's Disease. *Cells* **2020**, *9*, 2655. [CrossRef]

27. Prikhodko, V.; Chernyuk, D.; Sysoev, Y.; Zernov, N.; Okovityi, S.; Popugaeva, E. Potential Drug Candidates to Treat. TRPC6 Channel Deficiencies in the Pathophysiology of Alzheimer's Disease and Brain Ischemia. *Cells* **2020**, *9*, 2351. [CrossRef]

28. Pizzo, P.; Basso, E.; Filadi, R.; Greotti, E.; Leparulo, A.; Pendin, D.; Redolfi, N.; Rossini, M.; Vajente, N.; Pozzan, T.; et al. Presenilin-2 and Calcium Handling: Molecules, Organelles, Cells and Brain Networks. *Cells* **2020**, *9*, 2166. [CrossRef]

29. Lee, J.H.; McBrayer, M.K.; Wolfe, D.M.; Haslett, L.J.; Kumar, A.; Sato, Y.; Lie, P.P.Y.; Mohyuran, P.; Coffey, E.E.; Kompella, U.; et al. Presenilin 1 Maintains Lysosomal Ca^{2+} Homeostasis via TRPML1 by Regulating vATPase-Mediated Lysosome Acidification. *Cell Rep.* **2015**, *12*, 1430–1444. [CrossRef]

30. Checler, F. Processing of the beta-amyloid precursor protein and its regulation in Alzheimer's disease. *J. Neurochem.* **1995**, *65*, 1431–1444. [CrossRef]
31. Putney, J.W., Jr. New molecular players in capacitative Ca^{2+} entry. *J. Cell. Sci.* **2007**, *120*, 1959–1965. [CrossRef] [PubMed]
32. Chen, X.; Sooch, G.; Demaree, I.S.; White, F.A.; Obukhov, A.G. Transient Receptor Potential Canonical (TRPC) Channels: Then and Now. *Cells* **2020**, *9*, 1983. [CrossRef] [PubMed]
33. Zhang, H.; Sun, S.; Wu, L.; Pchitskaya, E.; Zakharova, O.; Tacer, K.F.; Bezprozvanny, I. Store-Operated Calcium Channel Complex. in Postsynaptic Spines: A New Therapeutic Target. for Alzheimer's Disease Treatment. *J. Neurosci.* **2016**, *36*, 11837–11850. [CrossRef]
34. Du, W.; Huang, J.; Yao, H.; Zhou, K.; Duan, B.; Wang, Y. Inhibition of TRPC6 degradation suppresses ischemic brain damage in rats. *J. Clin. Investig.* **2010**, *120*, 3480–3492. [CrossRef] [PubMed]
35. Corder, E.H.; Saunders, A.M.; Strittmatter, W.J.; Schmechel, D.E.; Gaskell, P.C.; Small, G.W.; Roses, A.D.; Haines, J.L.; Pericak-Vance, M.A. Gene dose of apolipoprotein E type 4 allele and the risk of Alzheimer's disease in late onset families. *Science* **1993**, *261*, 921–923. [CrossRef] [PubMed]

Publisher's Note: MDPI stays neutral with regard to jurisdictional claims in published maps and institutional affiliations.

Review

Alterations of the Endoplasmic Reticulum (ER) Calcium Signaling Molecular Components in Alzheimer's Disease

Mounia Chami * and Frédéric Checler

Team Labelled "Laboratory of Excellence (LABEX) DistAlz", INSERM, CNRS, IPMC, Université Côte d'Azur, 660 route des Lucioles, Sophia-Antipolis, 06560 Valbonne, France; checler@ipmc.cnrs.fr
* Correspondence: mchami@ipmc.cnrs.fr; Tel.: +33-4939-53457; Fax: +33-4939-53408

Received: 7 October 2020; Accepted: 30 November 2020; Published: 1 December 2020

Abstract: Sustained imbalance in intracellular calcium (Ca^{2+}) entry and clearance alters cellular integrity, ultimately leading to cellular homeostasis disequilibrium and cell death. Alzheimer's disease (AD) is the most common cause of dementia. Beside the major pathological features associated with AD-linked toxic amyloid beta ($A\beta$) and hyperphosphorylated tau (p-tau), several studies suggested the contribution of altered Ca^{2+} handling in AD development. These studies documented physical or functional interactions of $A\beta$ with several Ca^{2+} handling proteins located either at the plasma membrane or in intracellular organelles including the endoplasmic reticulum (ER), considered the major intracellular Ca^{2+} pool. In this review, we describe the cellular components of ER Ca^{2+} dysregulations likely responsible for AD. These include alterations of the inositol 1,4,5-trisphosphate receptors' (IP_3Rs) and ryanodine receptors' (RyRs) expression and function, dysfunction of the sarco-endoplasmic reticulum Ca^{2+} ATPase (SERCA) activity and upregulation of its truncated isoform (S1T), as well as presenilin (PS1, PS2)-mediated ER Ca^{2+} leak/ER Ca^{2+} release potentiation. Finally, we highlight the functional consequences of alterations of these ER Ca^{2+} components in AD pathology and unravel the potential benefit of targeting ER Ca^{2+} homeostasis as a tool to alleviate AD pathogenesis.

Keywords: calcium; Alzheimer's disease; endoplasmic reticulum; SERCA; IP_3R; RyR; S1T; presenilin

---

1. Introduction

1.1. Ca^{2+} Signaling

As a signal transduction molecule, calcium (Ca^{2+}) regulates a large number of neuronal processes including growth and differentiation, neurotransmitter release and synaptic function, activity-dependent changes in gene expression and apoptosis [1]. Cytosolic Ca^{2+} ([Ca^{2+}]cyt) signals are regulated in a spatiotemporal-dependent manner underlined by an intricate interplay between Ca^{2+} entry through the plasma membrane, storage in the internal stores (i.e., the endoplasmic reticulum (ER), considered the major dynamic Ca^{2+} intracellular pool), Ca^{2+} mobilization from the ER and its buffering by Ca^{2+}-binding proteins (CaBP) (Figure 1). Ca^{2+} entry through the plasma membrane occurs through ligand-dependent Ca^{2+} receptors (i.e., *N*-methyl-D-aspartate receptor (NMDA) and Alpha7 nicotinic acetylcholine receptors (nAChRs)) and through voltage-gated Ca^{2+} channels (VGCC) (Figure 1). Ca^{2+} mobilization from the ER occurs through activation of the inositol 1,4,5-trisphosphate receptors (IP_3R) downstream of metabotropic receptors (Figure 1), or through the activation of ryanodine receptors (RyRs) that are activated by a slight increase in [Ca^{2+}]cyt, a mechanism known as Ca^{2+}-induced Ca^{2+} release (CICR) (Figure 1). Elevations of cytosolic Ca^{2+} signals are "shut down" through the plasma membrane Na^+/Ca^{2+} exchanger (NCX) and two Ca^{2+} ATPases which consume ATP to actively extrude Ca^{2+} out of the cells (i.e., the plasma membrane

Ca^{2+} ATPase (PMCA)) or to actively sequester Ca^{2+} into the ER lumen (i.e., the sarco-endoplasmic reticulum Ca^{2+} ATPase (SERCA)) (Figure 1). Intriguingly, coupling between ER Ca^{2+} depletion and Ca^{2+} influx through the plasma membrane occurs through a canonical store-operated Ca^{2+} entry (SOCE) pathway [2] mainly consisting of a direct physical interaction between the Ca^{2+}-sensing stromal interacting molecules (STIM1/2) oligomers within the ER membrane and the pore-forming ORAI proteins in the plasma membrane [3–5] (Figure 1). Several lines of evidence indicate that Ca^{2+} homeostasis could be disrupted upon cellular challenges as well as in neurodegenerative conditions.

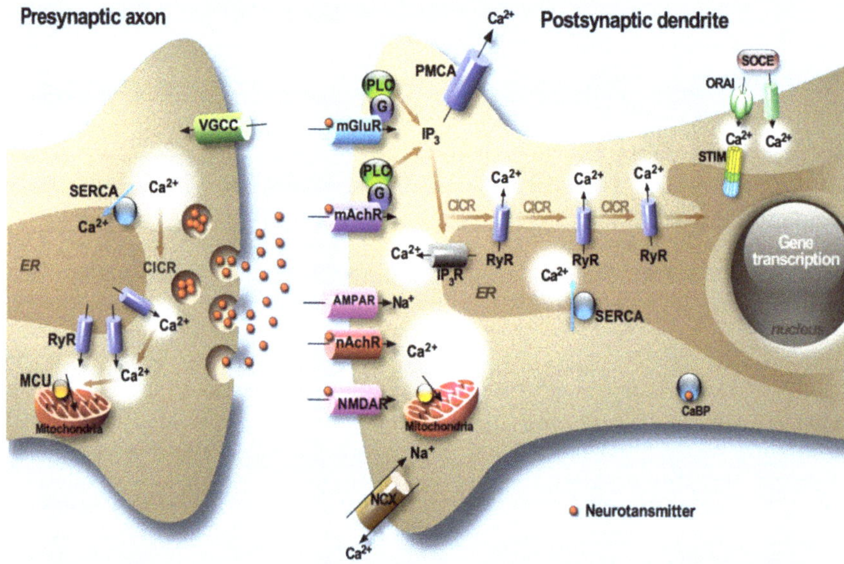

Figure 1. Elevations of intraneuronal [Ca^{2+}] are the result of an influx across the plasma membrane and the release from the ER through various channels and receptors. The low intraneuronal Ca^{2+} level is then maintained by the activity of Ca^{2+}-binding proteins (CaBP) and involves the sodium-Ca^{2+} exchanger (Na$^+$/Ca^{2+}) acting in concert with the ATP-dependent Ca^{2+} pumps located at the plasma membrane and the ER. Depletion of ER Ca^{2+} content activates the store-operated Ca^{2+} entry (SOCE) pathway.

1.2. Alzheimer's Disease

Alzheimer's disease (AD) is an age-associated dementia disorder characterized by the accumulation of extracellular amyloid-beta (Aβ) peptides in the senile plaques and by the hyperphosphorylation of tau (pTau) protein, leading to intracellular protein aggregation into bundles or filaments that are deposited as neurofibrillary tangles [6–8]. Notably, Aβ peptide derives from the sequential processing of the β-amyloid precursor protein (βAPP referred to as APP hereafter) [9,10] by the β-secretase (BACE1) and the γ–secretase complex (composed of presenilins (PSs: PS1 or PS2, the catalytic subunits of the enzyme), Nicastrin, anterior pharynx-defective-1 (APH-1) and presenilin enhancer-2 (PEN-2) [11,12])) (Figure 2). Importantly, a significant number of aggressive AD cases generally characterized by early onset are inherited in an autosomal-dominant manner (FAD: familial AD) and are caused by mutations on APP and on PS1 and PS2 [13,14] (Figure 2). These mutations either modify the nature of Aβ peptides and/or affect the levels of their production [15,16]. Besides the canonical disease-associated intracellular pTau and extracellular Aβ accumulations, recent studies unraveled additional processes that could contribute to AD progression, including: (i) the intracellular accumulation of Aβ [17,18] and other APP-derived fragments [18–24], and (ii) the spreading of both extracellular Tau and Aβ between neurons and between neurons and glial cells [25,26].

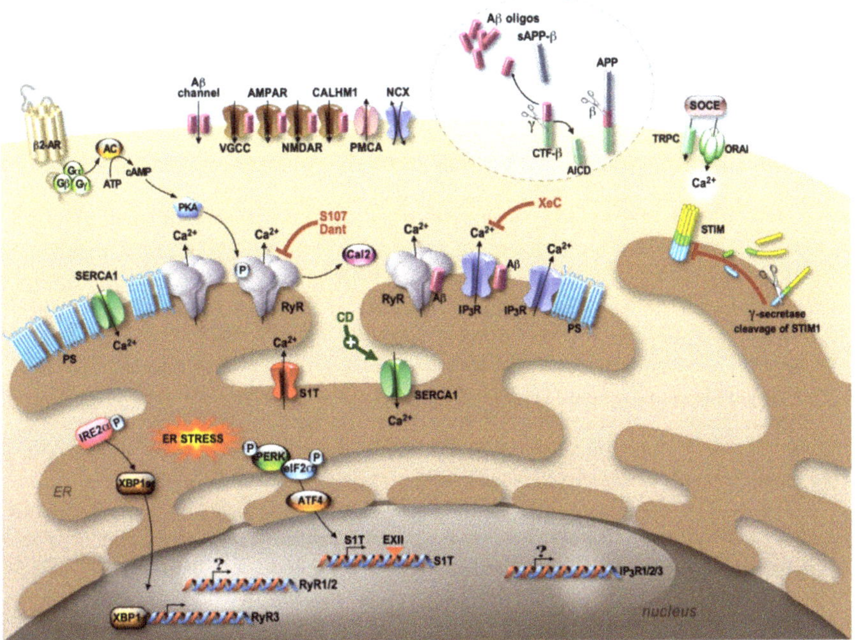

Figure 2. Aβ peptides are derived from the processing of the βAPP (APP) through the amyloidogenic pathway. APP is first cleaved by β–secretase (β), generating APP C-terminal fragment β (CTF-β), which is then cleaved by γ–secretase complex (γ) to produce Aβ and APP intracellular domain (AICD). At the plasma membrane, Aβ form a cation channel and modulate several Ca^{2+} channels (VGCC, AMPAR, NMDAR and CALHM1). ER Ca^{2+} deregulation occurs through presenilin (PS)-associated Ca^{2+} leak and/or enhanced IP₃R- and RyR-mediated Ca^{2+} release, dysfunctional Sarco-endoplasmic reticulum Ca^{2+} ATPase (SERCA) activity, enhanced expression of S1T driven by ER stress response and enhanced expression and dysfunctional IP₃Rs and RyRs. SOCE is also deregulated in AD and implicates STIM, ORAI and TRPC. RyR2 macromolecular complex destabilization (PKA phosphorylation and calstabin2 (Cal2) dissociation) is linked to β2-adrenergic receptor activation. Pharmacological stabilization of ER Ca^{2+} content by S107, Dantrolene (Dant) blocking RyRs-mediated Ca^{2+} release/leak, Xestospongin C (XeC) blocking IP₃R-mediated Ca^{2+} release and CD1163 (CD) activating SERCA provided beneficial effects in reversing several AD-related pathogenic paradigms in vitro and in vivo.

1.3. Physiology of ER Calcium Handling in Neurons

The ER forms a continuous and highly motile network distributed throughout the neuron. Within dendrites and dendritic spines, ER Ca^{2+} release is involved in modulating postsynaptic responses and synaptic plasticity [27]. In presynaptic nerve terminals, as well as in growth cones, ER is involved in vesicle fusion and neurotransmitter release [28,29]. In the soma, ER Ca^{2+} handling is coupled to the activation of Ca^{2+}-sensitive kinases and phosphatases [30]. In the perinuclear space, ER Ca^{2+} handling triggers gene transcription [31]. Ca^{2+} mobilization from the ER has been shown to be involved in growth cone activity and in the formation of new connections and/or the strengthening of preexisting connections that occur during learning and memory in the adult brain [32].

1.4. Calcium Deregulation in AD

As stated above, the tight but subtle control of intracellular Ca^{2+} homeostasis is required for neuronal health, development and function [29,30,33,34]. Therefore, persistent imbalance in Ca^{2+}

entry and clearance alters cellular integrity, leading to cellular homeostasis disequilibrium. These Ca^{2+} deregulations ultimately trigger excessive proliferation or cell death depending on the strength and the duration of the insult and in a cell-type-specific manner. Ca^{2+} signaling deregulation has a central role in AD pathophysiology [35]. The relevance of Ca^{2+} signaling in AD is supported by the fact that Ca^{2+} alterations were reported in both sporadic (SAD) and familial (FAD) forms of AD and that this can exacerbate Aβ formation and promote tau hyperphosphorylation [35–37]. As first evidence, in vitro studies have shown that Ca^{2+} may directly interact and enhance the proteolytic activity of BACE1 [38] and to stabilize γ–secretase and enhance its activity in reconstituted in vitro assay [39]. Moreover, tau hyperphosphorylation at disease-specific sites has been associated with abnormal intracellular Ca^{2+} signaling occurring upstream of Ca^{2+}/calmodulin (CaM)-dependent protein kinase II (CaMKII) and CDK5 activation [37,40–42]. The bulk of data gathered these last 30 years allows us to draw up a scenario where Ca^{2+} deregulation is not only a consequence of the disease but also participates in a feedback loop to disease progression and amplification [35,36,43–46]. These studies reported a Ca^{2+}-dependent enhancement of APP processing and the production of toxic APP-derived fragments, activation of signaling cascades through the modulation of kinases and phosphatases activities, thus affecting synaptic plasticity and cognitive function [34–36,47–49].

Several studies demonstrated a tight relationship between altered Ca^{2+} handling and the amyloidogenic cascade. These studies lead to identifying the physical or functional interaction of Aβ with several Ca^{2+} handling proteins in various AD models. At least four lines of evidence have emerged: (i) at the plasma membrane, Aβ has been shown to form a cation channel [50], or to act as a channel-modulator for the VGCCs, the nAChRs, the ionotropic glutamatereceptors NMDARs and AMPARs (α-amino-3-hydroxy-5-methyl-4-isoxazolepropionic acid receptors), the Ca^{2+} homeostasis modulator 1 (CALHM1), and more recently the store-operated Ca^{2+}channels (SOCE) (Figure 2) [51–60]; (ii) dysfunctional mitochondria were associated with Aβ-mediated Ca^{2+} toxicity [61,62] (discussed in this Special Issue [63]). Importantly, mitochondrial permeability transition pore, mitochondrial Ca^{2+} uniporter (MCU) dysfunctions and impaired mitochondrial Ca^{2+} efflux contribute to mitochondrial alteration in AD [63–65]; (iii) the autophagic failure in AD has been linked to lysosomal degradation defects [24,66] likely occurring upon lysosomal Ca^{2+} depletion [67,68]; (iv) a complex scenario of AD-associated ER Ca^{2+} dysregulation also emerged, where disturbances were linked to presenilin (PS1 and PS2)-associated ER Ca^{2+} leak and/or ER Ca^{2+} release potentiation functions [69–73], dysfunctional SERCA activity [74] and the upregulation of the recently described SERCA1 truncated isoform (S1T) [75], alterations of IP_3Rs function [56,69,70,72,76–81] and dysfunctional RyRs [44,80,82–94] (Figure 2).

Besides APP-derived amyloidogenic products, previous studies described a physiological role of APP in regulating Ca^{2+} signaling. Knockdown of endogenous APP increases the frequency and reduces the amplitude of neuronal Ca^{2+} oscillations [95]. In addition, a recent study specifically reported that APP-deficient cells exhibited elevated resting Ca^{2+} levels in the ER and reduced ER Ca^{2+} leakage rates [96]. Pathogenic tau has also been associated with nuclear Ca^{2+} deregulation [97], with increasing the ionic current of artificial membranes [98], with inducing spontaneous Ca^{2+} oscillations in the neurons [99] and with the inhibition of mitochondrial Ca^{2+} efflux via the mitochondrial Na^+/Ca^{2+} exchanger [99] (also discussed in this Special Issue [63]).

In this review, we will specifically present an update of the alterations of the molecular components controlling ER Ca^{2+} signaling in AD and discuss the potential benefit of targeting ER Ca^{2+} homeostasis as a tool to alleviate AD pathogenesis.

2. The Ryanodine Receptors: RyRs

RyRs are a family of three mammalian isoforms, RyR1, RyR2 and RyR3, mainly expressed in the skeletal muscle, heart and brain. All RyRs isoforms are expressed in the brain, with an abundance range of order as follows, RyR2 > RyR1 >> RyR3 [100,101]. RyRs activity is influenced on the one hand by Ca^{2+}, Mg^{2+} and ATP [102–105] and, on the other hand, by the integrated effects of co-proteins forming

RyR1 and RyR2 homotetramer macromolecular complexes [106–108]. These include calmodulin (CaM) [109,110], FKBP12 (12.0 kDa) and FKBP12.6 (12.6 kDa), known as Calstabin1 (Cal1) and Calstabin2 (Cal2), respectively [111]; PKA anchored to RyR1 and RyR2 via a kinase anchoring protein (mAKAP) [112,113], and Ca^{2+}/calmodulin-dependent protein kinase II (CaMKII) [114]. Other regulatory proteins were also described to interact with RYR1, thus controlling the channel gating activity [109]. RyR1/2 macromolecular complexes contain also the requisite molecular machinery allowing channel dephosphorylation (i.e., PP1 and PP2A) [113,115,116].

Enhanced RyR-mediated Ca^{2+} release was reported in primary cultured neurons derived from 3xTg-AD mice (knock in (KI) for the mutated PS1M146V and overexpressing mutated APP and microtubule-associated tau protein (PS1M146V/APPswe/tauP301L)) [85,87]. This was further confirmed in cellular models expressing wild-type or mutated APP, PS1 or PS2 [44,80,82–84,86–94,117]. Exacerbated IP_3R-evoked Ca^{2+} signals in AD mice (PS1KI and 3xTg-AD)-derived neurons were shown to be linked to RYR-associated CICR [85]. These findings were further supported by using the RyR blocker dantrolene (Dant), shown to reduce enhanced $[Ca^{2+}]$cyt level [92,93,118]. While some studies reported that RyR dysfunction in AD-related study models occurs independently of PS mutation or overexpression, namely in models expressing APP and overproducing Aβ [86,92,119–122], in many cases, PS mutation-mediated Ca^{2+} deregulation was associated with the alteration of the activity of RyRs (discussed beyond in PSs chapter). In addition, it was also reported that exogenous Aβ oligomers may directly stimulate RyR-mediated Ca^{2+} release [123] and that the application of soluble Aβ caused a marked increase in channel open probability [124].

RyR isoform expression is modified throughout AD progression and between different brain regions [125]. Exogenous application of Aβ peptide was also shown to specifically increase RyR3 isoform expression [86,123]. RyRs mRNAs increase throughout the lifetime of PS1-M146V transgenic mice and 3xTg-AD mice [84,85,87] as well as in cellular and mice AD models overexpressing wild-type or mutated APP (bearing the Swedish mutation APPswe) [92]. Conversely, neuronal conditional PS1/2 knockout (KO) (PScDKO) is associated with a downregulation of RyR2 expression, demonstrating that PS may regulate Ca^{2+} homeostasis and synaptic function via RyRs [126]. It has been proposed that the modulation of RyR expression may act as a disease promoter or a compensatory beneficial mechanism. In fact, while on the one hand, enhanced $[Ca^{2+}]$cyt response is associated with the increased expression of RyRs [127], the activation of the ER stress response factor X-box binding protein 1 spliced isoform (XBP1s) may occur upon Aβ oligomer treatment [128,129], triggering a reduction of $[Ca^{2+}]$cyt linked to the down-expression of the RyR3 isoform [130]. Accordingly, a dual role for endogenous RyR3 has been suggested in an AD mouse model. Thus, the deletion of RyR3 in young ($\leq$ 3 mo) APPPS1 mice increased hippocampal neuronal network excitability and accelerated AD pathology, leading to mushroom spine loss and increased Aβ accumulation. Meanwhile, deletion of RyR3 in older APPPS1 mice ($\geq$6 mo) rescued network excitability and mushroom spine loss, reduced Aβ load and reduced spontaneous seizure occurrence [131] (Figure 2).

RyRs mutations are liked to various pathologies affecting muscle and heart [132,133]. The development of transgenic mouse models (i.e., KO of RyR1, RyR2 or RyR3, or expressing RyR harboring disease mutations, or lacking exon sequence) [134] strengthens the fact that RyRs play a key role in physiology and pathophysiology. The viability of RyR3 KO mouse, in contrast to the RyR1 and RyR2 KO mice [135,136], led to the demonstration that RyR3-deficient mice exhibit decreased social behavior [137], greater locomotor activity [136,138], altered memory [138,139] associated with impaired maintenance of long-term potentiation (LTP) [140]. To date, no mutations have been reported in RYRs linked to brain disorders. Nevertheless, the role of leaky RyR2 in the pathogenesis of epilepsy has been described in the RyR2-R2474S mice model [101]. Interestingly, three single nuclear polymorphisms were significantly associated with risk for hypertension, diabetes and AD [141]. A meta-analysis based on four genome-wide association study (GWAS) also identified *RYR3* association with AD risk [142]. Another study observed a significant interaction between *RYR3* and *CACNA1C* (gene encoding for the

Ca^{2+} voltage-gated channel subunit Alpha1 C) in three independent datasets of AD Neuroimaging Initiative cohorts [143].

RyRs post-translational modifications (PTMs) shift the channel from a finely regulated state to a non-regulated Ca^{2+} leak channel. RyR PTMs were associated with different pathologies affecting skeletal muscle, heart and, recently, brain [133] [113,144–150]. Experimental transgenic mice expressing RyR harboring PKA-non-phosphorylated sites or phosphomimetic RyR mutants demonstrated the role of the PKA phosphorylation site in RyR macromolecular complex remodeling, Calstabin dissociation and ER Ca^{2+} leak [133]. In addition to phosphorylation sites, RyRs also contain a large number of amino acid residues that are potential targets for reactive oxygen species (ROS) and for reactive nitrogen species (RNS) [108,151,152]. Recently, we described a new molecular mechanism and signaling cascade underlying altered RyR-mediated intracellular Ca^{2+} release in AD [116,150,153]. We reported that the RyR2 channel undergoes PKA phosphorylation, oxidation/nitrosylation and depletion of the channel stabilizing subunit Calstabin2 in SH-SY5Y neuroblastoma cells expressing APP harboring the familial Swedish mutations (APPswe), in APP/PS1 (APPswe, PS1-M146V), as well as in 3xTg-AD, transgenic mice models and, most importantly, in human SAD brains [150,153]. We further reported that RyR2 macromolecular complex remodeling occurs through synergistic mitochondrial reactive oxygen species (ROS) production and β-adrenergic stimulation [150,153]. Notably, oxidative stress is considered a major contributor to AD pathogenesis [154,155], and β2-adrenergic receptors (β2-ARs) have also been implicated in the development of AD [51,156–161]. However, targeting β-adrenergic signaling is questionable, since both beneficial versus defective effects were described in AD mice [162–164]. In our study, we specifically targeted the downstream PKA-mediated RyR2 phosphorylation and macromolecular complex destabilization (Figure 2). We showed that pharmacological stabilization of calstabin2 on the RyR2 macromolecular complex by S107 (a benzothiazepine derivative molecule [101]) reduces elevated Ca^{2+} signals in AD cells [153], prevents ER Ca^{2+} leakage and reduces single channel open probabilities in AD mice brains [150]. Most importantly, S107 treatment reduces APP processing and Aβ production both in vitro and in vivo [150,153]. S107 administration also inhibited calpain activity and AMPK-dependent tau phosphorylation in an APP/PS1 mouse model [150]. These data agree well with previously reported studies demonstrating the beneficial effects of the pharmacological targeting of RyR with dantrolene [88,92,93,165]. In support of these findings, RyR macromolecular complex stabilization improved the hippocampal synaptic plasticity (LTP and LTD) and cognitive function of APP/PS1 and 3xTg-AD mice [150]. Importantly, we further showed that crossing APP/PS1 mice with RyR2-S2808A KI mice, harboring RyR2 channels that cannot be PKA-phosphorylated, resulted in improved cognitive function and decreased neuropathology. In contrast, phosphormimetic RyR2-S2808D KI mice exhibit early altered hippocampal synaptic plasticity (LTP and LTD) and cognitive dysfunction [150]. Overall, these results emphasize the broad implication of RyRs in ER Ca^{2+} signaling deregulation in AD occurring through the regulation of RYRs expression, CICR-dependent activity, macromolecular complex stability-linked to β2-AR signaling cascade, Aβ- and PS-mediated RyRs channel opening and likely RYR3 gene polymorphism.

3. The Inositol 1,4,5-Trisphosphate Receptors: IP$_3$Rs

Among the three IP$_3$Rs isoforms, the predominant one in neurons is IP$_3$R1 [166–169]. In addition to Ca^{2+} and IP$_3$, there are other allosteric IP$_3$R modulators, including ATP [170]. The activity of IP$_3$R can also be regulated by its phosphorylation by different kinases [166,170]. Among them are PKA, protein kinase C (PKC), cGMP-dependent protein kinase (PKG), CaMKII and different protein tyrosine kinases. Moreover, similarly to RyRs, the IP$_3$Rs can also be regulated by the redox status and by several interacting proteins (i.e., CaM-related Ca^{2+}-binding proteins (CaBPs), Bcl2 family members, proteases (Caspase-3 and calpain) and ER lumen-specific protein (ERp44) [170]).

IP$_3$Rs activity controls spine morphology, synaptic plasticity and memory consolidation [171–173]. Notably, alterations of IP$_3$Rs expression and function were reported to be implicated in Ca^{2+} signaling deregulation in several AD models [174]. Ca^{2+} imaging experiments demonstrated that orthologous expression of FAD PS1 mutants potentiates IP$_3$-mediated Ca^{2+} release [175]. These data were confirmed in cortical neurons isolated from PS1-M146V KI mice [79] and in cells expressing FAD PS1-DeltaE9 mutant [176]. PS1-DeltaE9 mutant cells harbored enhanced basal phosphoinositide hydrolysis and cyt[Ca^{2+}], which were both reversed by the PLC inhibitor neomycin. PS1-DeltaE9 mutant cells also showed high basal [Ca^{2+}] and agonist-evoked Ca^{2+} signals that were reversed by xestospongin C (XeC, a reversible IP$_3$R antagonist) [176]. The molecular mechanisms underlying enhanced IP$_3$R-mediated Ca^{2+} release have been described to be PS-dependent and/or PS-independent [73,177] (also discussed in PSs chapter below). The computational modeling of single IP$_3$R activity was used to analyze and quantify the pathological enhancement of IP$_3$R function by FAD-causing mutant PS [178]. This study revealed that the gain-of-function enhancement of IP$_3$R was sensitive to both IP$_3$ and Ca^{2+}, thus triggering a higher frequency of local Ca^{2+} signals, while enhancing the activity of the channel at extremely low ligand concentrations will lead to spontaneous Ca^{2+} signals in cells expressing FAD-causing mutant PS [178]. It has been consequently observed that the gain-of-function enhancement of IP$_3$R channels in cells expressing PS1-M146L leads to the opening of mitochondrial permeability transition pore (PTP) in high-conductance state, triggering a reduction in the inner mitochondrial membrane potential and in NADH and ATP levels [179]. Conversely, genetic reduction of IP$_3$R1 normalizes disturbed Ca^{2+} signaling in PS1-M146V KI mice and most importantly alleviates AD pathogenesis (i.e., rescues aberrant hippocampal long-term potentiation (LTP), attenuates Aβ accumulation and tau hyperphosphorylation and memory deficits) in both PS1-M146V KI and 3xTg-AD mice [72]. Accordingly, in vitro experiments showed that XeC effectively ameliorated Aβ42-induced apoptosis and intracellular Ca^{2+} overload in the primary hippocampal neurons [180]. Notably, intracerebroventricular injection of XeC reduced the number of Aβ plaques, alleviated ER stress response and significantly improved the cognitive behavior of APP/PS1 mice [180]. Exacerbated IP$_3$R-mediated Ca^{2+} release is also linked to Aβ, independently of PS overexpression/mutation. Jensen L.E., et al. reported that Aβ42 induced elevation of cytosolic Ca^{2+} in an IP$_3$R-dependent and -independent manner [91]. In addition, it was also shown that the treatment with Aβ42 significantly increased mRNA levels of IP$_3$R1/2 and mGluR5 [181]. Enhanced IP3R1 expression and ER Ca^{2+} release were also reported in astrocytes derived from the entorhinal cortex and from the hippocampus from WT mice and mice treated with Aβ42 oligomers [182]. Finally, as stated above, IP$_3$R function is regulated by several binding proteins. Thus, it is also conceivable that any alteration of the expression, localization, activity and binding affinity of these proteins may affect IP$_3$R structural/functional state, thus impacting AD development.

4. Presenilins 1 and 2: PS1/2

PS1 and PS2 are multispanning transmembrane (TM) proteins located in intracellular membranous organelles such as the ER, nuclear envelope and Golgi apparatus but also in multiple secretory and endocytic organelles as well as the plasma membrane [183]. PS1 was first cloned as a causative gene of FAD [184]. Its homologue PS2 gene was then identified, sharing an approximately 60% sequence homology as a whole and approximately 90% within the TM domains [185,186]. Accordingly, PS1 and PS2 were also shown to share similar predicted topology [187–189].

Both PS1 and PS2 are expressed in neurons [190] and are essential for embryonic development since PS1 KO mice die at birth [191] and PS1/PS2 double KO mice (PSDKO) mice die before embryonic day 9.5 [192]. Importantly, PSs were also shown to play key roles in neuronal function and survival. Therefore, conditional PSDKO mice show impaired spatial and associative memory, deficits in short- and long-term plasticity [193,194] and develop synaptic, dendritic and neuronal degeneration in an age-dependent manner [193]. Importantly, being the catalytic component of γ–secretase complex cleaving the APP [195], most of the PS mutations associated with early-onset FAD affect APP processing and, more particularly, the ratio Aβ40/42 by increasing the aggregation-prone Aβ42 species [196–198]. Several studies proposed the contribution of PSs to ER Ca^{2+} signaling deregulation in AD. It has been proposed that PSs act as ER Ca^{2+} leak channels, and FAD mutations in PSs disrupt this function, leading to ER Ca^{2+} overload [69,70,76]. Tu et al. also proposed that the full-length PSs function as ER Ca^{2+} leak channels independently of other γ–secretase components [69]. Cysteine point mutants combined with NMR studies revealed that TM7 and TM9, but not TM6, could play an important role in forming the conductance pore of mouse PS1 [77]. A recent study investigated the interaction of Ca^{2+} with both PS1 and PS2 using all-atom molecular dynamics (MD) simulations in realistic membrane models [199]. Although the Ca^{2+} leak event linked to PS1 or PS2 has been challenged in this study, the obtained data demonstrated the presence of four Ca^{2+} sites in membrane-bound PS1 and PS2 [199]. The authors speculated that Ca^{2+} may prevent PS maturation (i.e., "presenilinase" endoproteolysis generating PS N-and PS C-terminal derivatives [200]) by triggering conformational changes, thus preserving the immature Ca^{2+} regulation function. Meanwhile, conversely, PS maturation yielding a biologically active PS would abolish this Ca^{2+}-regulatory function [199]. Nevertheless, the PS-associated Ca^{2+} leak function was discussed in other studies proposing that FAD PSs directly potentiate the gating of IP_3R [81]. Exaggerated IP_3R-mediated Ca^{2+} responses were also reported in cells and neurons derived from transgenic mice expressing FAD-linked mutant PS1 or PS2 [84]. These findings agree well with data obtained in PS1-M146V KI mice neurons using whole-cell patch-clamp recording, flash photolysis and two-photon imaging [79]. Accordingly, genetic reduction of IP_3R 1 normalizes disturbed Ca^{2+} signaling in FAD PS1 mice and alleviates AD pathogenesis in PS1-M146V KI mice [72]. Other studies point to PS-linked disruptions in RyR signaling as an important ER molecular component associated with enhanced ER Ca^{2+} signals in both 3xTg-AD and PS1-M146V (KI) neurons [80]. PS1/PS2 were also shown to harbor a physical interaction with RyRs in the ER [83,117,201]. Specifically, PS2 interacts with RyR and with sorcin (a RyR regulator) in a Ca^{2+}-dependent manner in both cellular models and in the brain [201,202], thus increasing both mean currents and open probability of single brain RyR channels [203,204]. Discrepancies regarding the role of PSs in ER Ca^{2+} handling alterations were further highlighted in a recent study [205] showing that FAD PS2 mutants, but not FAD PS1, are able to partially block SERCA activity, thereby reducing ER Ca^{2+} content in either SH-SY5Y cells or FAD patient-derived fibroblasts [205]. Despite this incongruity concerning the exact molecular mechanism underlying PS-mediated ER Ca^{2+} deregulation, FAD PSs undoubtedly directly or indirectly contribute to the Ca^{2+} hypothesis in AD. However, whether PS-mediated ER Ca^{2+} deregulation is dependent on or independent of its endoproteolysis generating PS N-and PS C-terminal derivatives still remains an open question.

5. The Sarco-Endoplasmic Reticulum (SR/ER) Ca^{2+}-ATPase and Its Truncated Isoform: SERCA and S1T

SERCAs are integral ER proteins preserving low $[Ca^{2+}]cyt$ by pumping free Ca^{2+} ions into the ER lumen, utilizing ATP hydrolysis. The SERCA pumps are encoded by three distinct genes (SERCA1-3), resulting in 12 known protein isoforms, with tissue-specific expression patterns. SERCA2b is the most expressed isoform in neurons [206]. Despite the well-established structure and function of the SERCA pumps, their role in the central nervous system and whether it could be affected in brain diseases remain to be definitely established. Interestingly, SERCA-mediated Ca^{2+} dyshomeostasis has been associated with neuropathological conditions, such as bipolar disorder, schizophrenia, Parkinson's disease but also AD [207]. An initial study showed that SERCA activity is reduced in fibroblasts isolated from PSDKO. Immunoprecipitation analyses suggested a physical interaction between SERCA and PS1 and PS2 [74] and that modulation of SERCA expression regulates Aβ levels [74]. The interaction of PS1 holoprotein was further demonstrated in cells overexpressing PS1 and subjected to tunicamycin treatment [208]. It has also been shown that overexpressed wild-type or mutated PS2 triggered ER-passive leakage through IP_3R and RyR but also potently reduced ER Ca^{2+} uptake, an effect that has been counteracted by the overexpression of SERCA2b [71]. A recent study reported that the pharmacological SERCA activation by a quinoline derivative (CD1163), discovered via high-throughput screening of small molecules library, provides some beneficial effects in APP/PS1 mice [209] (Figure 2). Overall, these studies pinpointed the potential contribution of SERCA to ER Ca^{2+} dyshomeostsis in

AD cellular study models. However, dedicated studies in mice AD models and in human-derived samples are still needed to further support the beneficial versus pathogenic role of the modulation of SERCA expression and activity in AD development.

Accumulation of unfolded proteins into the ER as well as alteration of ER Ca^{2+} homeostasis induce ER stress, eliciting an unfolded protein response (UPR) [210,211]. Several studies have reported that UPR occurs in human AD brains [212,213] and in several AD study systems [214–216]. We previously demonstrated that the human SERCA1 truncated isoform (S1T) [217] is induced under pharmacological and physiopathological ER stress through the activation of the PERK-eIF2α-ATF4-CHOP pathway [218]. In turn, S1T expression induction triggered an amplification of ER stress and mitochondrial apoptosis [218]. UPR activation has been proposed to be linked to intracellular Aβ accumulation [219]. In a recent study, we revealed that S1T is upregulated in SH-SY5Y cells expressing APPswe [75]. Importantly, biochemical data indicate that enhanced human S1T expression correlates with Aβ load in human AD-affected brains and that S1T high neuronal immunostaining is selectively observed in human AD cases harboring focal Aβ. We further demonstrated that S1T expression is induced by exogenous application of Aβ oligomers in cells [75]. Interestingly, S1T overexpression in return enhances APP processing and the production of APP-derived toxic fragments (APP C-terminal fragments and Aβ) in cells and in 3xTgAD mice. Mechanistically, we find that S1T-mediated elevation of APP proteolysis occurs through the upregulation of BACE1 expression and enhanced activity [75]. In agreement with these findings, several lines of evidence indicated that enhanced phosphorylation of PERK and eIF2α in the AD brain is associated with increased amyloidogenic APP processing [214–216] through increased BACE1 expression [220,221]. We have also to consider that BACE1 upregulation occurring downstream of $[Ca^{2+}]$cyt elevation acts in a positive feedback loop with AD progression [222]. In addition, the induction of ER stress and the activation of UPR trigger neuroinflammation [223]. Accordingly, we demonstrated that S1T overexpression, as well as tunicamycin treatment, induce the expression of proinflammatory cytokines and increase the proliferation of active microglia [75]. Altogether, our data strengthen the molecular link between ER Ca^{2+} leak, ER stress and APP processing contributing to AD setting and/or progression.

6. The Molecular Bridge between ER Ca^{2+} Depletion and Plasma Membrane Ca^{2+} Entry: STIM/ORAI

The store-operated Ca^{2+} entry (SOCE) is an essential route for Ca^{2+} uptake to replenish intracellular Ca^{2+} stores [224]. The stromal interaction molecules STIM1 and STIM2 have been identified as essential components of SOCE and major sensors of the Ca^{2+} concentration located in the ER membrane (reviewed in [225]). Both STIM homologues are ubiquitously expressed in different cell types, with a higher STIM1 level in most tissues and a predominant expression of STIM2 in the brain [226]. A decrease in ER luminal Ca^{2+} concentration results in dissociation of Ca^{2+} from the STIM EF-hand domain, which, in turn, triggers oligomerization and activation of STIM1, a process that is reversed when luminal $[Ca^{2+}]$ returns to resting level [3,4]. Active STIM oligomers translocate to ER plasma membrane junctions and recruit and interact with ORAI channels located on the plasma membrane [227]. There are three ORAI isoforms displaying tissue-specific expression and activation patterns [225]. In addition, transient receptor potential channels (TRPC) can also be recruited by the ORAI/STIM complex, constituting an additional route for Ca^{2+} entry through the plasma membrane upon ER Ca^{2+} depletion [228] (Figure 1). Several TRPC isoforms were also identified and were described to harbor tissue/cell-specific expression and activation patterns [229].

Several studies pinpointed a role for STIM/ORAI in neuronal Ca^{2+} signaling-associated synaptic function [230,231]. Thus, the maturation of dendritic spines and the formation of functional synapses in immature hippocampal neurons is facilitated by the influx of Ca^{2+} through ORAI1 [232]. STIM also interact with/and or control the activity of several Ca^{2+} channels on the plasma membrane (L-type Ca^{2+} channels (Ca_V1.2), L-type VGCCs and mGluR [232]). Several studies suggested that the disruption of neuronal SOCE underlies AD pathogenesis. A direct connection between Aβ-induced synaptic mushroom spine loss and the neuronal SOCE pathway was reported in two studies. Popugaeva et al.

reported that the application of exogenous Aβ42 oligomers to hippocampal cultures or injection of Aβ42 oligomers directly into the hippocampal region resulted in the reduction of mushroom spines and activity of synaptic CaMKII, which were rescued by STIM2 overexpression [224]. Accordingly, similar findings were reported in APPKI hippocampal neurons accumulating extracellular Aβ42. Thus, it was shown that Aβ triggers mGluR5 receptor overactivation, leading to elevated ER Ca^{2+} levels, compensatory downregulation of STIM2 expression, impairment of synaptic SOCE and reduced CaMKII activity [58]. Inversely, overexpression of the constitutively active STIM1-D76A mutant and ORAI1 significantly reduced Aβ secretion [55]. A link between STIM1/2 and PS1 was also reported. The STIM2–SOCE–CaMKII pathway was downregulated in a PS1-M146V KI mouse model of AD, associated with loss of hippocampal mushroom spines [233], and conversely, STIM2 overexpression rescued synaptic SOCE and mushroom spine deficit in hippocampal neurons from PS1-M146V KI mice [233]. Intriguingly, even if STIM1 expression is not altered in the AD models cited above, it has been identified as a target of PS1-containing γ–secretase activity. In particular, FAD-linked PS1 mutations enhanced γ–secretase cleavage of STIM1, reducing the activation of ORAI1 and attenuating SOCE [234]. As a consequence, the inhibition of SOCE in hippocampal neurons triggered an alteration of the dendritic spine architecture [234]. A recent study showed that the hyperactivation of SOCE channels in neurons expressing PS1-DeltaE9 mutant is mediated by the STIM1 sensor and can be attenuated by pharmacological inhibition and genetic KO STIM1 [235]. Interestingly, SOCE in PS1-DeltaE9 mutant-expressing cells is not contributed by STIM2 but involves TRPC and ORAI subunits. Importantly, transgenic Drosophila flies expressing PS1-DeltaE9 in the cholinergic neuron system showed short-term memory loss, which was reversed upon pharmacological inhibition of STIM1 [235]. Accordingly, a recent study further supports the link between FAD PSs and altered SOCE, through the demonstration of reduced STIM1 expression in SH-SY5Y cells and in patient-derived fibroblasts expressing different FAD-PS mutations [205].

TRPC may also play a role in SOCE deregulation in AD. TRPC expression is not altered in mice and human AD brains [233]. However, reduced TRPC1 expression was observed in astrocytes derived from APP KO mice [181,236]. In addition, TRPC6 was shown to specifically interact with APP, thereby blocking its cleavage by γ–secretase and reducing Aβ production independently from its ion channel activity [237]. Conversely, PS2 mutations abolish agonist-induced TRPC6 activation [238]. Importantly, activation of TRPC6 stimulates the activity of the neuronal SOCE pathway in the spines and rescues mushroom spine loss and long-term potentiation impairment in APP KI mice [58]. A review by Prikhodko, V. et al. in this Special Issue addresses the potential use of TRPC modulators as drugs to treat AD [239].

7. Conclusions

Studies demonstrating the implication of ER Ca^{2+} deregulation in AD highlight a complex picture integrating several molecular ER Ca^{2+} components. This includes enhanced ER Ca^{2+} release through IP$_3$R and RyR, dysfunctional ER Ca^{2+} uptake by SERCA and upregulation of the S1T truncated isoform, gain- or loss-of-function of PS components of the γ–secretase complex. Disease-associated remodeling of this Ca^{2+} machinery toolkit is also coupled to specific cellular signaling cascades modulating the activity (i.e., post-translational modifications, interactions with regulatory proteins) and/or the expression of these Ca^{2+} channels and pump (i.e., linked to ER stress). Several studies also pinpointed the direct interaction of Aβ peptide with several members of the ER Ca^{2+} machinery, thus contributing to ER Ca^{2+} dyshomeostasis. In addition, recent studies demonstrated that the failure of the SOCE molecular bridge between the ER and the plasma membrane has to be seriously considered as a major molecular mechanism controlling ER Ca^{2+} content and consequently ER-mediated Ca^{2+} release. Finally, it becomes now evident that ER Ca^{2+} dyshomeostasis is significantly associated with AD development and/or progression. The treatment options for AD remain supportive and symptomatic, without attenuation of the ultimate prognosis; thus efforts have still to be made in defining therapeutic approaches targeting ER Ca^{2+} machinery to cure AD (Figure 2). The described

ER Ca^{2+} toolkits are enriched in ER–mitochondria contact sites known as mitochondria-associated membranes (MAMs) [240]. Importantly, besides Ca^{2+} tunneling from ER to mitochondria, MAMs impact various cellular housekeeping functions such as phospholipid, glucose, cholesterol and fatty acid metabolism, which are all altered in AD [240,241]. This may further highlight the potential relevance of targeting ER Ca^{2+} handling proteins as an attempt to alleviate both ER and mitochondria dysfunctions associated with AD.

Funding: This work has been developed and supported through the LABEX (excellence laboratory, program investment for the future) DISTALZ (Development of Innovative Strategies for a Transdisciplinary approach to Alzheimer's disease) to FC and by Fondation Vaincre Alzheimer to MC.

Acknowledgments: We thank Franck Aguila for figure drawing.

Conflicts of Interest: The authors declare no conflict of interest.

References

1. Berridge, M.J.; Bootman, M.D.; Roderick, H.L. Calcium signalling: Dynamics, homeostasis and remodelling. *Nat. Rev. Mol. Cell Biol.* **2003**, *4*, 517–529. [PubMed]
2. Bootman, M.D.; Collins, T.J.; Peppiatt, C.M.; Prothero, L.S.; MacKenzie, L.; De Smet, P.; Travers, M.; Tovey, S.C.; Seo, J.T.; Berridge, M.J.; et al. Calcium signalling—An overview. *Semin. Cell Dev. Biol.* **2001**, *12*, 3–10. [PubMed]
3. Cahalan, M.D. STIMulating store-operated Ca(2+) entry. *Nat. Cell Biol.* **2009**, *11*, 669–677. [PubMed]
4. Putney, J.W., Jr. New molecular players in capacitative Ca2+ entry. *J. Cell Sci.* **2007**, *120*, 1959–1965.
5. Smyth, J.T.; Hwang, S.Y.; Tomita, T.; DeHaven, W.I.; Mercer, J.C.; Putney, J.W. Activation and regulation of store-operated calcium entry. *J. Cell Mol. Med.* **2010**, *14*, 2337–2349.
6. Gulisano, W.; Maugeri, D.; Baltrons, M.A.; Fa, M.; Amato, A.; Palmeri, A.; D'Adamio, L.; Grassi, C.; Devanand, D.P.; Honig, L.S.; et al. Role of Amyloid-beta and Tau Proteins in Alzheimer's Disease: Confuting the Amyloid Cascade. *J. Alzheimer's Dis* **2018**, *64*, S611–S631. [CrossRef]
7. Selkoe, D.J.; Hardy, J. The amyloid hypothesis of Alzheimer's disease at 25 years. *EMBO Mol. Med.* **2016**, *8*, 595–608. [CrossRef]
8. Jones, D.T.; Graff-Radford, J.; Lowe, V.J.; Wiste, H.J.; Gunter, J.L.; Senjem, M.L.; Botha, H.; Kantarci, K.; Boeve, B.F.; Knopman, D.S.; et al. Tau, amyloid, and cascading network failure across the Alzheimer's disease spectrum. *Cortex* **2017**, *97*, 143–159. [CrossRef]
9. Checler, F. Processing of the beta-amyloid precursor protein and its regulation in Alzheimer's disease. *J. Neurochem.* **1995**, *65*, 1431–1444.
10. Zhang, Y.W.; Thompson, R.; Zhang, H.; Xu, H. APP processing in Alzheimer's disease. *Mol. Brain* **2011**, *4*, 3. [CrossRef]
11. Edbauer, D.; Winkler, E.; Regula, J.T.; Pesold, B.; Steiner, H.; Haass, C. Reconstitution of gamma-secretase activity. *Nat. Cell Biol.* **2003**, *5*, 486–488. [CrossRef] [PubMed]
12. Kimberly, W.T.; LaVoie, M.J.; Ostaszewski, B.L.; Ye, W.; Wolfe, M.S.; Selkoe, D.J. Gamma-secretase is a membrane protein complex comprised of presenilin, nicastrin, Aph-1, and Pen-2. *Proc. Natl. Acad. Sci. USA* **2003**, *100*, 6382–6387. [CrossRef] [PubMed]
13. Tanzi, R.E.; Bertram, L. Twenty years of the Alzheimer's disease amyloid hypothesis: A genetic perspective. *Cell* **2005**, *120*, 545–555. [PubMed]
14. Van Cauwenberghe, C.; Van Broeckhoven, C.; Sleegers, K. The genetic landscape of Alzheimer disease: Clinical implications and perspectives. *Genet. Med.* **2015**, *18*, 421–430. [CrossRef] [PubMed]
15. Golde, T.E.; Cai, X.D.; Shoji, M.; Younkin, S.G. Production of amyloid beta protein from normal amyloid beta-protein precursor (beta APP) and the mutated beta APPS linked to familial Alzheimer's disease. *Ann. N.Y. Acad. Sci.* **1993**, *695*, 103–108. [PubMed]
16. Wolfe, M.S. When loss is gain: Reduced presenilin proteolytic function leads to increased Abeta42/Abeta40. Talking Point on the role of presenilin mutations in Alzheimer disease. *EMBO Rep.* **2007**, *8*, 136–140.
17. Billings, L.M.; Oddo, S.; Green, K.N.; McGaugh, J.L.; LaFerla, F.M. Intraneuronal Abeta causes the onset of early Alzheimer's disease-related cognitive deficits in transgenic mice. *Neuron* **2005**, *45*, 675–688.

18. Del Prete, D.; Suski, J.M.; Oules, B.; Debayle, D.; Gay, A.S.; Lacas-Gervais, S.; Bussiere, R.; Bauer, C.; Pinton, P.; Paterlini-Brechot, P.; et al. Localization and Processing of the Amyloid-beta Protein Precursor in Mitochondria-Associated Membranes. *J. Alzheimer's Dis.* **2017**, *55*, 1549–1570. [CrossRef]

19. Pardossi-Piquard, R.; Petit, A.; Kawarai, T.; Sunyach, C.; Alves da Costa, C.; Vincent, B.; Ring, S.; D'Adamio, L.; Shen, J.; Muller, U.; et al. Presenilin-dependent transcriptional control of the Abeta-degrading enzyme neprilysin by intracellular domains of betaAPP and APLP. *Neuron* **2005**, *46*, 541–554. [CrossRef]

20. Nhan, H.S.; Chiang, K.; Koo, E.H. The multifaceted nature of amyloid precursor protein and its proteolytic fragments: Friends and foes. *Acta Neuropathol.* **2015**, *129*, 1–19. [CrossRef]

21. Baranger, K.; Marchalant, Y.; Bonnet, A.E.; Crouzin, N.; Carrete, A.; Paumier, J.M.; Py, N.A.; Bernard, A.; Bauer, C.; Charrat, E.; et al. MT5-MMP is a new pro-amyloidogenic proteinase that promotes amyloid pathology and cognitive decline in a transgenic mouse model of Alzheimer's disease. *Cell Mol. Life Sci.* **2016**, *73*, 217–236. [CrossRef] [PubMed]

22. Willem, M.; Tahirovic, S.; Busche, M.A.; Ovsepian, S.V.; Chafai, M.; Kootar, S.; Hornburg, D.; Evans, L.D.; Moore, S.; Daria, A.; et al. eta-Secretase processing of APP inhibits neuronal activity in the hippocampus. *Nature* **2015**, *526*, 443–447. [CrossRef] [PubMed]

23. Lauritzen, I.; Pardossi-Piquard, R.; Bauer, C.; Brigham, E.; Abraham, J.D.; Ranaldi, S.; Fraser, P.; St-George-Hyslop, P.; Le Thuc, O.; Espin, V.; et al. The beta-Secretase-Derived C-Terminal Fragment of betaAPP, C99, But Not Abeta, Is a Key Contributor to Early Intraneuronal Lesions in Triple-Transgenic Mouse Hippocampus. *J. Neurosci.* **2012**, *32*, 16243–16255. [CrossRef] [PubMed]

24. Lauritzen, I.; Pardossi-Piquard, R.; Bourgeois, A.; Pagnotta, S.; Biferi, M.G.; Barkats, M.; Lacor, P.; Klein, W.; Bauer, C.; Checler, F. Intraneuronal aggregation of the beta-CTF fragment of APP (C99) induces Abeta-independent lysosomal-autophagic pathology. *Acta Neuropathol.* **2016**, *132*, 257–276. [CrossRef]

25. Vergara, C.; Houben, S.; Suain, V.; Yilmaz, Z.; De Decker, R.; Vanden Dries, V.; Boom, A.; Mansour, S.; Leroy, K.; Ando, K.; et al. Amyloid-beta pathology enhances pathological fibrillary tau seeding induced by Alzheimer PHF in vivo. *Acta Neuropathol.* **2019**, *137*, 397–412. [CrossRef]

26. He, Z.; Guo, J.L.; McBride, J.D.; Narasimhan, S.; Kim, H.; Changolkar, L.; Zhang, B.; Gathagan, R.J.; Yue, C.; Dengler, C.; et al. Amyloid-beta plaques enhance Alzheimer's brain tau-seeded pathologies by facilitating neuritic plaque tau aggregation. *Nat. Med.* **2018**, *24*, 29–38. [CrossRef]

27. Holbro, N.; Grunditz, A.; Oertner, T.G. Differential distribution of endoplasmic reticulum controls metabotropic signaling and plasticity at hippocampal synapses. *Proc. Natl. Acad. Sci. USA* **2009**, *106*, 15055–15060. [CrossRef]

28. Dailey, M.E.; Bridgman, P.C. Dynamics of the endoplasmic reticulum and other membranous organelles in growth cones of cultured neurons. *J. Neurosci.* **1989**, *9*, 1897–1909. [CrossRef]

29. Emptage, N.J.; Reid, C.A.; Fine, A. Calcium stores in hippocampal synaptic boutons mediate short-term plasticity, store-operated Ca2+ entry, and spontaneous transmitter release. *Neuron* **2001**, *29*, 197–208. [CrossRef]

30. Berridge, M.J.; Bootman, M.D.; Lipp, P. Calcium—a life and death signal. *Nature* **1998**, *395*, 645–648. [CrossRef]

31. Li, W.; Llopis, J.; Whitney, M.; Zlokarnik, G.; Tsien, R.Y. Cell-permeant caged InsP3 ester shows that Ca2+ spike frequency can optimize gene expression. *Nature* **1998**, *392*, 936–941. [CrossRef] [PubMed]

32. Bandtlow, C.E.; Schmidt, M.F.; Hassinger, T.D.; Schwab, M.E.; Kater, S.B. Role of intracellular calcium in NI-35-evoked collapse of neuronal growth cones. *Science* **1993**, *259*, 80–83. [CrossRef] [PubMed]

33. Popugaeva, E.; Bezprozvanny, I. Can the calcium hypothesis explain synaptic loss in Alzheimer's disease? *Neurodegener. Dis.* **2014**, *13*, 139–141. [CrossRef] [PubMed]

34. Bojarski, L.; Herms, J.; Kuznicki, J. Calcium dysregulation in Alzheimer's disease. *Neurochem. Int.* **2008**, *52*, 621–633. [CrossRef] [PubMed]

35. Tong, B.C.; Wu, A.J.; Li, M.; Cheung, K.H. Calcium signaling in Alzheimer's disease & therapies. *Biochim. Biophys. Acta Mol. Cell Res.* **2018**, *1865*, 1745–1760. [CrossRef]

36. Del Prete, D.; Checler, F.; Chami, M. Ryanodine receptors: Physiological function and deregulation in Alzheimer disease. *Mol. Neurodegener.* **2014**, *9*, 21. [CrossRef]

37. Cao, L.L.; Guan, P.P.; Liang, Y.Y.; Huang, X.S.; Wang, P. Calcium Ions Stimulate the Hyperphosphorylation of Tau by Activating Microsomal Prostaglandin E Synthase 1. *Front. Aging Neurosci.* **2019**, *11*, 108. [CrossRef]

38. Hayley, M.; Perspicace, S.; Schulthess, T.; Seelig, J. Calcium enhances the proteolytic activity of BACE1: An in vitro biophysical and biochemical characterization of the BACE1-calcium interaction. *Biochim. Biophys. Acta* **2009**, *1788*, 1933–1938. [CrossRef]

39. Ho, M.; Hoke, D.E.; Chua, Y.J.; Li, Q.X.; Culvenor, J.G.; Masters, C.; White, A.R.; Evin, G. Effect of Metal Chelators on gamma-Secretase Indicates That Calcium and Magnesium Ions Facilitate Cleavage of Alzheimer Amyloid Precursor Substrate. *Int. J. Alzheimer's Dis.* **2010**, *2011*, 950932. [CrossRef]

40. Miller, S.G.; Kennedy, M.B. Regulation of brain type II Ca2+/calmodulin-dependent protein kinase by autophosphorylation: A Ca2+-triggered molecular switch. *Cell* **1986**, *44*, 861–870. [CrossRef]

41. Oka, M.; Fujisaki, N.; Maruko-Otake, A.; Ohtake, Y.; Shimizu, S.; Saito, T.; Hisanaga, S.I.; Iijima, K.M.; Ando, K. Ca2+/calmodulin-dependent protein kinase II promotes neurodegeneration caused by tau phosphorylated at Ser262/356 in a transgenic Drosophila model of tauopathy. *J. Biochem.* **2017**, *162*, 335–342. [CrossRef] [PubMed]

42. Litersky, J.M.; Johnson, G.V.; Jakes, R.; Goedert, M.; Lee, M.; Seubert, P. Tau protein is phosphorylated by cyclic AMP-dependent protein kinase and calcium/calmodulin-dependent protein kinase II within its microtubule-binding domains at Ser-262 and Ser-356. *Biochem. J.* **1996**, *316 (Pt 2)*, 655–660. [CrossRef]

43. Chakroborty, S.; Stutzmann, G.E. Early calcium dysregulation in Alzheimer's disease: Setting the stage for synaptic dysfunction. *Sci. China Life Sci.* **2011**, *54*, 752–762. [CrossRef] [PubMed]

44. Mattson, M.P. ER calcium and Alzheimer's disease: In a state of flux. *Sci. Signal.* **2010**, *3*, pe10. [CrossRef] [PubMed]

45. LaFerla, F.M. Calcium dyshomeostasis and intracellular signalling in Alzheimer's disease. *Nat. Rev. Neurosci.* **2002**, *3*, 862–872. [CrossRef] [PubMed]

46. Supnet, C.; Bezprozvanny, I. The dysregulation of intracellular calcium in Alzheimer disease. *Cell Calcium* **2010**, *47*, 183–189. [CrossRef] [PubMed]

47. Bezprozvanny, I.; Mattson, M.P. Neuronal calcium mishandling and the pathogenesis of Alzheimer's disease. *Trends Neurosci.* **2008**, *31*, 454–463. [CrossRef] [PubMed]

48. Berridge, M.J. Calcium signalling and Alzheimer's disease. *Neurochem. Res.* **2011**, *36*, 1149–1156. [CrossRef]

49. Chakroborty, S.; Stutzmann, G.E. Calcium channelopathies and Alzheimer's disease: Insight into therapeutic success and failures. *Eur. J. Pharmacol.* **2014**, *739*, 83–95. [CrossRef]

50. Arispe, N.; Rojas, E.; Pollard, H.B. Alzheimer disease amyloid beta protein forms calcium channels in bilayer membranes: Blockade by tromethamine and aluminum. *Proc. Natl. Acad. Sci. USA* **1993**, *90*, 567–571. [CrossRef]

51. Alberdi, E.; Sanchez-Gomez, M.V.; Cavaliere, F.; Perez-Samartin, A.; Zugaza, J.L.; Trullas, R.; Domercq, M.; Matute, C. Amyloid beta oligomers induce Ca2+ dysregulation and neuronal death through activation of ionotropic glutamate receptors. *Cell Calcium* **2010**, *47*, 264–272. [CrossRef] [PubMed]

52. Texido, L.; Martin-Satue, M.; Alberdi, E.; Solsona, C.; Matute, C. Amyloid beta peptide oligomers directly activate NMDA receptors. *Cell Calcium* **2011**, *49*, 184–190. [CrossRef] [PubMed]

53. Thibault, O.; Pancani, T.; Landfield, P.W.; Norris, C.M. Reduction in neuronal L-type calcium channel activity in a double knock-in mouse model of Alzheimer's disease. *Biochim. Biophys. Acta* **2012**, *1822*, 546–549. [CrossRef] [PubMed]

54. Ma, Z.; Siebert, A.P.; Cheung, K.H.; Lee, R.J.; Johnson, B.; Cohen, A.S.; Vingtdeux, V.; Marambaud, P.; Foskett, J.K. Calcium homeostasis modulator 1 (CALHM1) is the pore-forming subunit of an ion channel that mediates extracellular Ca2+ regulation of neuronal excitability. *Proc. Natl. Acad. Sci. USA* **2012**, *109*, E1963–E1971. [CrossRef]

55. Zeiger, W.; Vetrivel, K.S.; Buggia-Prevot, V.; Nguyen, P.D.; Wagner, S.L.; Villereal, M.L.; Thinakaran, G. Ca2+ influx through store-operated Ca2+ channels reduces Alzheimer disease beta-amyloid peptide secretion. *J. Biol. Chem.* **2013**, *288*, 26955–26966. [CrossRef]

56. Yoo, A.S.; Cheng, I.; Chung, S.; Grenfell, T.Z.; Lee, H.; Pack-Chung, E.; Handler, M.; Shen, J.; Xia, W.; Tesco, G.; et al. Presenilin-mediated modulation of capacitative calcium entry. *Neuron* **2000**, *27*, 561–572. [CrossRef]

57. Chernyuk, D.; Zernov, N.; Kabirova, M.; Bezprozvanny, I.; Popugaeva, E. Antagonist of neuronal store-operated calcium entry exerts beneficial effects in neurons expressing PSEN1DeltaE9 mutant linked to familial Alzheimer disease. *Neuroscience* **2019**, *410*, 118–127. [CrossRef]

58. Zhang, H.; Sun, S.; Wu, L.; Pchitskaya, E.; Zakharova, O.; Fon Tacer, K.; Bezprozvanny, I. Store-Operated Calcium Channel Complex in Postsynaptic Spines: A New Therapeutic Target for Alzheimer's Disease Treatment. *J. Neurosci.* **2016**, *36*, 11837–11850. [CrossRef]

59. Zhang, H.; Wu, L.; Pchitskaya, E.; Zakharova, O.; Saito, T.; Saido, T.; Bezprozvanny, I. Neuronal Store-Operated Calcium Entry and Mushroom Spine Loss in Amyloid Precursor Protein Knock-In Mouse Model of Alzheimer's Disease. *J. Neurosci.* **2015**, *35*, 13275–13286. [CrossRef]

60. Popugaeva, E.; Pchitskaya, E.; Speshilova, A.; Alexandrov, S.; Zhang, H.; Vlasova, O.; Bezprozvanny, I. STIM2 protects hippocampal mushroom spines from amyloid synaptotoxicity. *Mol. Neurodegener.* **2015**, *10*, 37. [CrossRef]

61. Tang, J.; Oliveros, A.; Jang, M.H. Dysfunctional Mitochondrial Bioenergetics and Synaptic Degeneration in Alzheimer Disease. *Int. Neurourol. J.* **2019**, *23*, S5–S10. [CrossRef] [PubMed]

62. Garcia-Escudero, V.; Martin-Maestro, P.; Perry, G.; Avila, J. Deconstructing mitochondrial dysfunction in Alzheimer disease. *Oxid. Med. Cell Longev.* **2013**, *2013*, 162152. [CrossRef] [PubMed]

63. Esteras, N.; Abramov, A.Y. Mitochondrial Calcium Deregulation in the Mechanism of Beta-Amyloid and Tau Pathology. *Cells* **2020**, *9*, 2135. [CrossRef] [PubMed]

64. Perez, M.J.; Ponce, D.P.; Aranguiz, A.; Behrens, M.I.; Quintanilla, R.A. Mitochondrial permeability transition pore contributes to mitochondrial dysfunction in fibroblasts of patients with sporadic Alzheimer's disease. *Redox Biol.* **2018**, *19*, 290–300. [CrossRef] [PubMed]

65. Jadiya, P.; Kolmetzky, D.W.; Tomar, D.; Di Meco, A.; Lombardi, A.A.; Lambert, J.P.; Luongo, T.S.; Ludtmann, M.H.; Pratico, D.; Elrod, J.W. Impaired mitochondrial calcium efflux contributes to disease progression in models of Alzheimer's disease. *Nat. Commun.* **2019**, *10*, 3885. [CrossRef] [PubMed]

66. Wolfe, D.M.; Lee, J.H.; Kumar, A.; Lee, S.; Orenstein, S.J.; Nixon, R.A. Autophagy failure in Alzheimer's disease and the role of defective lysosomal acidification. *Eur. J. Neurosci.* **2013**, *37*, 1949–1961. [CrossRef] [PubMed]

67. Lee, J.H.; Yu, W.H.; Kumar, A.; Lee, S.; Mohan, P.S.; Peterhoff, C.M.; Wolfe, D.M.; Martinez-Vicente, M.; Massey, A.C.; Sovak, G.; et al. Lysosomal proteolysis and autophagy require presenilin 1 and are disrupted by Alzheimer-related PS1 mutations. *Cell* **2010**, *141*, 1146–1158. [CrossRef]

68. Lee, J.H.; McBrayer, M.K.; Wolfe, D.M.; Haslett, L.J.; Kumar, A.; Sato, Y.; Lie, P.P.; Mohan, P.; Coffey, E.E.; Kompella, U.; et al. Presenilin 1 Maintains Lysosomal Ca(2+) Homeostasis via TRPML1 by Regulating vATPase-Mediated Lysosome Acidification. *Cell Rep.* **2015**, *12*, 1430–1444. [CrossRef]

69. Tu, H.; Nelson, O.; Bezprozvanny, A.; Wang, Z.; Lee, S.F.; Hao, Y.H.; Serneels, L.; De Strooper, B.; Yu, G.; Bezprozvanny, I. Presenilins form ER Ca2+ leak channels, a function disrupted by familial Alzheimer's disease-linked mutations. *Cell* **2006**, *126*, 981–993. [CrossRef]

70. Nelson, O.; Tu, H.; Lei, T.; Bentahir, M.; de Strooper, B.; Bezprozvanny, I. Familial Alzheimer disease-linked mutations specifically disrupt Ca2+ leak function of presenilin 1. *J. Clin. Invest.* **2007**, *117*, 1230–1239. [CrossRef]

71. Brunello, L.; Zampese, E.; Florean, C.; Pozzan, T.; Pizzo, P.; Fasolato, C. Presenilin-2 dampens intracellular Ca2+ stores by increasing Ca2+ leakage and reducing Ca2+ uptake. *J. Cell Mol. Med.* **2009**, *13*, 3358–3369. [CrossRef] [PubMed]

72. Shilling, D.; Muller, M.; Takano, H.; Mak, D.O.; Abel, T.; Coulter, D.A.; Foskett, J.K. Suppression of InsP3 receptor-mediated Ca2+ signaling alleviates mutant presenilin-linked familial Alzheimer's disease pathogenesis. *J. Neurosci.* **2014**, *34*, 6910–6923. [CrossRef] [PubMed]

73. Shilling, D.; Mak, D.O.; Kang, D.E.; Foskett, J.K. Lack of evidence for presenilins as endoplasmic reticulum Ca2+ leak channels. *J. Biol. Chem.* **2012**, *287*, 10933–10944. [CrossRef] [PubMed]

74. Green, K.N.; Demuro, A.; Akbari, Y.; Hitt, B.D.; Smith, I.F.; Parker, I.; LaFerla, F.M. SERCA pump activity is physiologically regulated by presenilin and regulates amyloid beta production. *J. Cell Biol.* **2008**, *181*, 1107–1116. [CrossRef]

75. Bussiere, R.; Oules, B.; Mary, A.; Vaillant-Beuchot, L.; Martin, C.; El Manaa, W.; Vallee, D.; Duplan, E.; Paterlini-Brechot, P.; Alves Da Costa, C.; et al. Upregulation of the Sarco-Endoplasmic Reticulum Calcium ATPase 1 Truncated Isoform Plays a Pathogenic Role in Alzheimer's Disease. *Cells* **2019**, *8*, 1539. [CrossRef]

76. Nelson, O.; Supnet, C.; Liu, H.; Bezprozvanny, I. Familial Alzheimer's disease mutations in presenilins: Effects on endoplasmic reticulum calcium homeostasis and correlation with clinical phenotypes. *J. Alzheimer's Dis.* **2010**, *21*, 781–793. [CrossRef]

77. Nelson, O.; Supnet, C.; Tolia, A.; Horre, K.; De Strooper, B.; Bezprozvanny, I. Mutagenesis mapping of the presenilin 1 calcium leak conductance pore. *J. Biol. Chem.* **2011**, *286*, 22339–22347. [CrossRef]

78. Zampese, E.; Fasolato, C.; Kipanyula, M.J.; Bortolozzi, M.; Pozzan, T.; Pizzo, P. Presenilin 2 modulates endoplasmic reticulum (ER)-mitochondria interactions and Ca2+ cross-talk. *Proc. Natl. Acad. Sci. USA* **2011**, *108*, 2777–2782. [CrossRef]

79. Stutzmann, G.E.; Caccamo, A.; LaFerla, F.M.; Parker, I. Dysregulated IP3 signaling in cortical neurons of knock-in mice expressing an Alzheimer's-linked mutation in presenilin1 results in exaggerated Ca2+ signals and altered membrane excitability. *J. Neurosci.* **2004**, *24*, 508–513. [CrossRef]

80. Stutzmann, G.E.; Smith, I.; Caccamo, A.; Oddo, S.; Parker, I.; Laferla, F. Enhanced ryanodine-mediated calcium release in mutant PS1-expressing Alzheimer's mouse models. *Ann. N.Y. Acad. Sci.* **2007**, *1097*, 265–277. [CrossRef]

81. Cheung, K.H.; Shineman, D.; Muller, M.; Cardenas, C.; Mei, L.; Yang, J.; Tomita, T.; Iwatsubo, T.; Lee, V.M.; Foskett, J.K. Mechanism of Ca2+ disruption in Alzheimer's disease by presenilin regulation of InsP(3) receptor channel gating. *Neuron* **2008**, *58*, 871–883. [CrossRef] [PubMed]

82. Querfurth, H.W.; Jiang, J.; Geiger, J.D.; Selkoe, D.J. Caffeine stimulates amyloid beta-peptide release from beta-amyloid precursor protein-transfected HEK293 cells. *J. Neurochem.* **1997**, *69*, 1580–1591. [CrossRef] [PubMed]

83. Chan, S.L.; Mayne, M.; Holden, C.P.; Geiger, J.D.; Mattson, M.P. Presenilin-1 mutations increase levels of ryanodine receptors and calcium release in PC12 cells and cortical neurons. *J. Biol. Chem.* **2000**, *275*, 18195–18200. [CrossRef] [PubMed]

84. Smith, I.F.; Hitt, B.; Green, K.N.; Oddo, S.; LaFerla, F.M. Enhanced caffeine-induced Ca2+ release in the 3xTg-AD mouse model of Alzheimer's disease. *J. Neurochem.* **2005**, *94*, 1711–1718. [CrossRef]

85. Stutzmann, G.E.; Smith, I.; Caccamo, A.; Oddo, S.; Laferla, F.M.; Parker, I. Enhanced ryanodine receptor recruitment contributes to Ca2+ disruptions in young, adult, and aged Alzheimer's disease mice. *J. Neurosci.* **2006**, *26*, 5180–5189. [CrossRef]

86. Supnet, C.; Grant, J.; Kong, H.; Westaway, D.; Mayne, M. Amyloid-beta-(1-42) increases ryanodine receptor-3 expression and function in neurons of TgCRND8 mice. *J. Biol. Chem.* **2006**, *281*, 38440–38447. [CrossRef]

87. Chakroborty, S.; Goussakov, I.; Miller, M.B.; Stutzmann, G.E. Deviant ryanodine receptor-mediated calcium release resets synaptic homeostasis in presymptomatic 3xTg-AD mice. *J. Neurosci.* **2009**, *29*, 9458–9470. [CrossRef]

88. Paula-Lima, A.C.; Adasme, T.; SanMartin, C.; Sebollela, A.; Hetz, C.; Carrasco, M.A.; Ferreira, S.T.; Hidalgo, C. Amyloid beta-peptide oligomers stimulate RyR-mediated Ca2+ release inducing mitochondrial fragmentation in hippocampal neurons and prevent RyR-mediated dendritic spine remodeling produced by BDNF. *Antioxid. Redox Signal.* **2011**, *14*, 1209–1223. [CrossRef]

89. Ito, E.; Oka, K.; Etcheberrigaray, R.; Nelson, T.J.; McPhie, D.L.; Tofel-Grehl, B.; Gibson, G.E.; Alkon, D.L. Internal Ca2+ mobilization is altered in fibroblasts from patients with Alzheimer disease. *Proc. Natl. Acad. Sci. USA* **1994**, *91*, 534–538. [CrossRef]

90. Goussakov, I.; Miller, M.B.; Stutzmann, G.E. NMDA-mediated Ca(2+) influx drives aberrant ryanodine receptor activation in dendrites of young Alzheimer's disease mice. *J. Neurosci.* **2010**, *30*, 12128–12137. [CrossRef]

91. Jensen, L.E.; Bultynck, G.; Luyten, T.; Amijee, H.; Bootman, M.D.; Roderick, H.L. Alzheimer's disease-associated peptide Abeta42 mobilizes ER Ca(2+) via InsP3R-dependent and -independent mechanisms. *Front. Mol. Neurosci.* **2013**, *6*, 36. [CrossRef] [PubMed]

92. Oules, B.; Del Prete, D.; Greco, B.; Zhang, X.; Lauritzen, I.; Sevalle, J.; Moreno, S.; Paterlini-Brechot, P.; Trebak, M.; Checler, F.; et al. Ryanodine receptor blockade reduces amyloid-beta load and memory impairments in Tg2576 mouse model of Alzheimer disease. *J. Neurosci.* **2012**, *32*, 11820–11834. [CrossRef] [PubMed]

93. Chakroborty, S.; Briggs, C.; Miller, M.B.; Goussakov, I.; Schneider, C.; Kim, J.; Wicks, J.; Richardson, J.C.; Conklin, V.; Cameransi, B.G.; et al. Stabilizing ER Ca2+ channel function as an early preventative strategy for Alzheimer's disease. *PLoS ONE* **2012**, *7*, e52056. [CrossRef] [PubMed]

94. Chakroborty, S.; Kim, J.; Schneider, C.; Jacobson, C.; Molgo, J.; Stutzmann, G.E. Early presynaptic and postsynaptic calcium signaling abnormalities mask underlying synaptic depression in presymptomatic Alzheimer's disease mice. *J. Neurosci.* **2012**, *32*, 8341–8353. [CrossRef] [PubMed]

95. Santos, S.F.; Pierrot, N.; Morel, N.; Gailly, P.; Sindic, C.; Octave, J.N. Expression of human amyloid precursor protein in rat cortical neurons inhibits calcium oscillations. *J. Neurosci.* **2009**, *29*, 4708–4718. [CrossRef] [PubMed]
96. Gazda, K.; Kuznicki, J.; Wegierski, T. Knockdown of amyloid precursor protein increases calcium levels in the endoplasmic reticulum. *Sci. Rep.* **2017**, *7*, 14512. [CrossRef]
97. Mahoney, R.; Ochoa Thomas, E.; Ramirez, P.; Miller, H.E.; Beckmann, A.; Zuniga, G.; Dobrowolski, R.; Frost, B. Pathogenic Tau Causes a Toxic Depletion of Nuclear Calcium. *Cell Rep.* **2020**, *32*, 107900. [CrossRef]
98. Esteras, N.; Kundel, F.; Amodeo, G.F.; Pavlov, E.V.; Klenerman, D.; Abramov, A.Y. Insoluble tau aggregates induce neuronal death through modification of membrane ion conductance, activation of voltage-gated calcium channels and NADPH oxidase. *FEBS J.* **2020**. [CrossRef]
99. Britti, E.; Ros, J.; Esteras, N.; Abramov, A.Y. Tau inhibits mitochondrial calcium efflux and makes neurons vulnerable to calcium-induced cell death. *Cell Calcium* **2020**, *86*, 102150. [CrossRef]
100. Furuichi, T.; Furutama, D.; Hakamata, Y.; Nakai, J.; Takeshima, H.; Mikoshiba, K. Multiple types of ryanodine receptor/Ca2+ release channels are differentially expressed in rabbit brain. *J. Neurosci.* **1994**, *14*, 4794–4805. [CrossRef]
101. Lehnart, S.E.; Mongillo, M.; Bellinger, A.; Lindegger, N.; Chen, B.X.; Hsueh, W.; Reiken, S.; Wronska, A.; Drew, L.J.; Ward, C.W.; et al. Leaky Ca2+ release channel/ryanodine receptor 2 causes seizures and sudden cardiac death in mice. *J. Clin. Investig.* **2008**, *118*, 2230–2245. [CrossRef] [PubMed]
102. Meissner, G.; Darling, E.; Eveleth, J. Kinetics of rapid Ca2+ release by sarcoplasmic reticulum. Effects of Ca2+, Mg2+, and adenine nucleotides. *Biochemistry* **1986**, *25*, 236–244. [CrossRef] [PubMed]
103. Meissner, G. The structural basis of ryanodine receptor ion channel function. *J. Gen. Physiol.* **2017**, *149*, 1065–1089. [CrossRef] [PubMed]
104. Laver, D.R.; Honen, B.N. Luminal Mg2+, a key factor controlling RYR2-mediated Ca2+ release: Cytoplasmic and luminal regulation modeled in a tetrameric channel. *J. Gen. Physiol.* **2008**, *132*, 429–446. [CrossRef] [PubMed]
105. Tencerova, B.; Zahradnikova, A.; Gaburjakova, J.; Gaburjakova, M. Luminal Ca2+ controls activation of the cardiac ryanodine receptor by ATP. *J. Gen. Physiol.* **2012**, *140*, 93–108. [CrossRef] [PubMed]
106. Yan, Z.; Bai, X.; Yan, C.; Wu, J.; Li, Z.; Xie, T.; Peng, W.; Yin, C.; Li, X.; Scheres, S.H.W.; et al. Structure of the rabbit ryanodine receptor RyR1 at near-atomic resolution. *Nature* **2015**, *517*, 50–55. [CrossRef]
107. Peng, W.; Shen, H.; Wu, J.; Guo, W.; Pan, X.; Wang, R.; Chen, S.R.; Yan, N. Structural basis for the gating mechanism of the type 2 ryanodine receptor RyR2. *Science* **2016**, *354*. [CrossRef]
108. Denniss, A.; Dulhunty, A.F.; Beard, N.A. Ryanodine receptor Ca(2+) release channel post-translational modification: Central player in cardiac and skeletal muscle disease. *Int. J. Biochem. Cell Biol.* **2018**, *101*, 49–53. [CrossRef]
109. Lanner, J.T.; Georgiou, D.K.; Joshi, A.D.; Hamilton, S.L. Ryanodine receptors: Structure, expression, molecular details, and function in calcium release. *Cold Spring Harb. Perspect. Biol.* **2010**, *2*, a003996. [CrossRef]
110. Balshaw, D.M.; Yamaguchi, N.; Meissner, G. Modulation of intracellular calcium-release channels by calmodulin. *J. Membr. Biol.* **2002**, *185*, 1–8. [CrossRef]
111. MacMillan, D. FK506 binding proteins: Cellular regulators of intracellular Ca2+ signalling. *Eur. J. Pharmacol.* **2013**, *700*, 181–193. [CrossRef] [PubMed]
112. Zalk, R.; Lehnart, S.E.; Marks, A.R. Modulation of the ryanodine receptor and intracellular calcium. *Annu. Rev. Biochem.* **2007**, *76*, 367–385. [CrossRef] [PubMed]
113. Marx, S.O.; Reiken, S.; Hisamatsu, Y.; Jayaraman, T.; Burkhoff, D.; Rosemblit, N.; Marks, A.R. PKA phosphorylation dissociates FKBP12.6 from the calcium release channel (ryanodine receptor): Defective regulation in failing hearts. *Cell* **2000**, *101*, 365–376. [CrossRef]
114. Currie, S. Cardiac ryanodine receptor phosphorylation by CaM Kinase II: Keeping the balance right. *Front. Biosci.* **2009**, *14*, 5134–5156. [CrossRef]
115. Allen, P.B.; Ouimet, C.C.; Greengard, P. Spinophilin, a novel protein phosphatase 1 binding protein localized to dendritic spines. *Proc. Natl. Acad. Sci. USA* **1997**, *94*, 9956–9961. [CrossRef]
116. Chami, M.; Checler, F. Targeting Post-Translational Remodeling of Ryanodine Receptor: A New Track for Alzheimer's Disease Therapy? *Curr. Alzheimer Res.* **2020**, *17*, 313–323. [CrossRef]

117. Lee, S.Y.; Hwang, D.Y.; Kim, Y.K.; Lee, J.W.; Shin, I.C.; Oh, K.W.; Lee, M.K.; Lim, J.S.; Yoon, D.Y.; Hwang, S.J.; et al. PS2 mutation increases neuronal cell vulnerability to neurotoxicants through activation of caspase-3 by enhancing of ryanodine receptor-mediated calcium release. *FASEB J.* **2006**, *20*, 151–153.
118. Guo, Q.; Sopher, B.L.; Furukawa, K.; Pham, D.G.; Robinson, N.; Martin, G.M.; Mattson, M.P. Alzheimer's presenilin mutation sensitizes neural cells to apoptosis induced by trophic factor withdrawal and amyloid beta-peptide: Involvement of calcium and oxyradicals. *J. Neurosci.* **1997**, *17*, 4212–4222. [CrossRef]
119. Leissring, M.A.; Murphy, M.P.; Mead, T.R.; Akbari, Y.; Sugarman, M.C.; Jannatipour, M.; Anliker, B.; Muller, U.; Saftig, P.; De Strooper, B.; et al. A physiologic signaling role for the gamma -secretase-derived intracellular fragment of APP. *Proc. Natl. Acad. Sci. USA* **2002**, *99*, 4697–4702. [CrossRef]
120. Lopez, J.R.; Lyckman, A.; Oddo, S.; Laferla, F.M.; Querfurth, H.W.; Shtifman, A. Increased intraneuronal resting [Ca(2+)] in adult Alzheimer's disease mice. *J. Neurochem.* **2008**, *105*, 262–271. [CrossRef]
121. Rojas, G.; Cardenas, A.M.; Fernandez-Olivares, P.; Shimahara, T.; Segura-Aguilar, J.; Caviedes, R.; Caviedes, P. Effect of the knockdown of amyloid precursor protein on intracellular calcium increases in a neuronal cell line derived from the cerebral cortex of a trisomy 16 mouse. *Exp. Neurol.* **2008**, *209*, 234–242. [CrossRef] [PubMed]
122. Niu, Y.; Su, Z.; Zhao, C.; Song, B.; Zhang, X.; Zhao, N.; Shen, X.; Gong, Y. Effect of amyloid beta on capacitive calcium entry in neural 2a cells. *Brain Res. Bull.* **2009**, *78*, 152–157. [CrossRef] [PubMed]
123. Paula-Lima, A.C.; Hidalgo, C. Amyloid beta-peptide oligomers, ryanodine receptor-mediated Ca(2+) release, and Wnt-5a/Ca(2+) signaling: Opposing roles in neuronal mitochondrial dynamics? *Front. Cell Neurosci.* **2013**, *7*, 120. [CrossRef] [PubMed]
124. Shtifman, A.; Ward, C.W.; Laver, D.R.; Bannister, M.L.; Lopez, J.R.; Kitazawa, M.; LaFerla, F.M.; Ikemoto, N.; Querfurth, H.W. Amyloid-beta protein impairs Ca2+ release and contractility in skeletal muscle. *Neurobiol. Aging* **2010**, *31*, 2080–2090. [CrossRef] [PubMed]
125. Kelliher, M.; Fastbom, J.; Cowburn, R.F.; Bonkale, W.; Ohm, T.G.; Ravid, R.; Sorrentino, V.; O'Neill, C. Alterations in the ryanodine receptor calcium release channel correlate with Alzheimer's disease neurofibrillary and beta-amyloid pathologies. *Neuroscience* **1999**, *92*, 499–513. [CrossRef]
126. Wu, B.; Yamaguchi, H.; Lai, F.A.; Shen, J. Presenilins regulate calcium homeostasis and presynaptic function via ryanodine receptors in hippocampal neurons. *Proc. Natl. Acad. Sci. USA* **2013**, *110*, 15091–15096. [CrossRef]
127. Kipanyula, M.J.; Contreras, L.; Zampese, E.; Lazzari, C.; Wong, A.K.; Pizzo, P.; Fasolato, C.; Pozzan, T. Ca2+ dysregulation in neurons from transgenic mice expressing mutant presenilin 2. *Aging Cell* **2012**, *11*, 885–893. [CrossRef]
128. Cisse, M.; Duplan, E.; Lorivel, T.; Dunys, J.; Bauer, C.; Meckler, X.; Gerakis, Y.; Lauritzen, I.; Checler, F. The transcription factor XBP1s restores hippocampal synaptic plasticity and memory by control of the Kalirin-7 pathway in Alzheimer model. *Mol. Psychiatry* **2017**, *22*, 1562–1575. [CrossRef]
129. Gerakis, Y.; Dunys, J.; Bauer, C.; Checler, F. Abeta42 oligomers modulate beta-secretase through an XBP-1s-dependent pathway involving HRD1. *Sci. Rep.* **2016**, *6*, 37436. [CrossRef]
130. Casas-Tinto, S.; Zhang, Y.; Sanchez-Garcia, J.; Gomez-Velazquez, M.; Rincon-Limas, D.E.; Fernandez-Funez, P. The ER stress factor XBP1s prevents amyloid-beta neurotoxicity. *Hum. Mol. Genet.* **2011**, *20*, 2144–2160. [CrossRef]
131. Liu, J.; Supnet, C.; Sun, S.; Zhang, H.; Good, L.; Popugaeva, E.; Bezprozvanny, I. The role of ryanodine receptor type 3 in a mouse model of Alzheimer disease. *Channels* **2014**, *8*, 230–242. [CrossRef] [PubMed]
132. Mickelson, J.R.; Gallant, E.M.; Litterer, L.A.; Johnson, K.M.; Rempel, W.E.; Louis, C.F. Abnormal sarcoplasmic reticulum ryanodine receptor in malignant hyperthermia. *J. Biol. Chem.* **1988**, *263*, 9310–9315. [PubMed]
133. Kushnir, A.; Wajsberg, B.; Marks, A.R. Ryanodine receptor dysfunction in human disorders. *Biochim. Biophys. Acta Mol. Cell Res.* **2018**, *1865*, 1687–1697. [CrossRef] [PubMed]
134. Kushnir, A.; Betzenhauser, M.J.; Marks, A.R. Ryanodine receptor studies using genetically engineered mice. *FEBS Lett.* **2010**, *584*, 1956–1965. [CrossRef]
135. Takeshima, H.; Iino, M.; Takekura, H.; Nishi, M.; Kuno, J.; Minowa, O.; Takano, H.; Noda, T. Excitation-contraction uncoupling and muscular degeneration in mice lacking functional skeletal muscle ryanodine-receptor gene. *Nature* **1994**, *369*, 556–559. [CrossRef]

136. Takeshima, H.; Ikemoto, T.; Nishi, M.; Nishiyama, N.; Shimuta, M.; Sugitani, Y.; Kuno, J.; Saito, I.; Saito, H.; Endo, M.; et al. Generation and characterization of mutant mice lacking ryanodine receptor type 3. *J. Biol. Chem.* **1996**, *271*, 19649–19652. [CrossRef]

137. Matsuo, N.; Tanda, K.; Nakanishi, K.; Yamasaki, N.; Toyama, K.; Takao, K.; Takeshima, H.; Miyakawa, T. Comprehensive behavioral phenotyping of ryanodine receptor type 3 (RyR3) knockout mice: Decreased social contact duration in two social interaction tests. *Front. Behav. Neurosci.* **2009**, *3*, 3. [CrossRef]

138. Balschun, D.; Wolfer, D.P.; Bertocchini, F.; Barone, V.; Conti, A.; Zuschratter, W.; Missiaen, L.; Lipp, H.P.; Frey, J.U.; Sorrentino, V. Deletion of the ryanodine receptor type 3 (RyR3) impairs forms of synaptic plasticity and spatial learning. *EMBO J.* **1999**, *18*, 5264–5273. [CrossRef]

139. Futatsugi, A.; Kato, K.; Ogura, H.; Li, S.T.; Nagata, E.; Kuwajima, G.; Tanaka, K.; Itohara, S.; Mikoshiba, K. Facilitation of NMDAR-independent LTP and spatial learning in mutant mice lacking ryanodine receptor type 3. *Neuron* **1999**, *24*, 701–713. [CrossRef]

140. Shimuta, M.; Yoshikawa, M.; Fukaya, M.; Watanabe, M.; Takeshima, H.; Manabe, T. Postsynaptic modulation of AMPA receptor-mediated synaptic responses and LTP by the type 3 ryanodine receptor. *Mol. Cell Neurosci.* **2001**, *17*, 921–930. [CrossRef]

141. Gong, S.; Su, B.B.; Tovar, H.; Mao, C.; Gonzalez, V.; Liu, Y.; Lu, Y.; Wang, K.S.; Xu, C. Polymorphisms Within RYR3 Gene Are Associated With Risk and Age at Onset of Hypertension, Diabetes, and Alzheimer's Disease. *Am. J. Hypertens.* **2018**, *31*, 818–826. [CrossRef] [PubMed]

142. Sun, J.; Song, F.; Wang, J.; Han, G.; Bai, Z.; Xie, B.; Feng, X.; Jia, J.; Duan, Y.; Lei, H. Hidden risk genes with high-order intragenic epistasis in Alzheimer's disease. *J. Alzheimer's Dis.* **2014**, *41*, 1039–1056. [CrossRef] [PubMed]

143. Koran, M.E.; Hohman, T.J.; Thornton-Wells, T.A. Genetic interactions found between calcium channel genes modulate amyloid load measured by positron emission tomography. *Hum. Genet.* **2014**, *133*, 85–93. [CrossRef] [PubMed]

144. Wehrens, X.H.; Lehnart, S.E.; Reiken, S.R.; Deng, S.X.; Vest, J.A.; Cervantes, D.; Coromilas, J.; Landry, D.W.; Marks, A.R. Protection from cardiac arrhythmia through ryanodine receptor-stabilizing protein calstabin2. *Science* **2004**, *304*, 292–296. [CrossRef]

145. Andersson, D.C.; Betzenhauser, M.J.; Reiken, S.; Meli, A.C.; Umanskaya, A.; Xie, W.; Shiomi, T.; Zalk, R.; Lacampagne, A.; Marks, A.R. Ryanodine receptor oxidation causes intracellular calcium leak and muscle weakness in aging. *Cell Metab.* **2011**, *14*, 196–207. [CrossRef] [PubMed]

146. Bellinger, A.M.; Reiken, S.; Carlson, C.; Mongillo, M.; Liu, X.; Rothman, L.; Matecki, S.; Lacampagne, A.; Marks, A.R. Hypernitrosylated ryanodine receptor calcium release channels are leaky in dystrophic muscle. *Nat. Med.* **2009**, *15*, 325–330. [CrossRef]

147. Bellinger, A.M.; Reiken, S.; Dura, M.; Murphy, P.W.; Deng, S.X.; Landry, D.W.; Nieman, D.; Lehnart, S.E.; Samaru, M.; LaCampagne, A.; et al. Remodeling of ryanodine receptor complex causes "leaky" channels: A molecular mechanism for decreased exercise capacity. *Proc. Natl. Acad. Sci. USA* **2008**, *105*, 2198–2202. [CrossRef]

148. Fauconnier, J.; Meli, A.C.; Thireau, J.; Roberge, S.; Shan, J.; Sassi, Y.; Reiken, S.R.; Rauzier, J.M.; Marchand, A.; Chauvier, D.; et al. Ryanodine receptor leak mediated by caspase-8 activation leads to left ventricular injury after myocardial ischemia-reperfusion. *Proc. Natl. Acad. Sci. USA* **2011**, *108*, 13258–13263. [CrossRef]

149. Liu, X.; Betzenhauser, M.J.; Reiken, S.; Meli, A.C.; Xie, W.; Chen, B.X.; Arancio, O.; Marks, A.R. Role of leaky neuronal ryanodine receptors in stress-induced cognitive dysfunction. *Cell* **2012**, *150*, 1055–1067. [CrossRef]

150. Lacampagne, A.; Liu, X.; Reiken, S.; Bussiere, R.; Meli, A.C.; Lauritzen, I.; Teich, A.F.; Zalk, R.; Saint, N.; Arancio, O.; et al. Post-translational remodeling of ryanodine receptor induces calcium leak leading to Alzheimer's disease-like pathologies and cognitive deficits. *Acta Neuropathol.* **2017**, *134*, 749–767. [CrossRef]

151. Petrotchenko, E.V.; Yamaguchi, N.; Pasek, D.A.; Borchers, C.H.; Meissner, G. Mass spectrometric analysis and mutagenesis predict involvement of multiple cysteines in redox regulation of the skeletal muscle ryanodine receptor ion channel complex. *Res. Rep. Biol.* **2011**, *2011*, 13–21. [CrossRef] [PubMed]

152. Sun, J.; Yamaguchi, N.; Xu, L.; Eu, J.P.; Stamler, J.S.; Meissner, G. Regulation of the cardiac muscle ryanodine receptor by O(2) tension and S-nitrosoglutathione. *Biochemistry* **2008**, *47*, 13985–13990. [CrossRef] [PubMed]

153. Bussiere, R.; Lacampagne, A.; Reiken, S.; Liu, X.; Scheuerman, V.; Zalk, R.; Martin, C.; Checler, F.; Marks, A.R.; Chami, M. Amyloid beta production is regulated by beta2-adrenergic signaling-mediated post-translational modifications of the ryanodine receptor. *J. Biol. Chem.* **2017**, *292*, 10153–10168. [CrossRef]

154. Hidalgo, C.; Carrasco, M.A. Redox control of brain calcium in health and disease. *Antioxid. Redox Signal.* **2011**, *14*, 1203–1207. [CrossRef] [PubMed]

155. Von Bernhardi, R.; Eugenin, J. Alzheimer's disease: Redox dysregulation as a common denominator for diverse pathogenic mechanisms. *Antioxid. Redox Signal.* **2012**, *16*, 974–1031. [CrossRef]

156. Echeverria, V.; Ducatenzeiler, A.; Chen, C.H.; Cuello, A.C. Endogenous beta-amyloid peptide synthesis modulates cAMP response element-regulated gene expression in PC12 cells. *Neuroscience* **2005**, *135*, 1193–1202. [CrossRef] [PubMed]

157. Igbavboa, U.; Johnson-Anuna, L.N.; Rossello, X.; Butterick, T.A.; Sun, G.Y.; Wood, W.G. Amyloid beta-protein1-42 increases cAMP and apolipoprotein E levels which are inhibited by beta1 and beta2-adrenergic receptor antagonists in mouse primary astrocytes. *Neuroscience* **2006**, *142*, 655–660. [CrossRef] [PubMed]

158. Prapong, T.; Uemura, E.; Hsu, W.H. G protein and cAMP-dependent protein kinase mediate amyloid beta-peptide inhibition of neuronal glucose uptake. *Exp. Neurol.* **2001**, *167*, 59–64. [CrossRef]

159. Palavicini, J.P.; Wang, H.; Bianchi, E.; Xu, S.; Rao, J.S.; Kang, D.E.; Lakshmana, M.K. RanBP9 aggravates synaptic damage in the mouse brain and is inversely correlated to spinophilin levels in Alzheimer's brain synaptosomes. *Cell Death Dis.* **2013**, *4*, e667. [CrossRef]

160. Marambaud, P.; Ancolio, K.; Alves da Costa, C.; Checler, F. Effect of protein kinase A inhibitors on the production of Abeta40 and Abeta42 by human cells expressing normal and Alzheimer's disease-linked mutated betaAPP and presenilin 1. *Br. J. Pharmacol.* **1999**, *126*, 1186–1190. [CrossRef]

161. Bekris, L.M.; Yu, C.E.; Bird, T.D.; Tsuang, D.W. Genetics of Alzheimer disease. *J. Geriatr. Psychiatry Neurol.* **2010**, *23*, 213–227. [CrossRef] [PubMed]

162. Branca, C.; Wisely, E.V.; Hartman, L.K.; Caccamo, A.; Oddo, S. Administration of a selective beta2 adrenergic receptor antagonist exacerbates neuropathology and cognitive deficits in a mouse model of Alzheimer's disease. *Neurobiol. Aging* **2014**, *35*, 2726–2735. [CrossRef] [PubMed]

163. Dobarro, M.; Gerenu, G.; Ramirez, M.J. Propranolol reduces cognitive deficits, amyloid and tau pathology in Alzheimer's transgenic mice. *Int. J. Neuropsychopharmacol.* **2013**, *16*, 2245–2257. [CrossRef] [PubMed]

164. Dang, V.; Medina, B.; Das, D.; Moghadam, S.; Martin, K.J.; Lin, B.; Naik, P.; Patel, D.; Nosheny, R.; Wesson Ashford, J.; et al. Formoterol, a long-acting beta2 adrenergic agonist, improves cognitive function and promotes dendritic complexity in a mouse model of Down syndrome. *Biol. Psychiatry* **2014**, *75*, 179–188. [CrossRef]

165. Peng, J.; Liang, G.; Inan, S.; Wu, Z.; Joseph, D.J.; Meng, Q.; Peng, Y.; Eckenhoff, M.F.; Wei, H. Dantrolene ameliorates cognitive decline and neuropathology in Alzheimer triple transgenic mice. *Neurosci. Lett.* **2012**, *516*, 274–279. [CrossRef]

166. Mikoshiba, K. IP3 receptor/Ca2+ channel: From discovery to new signaling concepts. *J. Neurochem.* **2007**, *102*, 1426–1446. [CrossRef]

167. Ross, C.A.; Meldolesi, J.; Milner, T.A.; Satoh, T.; Supattapone, S.; Snyder, S.H. Inositol 1,4,5-trisphosphate receptor localized to endoplasmic reticulum in cerebellar Purkinje neurons. *Nature* **1989**, *339*, 468–470. [CrossRef]

168. Furuichi, T.; Yoshikawa, S.; Miyawaki, A.; Wada, K.; Maeda, N.; Mikoshiba, K. Primary structure and functional expression of the inositol 1,4,5-trisphosphate-binding protein P400. *Nature* **1989**, *342*, 32–38. [CrossRef]

169. Mignery, G.A.; Sudhof, T.C.; Takei, K.; De Camilli, P. Putative receptor for inositol 1,4,5-trisphosphate similar to ryanodine receptor. *Nature* **1989**, *342*, 192–195. [CrossRef]

170. Foskett, J.K.; White, C.; Cheung, K.H.; Mak, D.O. Inositol trisphosphate receptor Ca2+ release channels. *Physiol. Rev.* **2007**, *87*, 593–658. [CrossRef]

171. Inoue, T.; Kato, K.; Kohda, K.; Mikoshiba, K. Type 1 inositol 1,4,5-trisphosphate receptor is required for induction of long-term depression in cerebellar Purkinje neurons. *J. Neurosci.* **1998**, *18*, 5366–5373. [CrossRef] [PubMed]

172. Baker, K.D.; Edwards, T.M.; Rickard, N.S. The role of intracellular calcium stores in synaptic plasticity and memory consolidation. *Neurosci. Biobehav. Rev.* **2013**, *37*, 1211–1239. [CrossRef] [PubMed]

173. Sugawara, T.; Hisatsune, C.; Le, T.D.; Hashikawa, T.; Hirono, M.; Hattori, M.; Nagao, S.; Mikoshiba, K. Type 1 inositol trisphosphate receptor regulates cerebellar circuits by maintaining the spine morphology of purkinje cells in adult mice. *J. Neurosci.* **2013**, *33*, 12186–12196. [CrossRef] [PubMed]

174. Egorova, P.A.; Bezprozvanny, I.B. Inositol 1,4,5-trisphosphate receptors and neurodegenerative disorders. *FEBS J.* **2018**, *285*, 3547–3565. [CrossRef]

175. Leissring, M.A.; Paul, B.A.; Parker, I.; Cotman, C.W.; LaFerla, F.M. Alzheimer's presenilin-1 mutation potentiates inositol 1,4,5-trisphosphate-mediated calcium signaling in Xenopus oocytes. *J. Neurochem.* **1999**, *72*, 1061–1068. [CrossRef]

176. Cedazo-Minguez, A.; Popescu, B.O.; Ankarcrona, M.; Nishimura, T.; Cowburn, R.F. The presenilin 1 deltaE9 mutation gives enhanced basal phospholipase C activity and a resultant increase in intracellular calcium concentrations. *J. Biol. Chem.* **2002**, *277*, 36646–36655. [CrossRef]

177. Bezprozvanny, I.; Supnet, C.; Sun, S.; Zhang, H.; De Strooper, B. Response to Shilling et al. (10.1074/jbc.M111.300491). *J. Biol. Chem.* **2012**, *287*, 20469. [CrossRef]

178. Mak, D.O.; Cheung, K.H.; Toglia, P.; Foskett, J.K.; Ullah, G. Analyzing and Quantifying the Gain-of-Function Enhancement of IP3 Receptor Gating by Familial Alzheimer's Disease-Causing Mutants in Presenilins. *PLoS Comput. Biol.* **2015**, *11*, e1004529. [CrossRef]

179. Toglia, P.; Ullah, G. The gain-of-function enhancement of IP3-receptor channel gating by familial Alzheimer's disease-linked presenilin mutants increases the open probability of mitochondrial permeability transition pore. *Cell Calcium* **2016**, *60*, 13–24. [CrossRef]

180. Wang, Z.J.; Zhao, F.; Wang, C.F.; Zhang, X.M.; Xiao, Y.; Zhou, F.; Wu, M.N.; Zhang, J.; Qi, J.S.; Yang, W. Xestospongin C, a Reversible IP3 Receptor Antagonist, Alleviates the Cognitive and Pathological Impairments in APP/PS1 Mice of Alzheimer's Disease. *J. Alzheimer's Dis.* **2019**, *72*, 1217–1231. [CrossRef]

181. Ronco, V.; Grolla, A.A.; Glasnov, T.N.; Canonico, P.L.; Verkhratsky, A.; Genazzani, A.A.; Lim, D. Differential deregulation of astrocytic calcium signalling by amyloid-beta, TNFalpha, IL-1beta and LPS. *Cell Calcium* **2014**, *55*, 219–229. [CrossRef] [PubMed]

182. Grolla, A.A.; Sim, J.A.; Lim, D.; Rodriguez, J.J.; Genazzani, A.A.; Verkhratsky, A. Amyloid-beta and Alzheimer's disease type pathology differentially affects the calcium signalling toolkit in astrocytes from different brain regions. *Cell Death Dis.* **2013**, *4*, e623. [CrossRef] [PubMed]

183. De Strooper, B.; Beullens, M.; Contreras, B.; Levesque, L.; Craessaerts, K.; Cordell, B.; Moechars, D.; Bollen, M.; Fraser, P.; George-Hyslop, P.S.; et al. Phosphorylation, subcellular localization, and membrane orientation of the Alzheimer's disease-associated presenilins. *J. Biol. Chem.* **1997**, *272*, 3590–3598. [CrossRef] [PubMed]

184. Sherrington, R.; Rogaev, E.I.; Liang, Y.; Rogaeva, E.A.; Levesque, G.; Ikeda, M.; Chi, H.; Lin, C.; Li, G.; Holman, K.; et al. Cloning of a gene bearing missense mutations in early-onset familial Alzheimer's disease. *Nature* **1995**, *375*, 754–760. [CrossRef]

185. Rogaev, E.I.; Sherrington, R.; Rogaeva, E.A.; Levesque, G.; Ikeda, M.; Liang, Y.; Chi, H.; Lin, C.; Holman, K.; Tsuda, T.; et al. Familial Alzheimer's disease in kindreds with missense mutations in a gene on chromosome 1 related to the Alzheimer's disease type 3 gene. *Nature* **1995**, *376*, 775–778. [CrossRef]

186. Levy-Lahad, E.; Wasco, W.; Poorkaj, P.; Romano, D.M.; Oshima, J.; Pettingell, W.H.; Yu, C.E.; Jondro, P.D.; Schmidt, S.D.; Wang, K.; et al. Candidate gene for the chromosome 1 familial Alzheimer's disease locus. *Science* **1995**, *269*, 973–977. [CrossRef]

187. Lu, P.; Bai, X.C.; Ma, D.; Xie, T.; Yan, C.; Sun, L.; Yang, G.; Zhao, Y.; Zhou, R.; Scheres, S.H.W.; et al. Three-dimensional structure of human gamma-secretase. *Nature* **2014**, *512*, 166–170. [CrossRef]

188. Bai, X.C.; Yan, C.; Yang, G.; Lu, P.; Ma, D.; Sun, L.; Zhou, R.; Scheres, S.H.W.; Shi, Y. An atomic structure of human gamma-secretase. *Nature* **2015**, *525*, 212–217. [CrossRef]

189. Yang, G.; Zhou, R.; Shi, Y. Cryo-EM structures of human gamma-secretase. *Curr. Opin. Struct. Biol.* **2017**, *46*, 55–64. [CrossRef]

190. Lee, M.K.; Slunt, H.H.; Martin, L.J.; Thinakaran, G.; Kim, G.; Gandy, S.E.; Seeger, M.; Koo, E.; Price, D.L.; Sisodia, S.S. Expression of presenilin 1 and 2 (PS1 and PS2) in human and murine tissues. *J. Neurosci.* **1996**, *16*, 7513–7525. [CrossRef]

191. Wong, P.C.; Zheng, H.; Chen, H.; Becher, M.W.; Sirinathsinghji, D.J.; Trumbauer, M.E.; Chen, H.Y.; Price, D.L.; Van der Ploeg, L.H.; Sisodia, S.S. Presenilin 1 is required for Notch1 and DII1 expression in the paraxial mesoderm. *Nature* **1997**, *387*, 288–292. [CrossRef] [PubMed]

192. Donoviel, D.B.; Hadjantonakis, A.K.; Ikeda, M.; Zheng, H.; Hyslop, P.S.; Bernstein, A. Mice lacking both presenilin genes exhibit early embryonic patterning defects. *Genes Dev.* **1999**, *13*, 2801–2810. [CrossRef] [PubMed]

193. Beglopoulos, V.; Sun, X.; Saura, C.A.; Lemere, C.A.; Kim, R.D.; Shen, J. Reduced beta-amyloid production and increased inflammatory responses in presenilin conditional knock-out mice. *J. Biol. Chem.* **2004**, *279*, 46907–46914. [CrossRef] [PubMed]

194. Zhang, C.; Wu, B.; Beglopoulos, V.; Wines-Samuelson, M.; Zhang, D.; Dragatsis, I.; Sudhof, T.C.; Shen, J. Presenilins are essential for regulating neurotransmitter release. *Nature* **2009**, *460*, 632–636. [CrossRef]

195. Wolfe, M.S.; Xia, W.; Ostaszewski, B.L.; Diehl, T.S.; Kimberly, W.T.; Selkoe, D.J. Two transmembrane aspartates in presenilin-1 required for presenilin endoproteolysis and gamma-secretase activity. *Nature* **1999**, *398*, 513–517. [CrossRef]

196. Borchelt, D.R.; Thinakaran, G.; Eckman, C.B.; Lee, M.K.; Davenport, F.; Ratovitsky, T.; Prada, C.M.; Kim, G.; Seekins, S.; Yager, D.; et al. Familial Alzheimer's disease-linked presenilin 1 variants elevate Abeta1-42/1-40 ratio in vitro and in vivo. *Neuron* **1996**, *17*, 1005–1013. [CrossRef]

197. Tomita, T.; Maruyama, K.; Saido, T.C.; Kume, H.; Shinozaki, K.; Tokuhiro, S.; Capell, A.; Walter, J.; Grunberg, J.; Haass, C.; et al. The presenilin 2 mutation (N141I) linked to familial Alzheimer disease (Volga German families) increases the secretion of amyloid beta protein ending at the 42nd (or 43rd) residue. *Proc. Natl. Acad. Sci. USA* **1997**, *94*, 2025–2030. [CrossRef]

198. Scheuner, D.; Eckman, C.; Jensen, M.; Song, X.; Citron, M.; Suzuki, N.; Bird, T.D.; Hardy, J.; Hutton, M.; Kukull, W.; et al. Secreted amyloid beta-protein similar to that in the senile plaques of Alzheimer's disease is increased in vivo by the presenilin 1 and 2 and APP mutations linked to familial Alzheimer's disease. *Nat. Med.* **1996**, *2*, 864–870. [CrossRef]

199. Mehra, R.; Kepp, K.P. Identification of Structural Calcium Binding Sites in Membrane-Bound Presenilin 1 and 2. *J. Phys. Chem. B* **2020**, *124*, 4697–4711. [CrossRef]

200. Thinakaran, G.; Borchelt, D.R.; Lee, M.K.; Slunt, H.H.; Spitzer, L.; Kim, G.; Ratovitsky, T.; Davenport, F.; Nordstedt, C.; Seeger, M.; et al. Endoproteolysis of presenilin 1 and accumulation of processed derivatives in vivo. *Neuron* **1996**, *17*, 181–190. [CrossRef]

201. Takeda, T.; Asahi, M.; Yamaguchi, O.; Hikoso, S.; Nakayama, H.; Kusakari, Y.; Kawai, M.; Hongo, K.; Higuchi, Y.; Kashiwase, K.; et al. Presenilin 2 regulates the systolic function of heart by modulating Ca2+ signaling. *FASEB J.* **2005**, *19*, 2069–2071. [CrossRef] [PubMed]

202. Pack-Chung, E.; Meyers, M.B.; Pettingell, W.P.; Moir, R.D.; Brownawell, A.M.; Cheng, I.; Tanzi, R.E.; Kim, T.W. Presenilin 2 interacts with sorcin, a modulator of the ryanodine receptor. *J. Biol. Chem.* **2000**, *275*, 14440–14445. [CrossRef] [PubMed]

203. Hayrapetyan, V.; Rybalchenko, V.; Rybalchenko, N.; Koulen, P. The N-terminus of presenilin-2 increases single channel activity of brain ryanodine receptors through direct protein-protein interaction. *Cell Calcium* **2008**, *44*, 507–518. [CrossRef] [PubMed]

204. Rybalchenko, V.; Hwang, S.Y.; Rybalchenko, N.; Koulen, P. The cytosolic N-terminus of presenilin-1 potentiates mouse ryanodine receptor single channel activity. *Int. J. Biochem. Cell Biol.* **2008**, *40*, 84–97. [CrossRef] [PubMed]

205. Greotti, E.; Capitanio, P.; Wong, A.; Pozzan, T.; Pizzo, P.; Pendin, D. Familial Alzheimer's disease-linked presenilin mutants and intracellular Ca(2+) handling: A single-organelle, FRET-based analysis. *Cell Calcium* **2019**, *79*, 44–56. [CrossRef] [PubMed]

206. Baba-Aissa, F.; Raeymaekers, L.; Wuytack, F.; Dode, L.; Casteels, R. Distribution and isoform diversity of the organellar Ca2+ pumps in the brain. *Mol. Chem. Neuropathol.* **1998**, *33*, 199–208. [CrossRef] [PubMed]

207. Britzolaki, A.; Saurine, J.; Klocke, B.; Pitychoutis, P.M. A Role for SERCA Pumps in the Neurobiology of Neuropsychiatric and Neurodegenerative Disorders. *Adv. Exp. Med. Biol.* **2020**, *1131*, 131–161. [CrossRef]

208. Jin, H.; Sanjo, N.; Uchihara, T.; Watabe, K.; St George-Hyslop, P.; Fraser, P.E.; Mizusawa, H. Presenilin-1 holoprotein is an interacting partner of sarco endoplasmic reticulum calcium-ATPase and confers resistance to endoplasmic reticulum stress. *J. Alzheimer's Dis.* **2010**, *20*, 261–273. [CrossRef]

209. Krajnak, K.; Dahl, R. A new target for Alzheimer's disease: A small molecule SERCA activator is neuroprotective in vitro and improves memory and cognition in APP/PS1 mice. *Bioorg. Med. Chem. Lett.* **2018**, *28*, 1591–1594. [CrossRef]

210. Zhao, L.; Ackerman, S.L. Endoplasmic reticulum stress in health and disease. *Curr. Opin. Cell Biol.* **2006**, *18*, 444–452. [CrossRef]

211. Hetz, C.; Papa, F.R. The Unfolded Protein Response and Cell Fate Control. *Mol. Cell* **2018**, *69*, 169–181. [CrossRef] [PubMed]

212. Hoozemans, J.J.; Veerhuis, R.; Van Haastert, E.S.; Rozemuller, J.M.; Baas, F.; Eikelenboom, P.; Scheper, W. The unfolded protein response is activated in Alzheimer's disease. *Acta Neuropathol.* **2005**, *110*, 165–172. [CrossRef] [PubMed]

213. Hoozemans, J.J.; van Haastert, E.S.; Nijholt, D.A.; Rozemuller, A.J.; Eikelenboom, P.; Scheper, W. The unfolded protein response is activated in pretangle neurons in Alzheimer's disease hippocampus. *Am. J. Pathol.* **2009**, *174*, 1241–1251. [CrossRef] [PubMed]

214. Chang, R.C.; Suen, K.C.; Ma, C.H.; Elyaman, W.; Ng, H.K.; Hugon, J. Involvement of double-stranded RNA-dependent protein kinase and phosphorylation of eukaryotic initiation factor-2alpha in neuronal degeneration. *J. Neurochem.* **2002**, *83*, 1215–1225. [CrossRef]

215. Page, G.; Rioux Bilan, A.; Ingrand, S.; Lafay-Chebassier, C.; Pain, S.; Perault Pochat, M.C.; Bouras, C.; Bayer, T.; Hugon, J. Activated double-stranded RNA-dependent protein kinase and neuronal death in models of Alzheimer's disease. *Neuroscience* **2006**, *139*, 1343–1354. [CrossRef]

216. Kim, H.S.; Choi, Y.; Shin, K.Y.; Joo, Y.; Lee, Y.K.; Jung, S.Y.; Suh, Y.H.; Kim, J.H. Swedish amyloid precursor protein mutation increases phosphorylation of eIF2alpha in vitro and in vivo. *J. Neurosci. Res.* **2007**, *85*, 1528–1537. [CrossRef]

217. Chami, M.; Gozuacik, D.; Lagorce, D.; Brini, M.; Falson, P.; Peaucellier, G.; Pinton, P.; Lecoeur, H.; Gougeon, M.L.; le Maire, M.; et al. SERCA1 truncated proteins unable to pump calcium reduce the endoplasmic reticulum calcium concentration and induce apoptosis. *J. Cell Biol.* **2001**, *153*, 1301–1314. [CrossRef]

218. Chami, M.; Oules, B.; Szabadkai, G.; Tacine, R.; Rizzuto, R.; Paterlini-Brechot, P. Role of SERCA1 truncated isoform in the proapoptotic calcium transfer from ER to mitochondria during ER stress. *Mol. Cell* **2008**, *32*, 641–651. [CrossRef]

219. Nishitsuji, K.; Tomiyama, T.; Ishibashi, K.; Ito, K.; Teraoka, R.; Lambert, M.P.; Klein, W.L.; Mori, H. The E693Delta mutation in amyloid precursor protein increases intracellular accumulation of amyloid beta oligomers and causes endoplasmic reticulum stress-induced apoptosis in cultured cells. *Am. J. Pathol.* **2009**, *174*, 957–969. [CrossRef]

220. O'Connor, T.; Sadleir, K.R.; Maus, E.; Velliquette, R.A.; Zhao, J.; Cole, S.L.; Eimer, W.A.; Hitt, B.; Bembinster, L.A.; Lammich, S.; et al. Phosphorylation of the translation initiation factor eIF2alpha increases BACE1 levels and promotes amyloidogenesis. *Neuron* **2008**, *60*, 988–1009. [CrossRef]

221. Devi, L.; Ohno, M. PERK mediates eIF2alpha phosphorylation responsible for BACE1 elevation, CREB dysfunction and neurodegeneration in a mouse model of Alzheimer's disease. *Neurobiol. Aging* **2014**, *35*, 2272–2281. [CrossRef] [PubMed]

222. Chami, L.; Checler, F. BACE1 is at the crossroad of a toxic vicious cycle involving cellular stress and beta-amyloid production in Alzheimer's disease. *Mol. Neurodegener.* **2012**, *7*, 52. [CrossRef] [PubMed]

223. Salminen, A.; Kauppinen, A.; Suuronen, T.; Kaarniranta, K.; Ojala, J. ER stress in Alzheimer's disease: A novel neuronal trigger for inflammation and Alzheimer's pathology. *J. Neuroinflamm.* **2009**, *6*, 41. [CrossRef] [PubMed]

224. Popugaeva, E.; Bezprozvanny, I. STIM proteins as regulators of neuronal store-operated calcium influx. *Neurodegener. Dis. Manag.* **2018**, *8*, 5–7. [CrossRef] [PubMed]

225. Secondo, A.; Bagetta, G.; Amantea, D. On the Role of Store-Operated Calcium Entry in Acute and Chronic Neurodegenerative Diseases. *Front. Mol. Neurosci.* **2018**, *11*, 87. [CrossRef]

226. Kraft, R. STIM and ORAI proteins in the nervous system. *Channels* **2015**, *9*, 245–252. [CrossRef]

227. Luik, R.M.; Wu, M.M.; Buchanan, J.; Lewis, R.S. The elementary unit of store-operated Ca2+ entry: Local activation of CRAC channels by STIM1 at ER-plasma membrane junctions. *J. Cell Biol.* **2006**, *174*, 815–825. [CrossRef]

228. Ambudkar, I.S.; Ong, H.L.; Liu, X.; Bandyopadhyay, B.C.; Cheng, K.T. TRPC1: The link between functionally distinct store-operated calcium channels. *Cell Calcium* **2007**, *42*, 213–223. [CrossRef]

229. Chen, X.; Sooch, G.; Demaree, I.S.; White, F.A.; Obukhov, A.G. Transient Receptor Potential Canonical (TRPC) Channels: Then and Now. *Cells* **2020**, *9*, 1983. [CrossRef]

230. Moccia, F.; Zuccolo, E.; Soda, T.; Tanzi, F.; Guerra, G.; Mapelli, L.; Lodola, F.; D'Angelo, E. Stim and Orai proteins in neuronal Ca(2+) signaling and excitability. *Front. Cell Neurosci.* **2015**, *9*, 153. [CrossRef]

231. Wegierski, T.; Kuznicki, J. Neuronal calcium signaling via store-operated channels in health and disease. *Cell Calcium* **2018**, *74*, 102–111. [CrossRef] [PubMed]

232. Heine, M.; Heck, J.; Ciuraszkiewicz, A.; Bikbaev, A. Dynamic compartmentalization of calcium channel signalling in neurons. *Neuropharmacology* **2020**, *169*, 107556. [CrossRef] [PubMed]
233. Sun, S.; Zhang, H.; Liu, J.; Popugaeva, E.; Xu, N.J.; Feske, S.; White, C.L., 3rd; Bezprozvanny, I. Reduced synaptic STIM2 expression and impaired store-operated calcium entry cause destabilization of mature spines in mutant presenilin mice. *Neuron* **2014**, *82*, 79–93. [CrossRef] [PubMed]
234. Tong, B.C.; Lee, C.S.; Cheng, W.H.; Lai, K.O.; Foskett, J.K.; Cheung, K.H. Familial Alzheimer's disease-associated presenilin 1 mutants promote gamma-secretase cleavage of STIM1 to impair store-operated Ca2+ entry. *Sci. Signal.* **2016**, *9*, ra89. [CrossRef]
235. Ryazantseva, M.; Goncharova, A.; Skobeleva, K.; Erokhin, M.; Methner, A.; Georgiev, P.; Kaznacheyeva, E. Presenilin-1 Delta E9 Mutant Induces STIM1-Driven Store-Operated Calcium Channel Hyperactivation in Hippocampal Neurons. *Mol. Neurobiol.* **2018**, *55*, 4667–4680. [CrossRef]
236. Linde, C.I.; Baryshnikov, S.G.; Mazzocco-Spezzia, A.; Golovina, V.A. Dysregulation of Ca2+ signaling in astrocytes from mice lacking amyloid precursor protein. *Am. J. Physiol. Cell Physiol.* **2011**, *300*, C1502–C1512. [CrossRef]
237. Wang, J.; Lu, R.; Yang, J.; Li, H.; He, Z.; Jing, N.; Wang, X.; Wang, Y. TRPC6 specifically interacts with APP to inhibit its cleavage by gamma-secretase and reduce Abeta production. *Nat. Commun.* **2015**, *6*, 8876. [CrossRef]
238. Lessard, C.B.; Lussier, M.P.; Cayouette, S.; Bourque, G.; Boulay, G. The overexpression of presenilin2 and Alzheimer's-disease-linked presenilin2 variants influences TRPC6-enhanced Ca2+ entry into HEK293 cells. *Cell Signal.* **2005**, *17*, 437–445. [CrossRef]
239. Prikhodko, V.; Chernyuk, D.; Sysoev, Y.; Zernov, N.; Okovityi, S.; Popugaeva, E. Potential Drug Candidates to Treat TRPC6 Channel Deficiencies in the Pathophysiology of Alzheimer's Disease and Brain Ischemia. *Cells* **2020**, *9*, 2351. [CrossRef]
240. Giorgi, C.; Missiroli, S.; Patergnani, S.; Duszynski, J.; Wieckowski, M.R.; Pinton, P. Mitochondria-associated membranes: Composition, molecular mechanisms, and physiopathological implications. *Antioxid. Redox Signal.* **2015**, *22*, 995–1019. [CrossRef]
241. Area-Gomez, E.; de Groof, A.; Bonilla, E.; Montesinos, J.; Tanji, K.; Boldogh, I.; Pon, L.; Schon, E.A. A key role for MAM in mediating mitochondrial dysfunction in Alzheimer disease. *Cell Death Dis.* **2018**, *9*, 335. [CrossRef] [PubMed]

Publisher's Note: MDPI stays neutral with regard to jurisdictional claims in published maps and institutional affiliations.

Review

Ca^{2+} Dyshomeostasis Disrupts Neuronal and Synaptic Function in Alzheimer's Disease

John McDaid [1,2], Sarah Mustaly-Kalimi [1,2] and Grace E. Stutzmann [1,2,3,*]

1   Center for Neurodegenerative Disease and Therapeutics, Rosalind Franklin University of Medicine and Science, 3333 Green Bay Rd., North Chicago, IL 60064, USA; john.mcdaid@rosalindfranklin.edu (J.M.); sarah.mustaly@my.rfums.org (S.M.-K.)

2   School of Graduate and Postdoctoral Studies, Rosalind Franklin University of Medicine and Science, 3333 Green Bay Rd., North Chicago, IL 60064, USA

3   Chicago Medical School, Rosalind Franklin University of Medicine and Science, 3333 Green Bay Rd., North Chicago, IL 60064, USA

*   Correspondence: grace.stutzmann@rosalindfranklin.edu; Tel.: +1-847-578-8540

Received: 3 November 2020; Accepted: 7 December 2020; Published: 10 December 2020

Abstract: Ca^{2+} homeostasis is essential for multiple neuronal functions and thus, Ca^{2+} dyshomeostasis can lead to widespread impairment of cellular and synaptic signaling, subsequently contributing to dementia and Alzheimer's disease (AD). While numerous studies implicate Ca^{2+} mishandling in AD, the cellular basis for loss of cognitive function remains under investigation. The process of synaptic degradation and degeneration in AD is slow, and constitutes a series of maladaptive processes each contributing to a further destabilization of the Ca^{2+} homeostatic machinery. Ca^{2+} homeostasis involves precise maintenance of cytosolic Ca^{2+} levels, despite extracellular influx via multiple synaptic Ca^{2+} channels, and intracellular release via organelles such as the endoplasmic reticulum (ER) via ryanodine receptor (RyRs) and IP$_3$R, lysosomes via transient receptor potential mucolipin channel (TRPML) and two pore channel (TPC), and mitochondria via the permeability transition pore (PTP). Furthermore, functioning of these organelles relies upon regulated inter-organelle Ca^{2+} handling, with aberrant signaling resulting in synaptic dysfunction, protein mishandling, oxidative stress and defective bioenergetics, among other consequences consistent with AD. With few effective treatments currently available to mitigate AD, the past few years have seen a significant increase in the study of synaptic and cellular mechanisms as drivers of AD, including Ca^{2+} dyshomeostasis. Here, we detail some key findings and discuss implications for future AD treatments.

Keywords: calcium; synaptic; glutamate; nicotinic receptors; mitochondria; autophagy; lysosome

---

1. Ca^{2+} Dysregulation and Synaptic Defects in AD

The synapse, as the primary site of communication between neurons, plays a vital role in the transmission of neuronal impulses and information, and for encoding of learning and memory, all of which are affected in Alzheimer's disease (AD). AD, as a progressive neurodegenerative disease, is characterized by Ca^{2+} dysregulation i.e., a "calciumopathy" [1] and synapse loss, i.e., a "synaptopathy" [2], with emerging evidence for a causal link between the two. Synaptic density is decreased in post-mortem brain tissue from AD patients [3,4], and while amyloid plaques have been implicated in AD related synaptic loss, synaptic deficits occur prior to and in the absence of amyloid plaques [5], and may also be due to Ca^{2+} dysregulation, an effect which is exhibited in presymptomatic AD mouse models [6–9]. Ca^{2+} dysregulation is characterized by exaggerated Ca^{2+} responses to synaptic and other stimuli, as well as abnormal Ca^{2+} homeostasis [10,11], both of which may result in elevated resting cytosolic Ca^{2+}, an effect which is observed in AD and older non-AD rodent models [12–16].

1.1. Ca^{2+} Dysregulation Disrupts Synaptic Networks in AD

Synapses are unique Ca^{2+} entry points in the neuronal architecture, expressing both pre- and postsynaptic Ca^{2+} channels/receptors, including presynaptic RyRs, N/P/Q voltage gated Ca^{2+} channels (VGCCs), α7 nicotinic acetylcholine receptors (α7 nAChRs), and postsynaptic L-type VGCCs, RyRs and NMDA receptors (NMDARs). NMDARs in particular are one of the most well characterized postsynaptic glutamate receptors, with a high Ca^{2+} permeability and an established role in hippocampal synaptic plasticity [17]. At hyperpolarized potentials, NMDARs are blocked by Mg^{2+}, but postsynaptic depolarization results in removal of Mg^{2+} block, and receptor disinhibition. Repeated NMDAR activation enhances postsynaptic Ca^{2+} entry, an effect which is facilitated by RyRs through Ca^{2+}-induced-Ca^{2+} release (CICR), thus driving increased postsynaptic AMPA receptor (AMPAR) expression and subsequent synaptic long-term potentiation (LTP). This dual role of the NMDAR as coincidence detector and postsynaptic Ca^{2+} entry channel makes it uniquely positioned to mediate synaptic potentiation resulting from concurrent pre- and postsynaptic activation, thus forming a mechanistic basis for Hebbian plasticity and associative learning.

Paradoxically, NMDARS, as well as playing a role in synaptic plasticity, may also play a role in synaptic loss [18] and cell death [19]. The role of NMDARs in the deleterious effects of AD is further illustrated by the efficacy of the NMDAR antagonist memantine as a treatment for moderate to severe AD [20,21]. Interestingly, the clinical efficacy, or lack thereof, of specific Ca^{2+} channel antagonists could serve as a useful pointer for a role for those Ca^{2+} channels in the pathophysiology of AD, with the failure of large scale clinical trials for L-type VGCC antagonists in particular, contrasting with the positive effects of memantine [22]. AD is characterized by synaptic loss [2,4,23], including loss of synaptic terminals and dendritic spines [4], and similar dendritic spine loss is accompanied by impaired synaptic transmission and plasticity in animal models of AD [8,24–29]. Although the cause of synaptic loss in AD is not fully understood, it is thought to be associated with increased ER- Ca^{2+} release within spines [8,9,30,31], and at later disease states, toxic soluble Aβ species [32], resulting in hippocampal dendritic spine loss via NMDAR activation [33–35]. In contrast to Aβ, synaptic effects of abnormal tau expression have not been as extensively studied, however, a few recent studies have implicated effects of tau on VGCC function and synaptic signaling [36–38]. Specifically, tau accumulation may lead to synaptic loss and impairment of synaptic function via activation of calcineurin [39].

Hippocampal and cortical neurites in close proximity to amyloid plaques demonstrate Ca^{2+} hyperactivity in vivo, in presymptomatic AD mice, an effect which is blocked by AMPA receptor and NMDAR antagonists, suggesting that this hyperactivity is synaptically driven [6,7]. Although Ca^{2+} hyperactivity occurred mainly in the vicinity of insoluble dense core plaques, these plaques are also surrounded by soluble Aβ [40,41], which causes similar hyperactivity in WT mice [6]. Furthermore, plaque proximity has been reported to have no effect on evoked dendritic RyR and VGCC mediated Ca^{2+} signaling in AD mice [42], raising the possibility that some of the hyperactivity observed may be due to a presynaptic mechanism. Indeed, the presynaptic Ca^{2+} hyperactivity observed in an AD mouse model was inhibited by the sarcoplasmic endoplasmic Ca^{2+}-ATPase (SERCA) pump inhibitor cyclopiazonic acid, indicating that this Ca^{2+} hyperactivity is driven by activation of presynaptic Ca^{2+} stores [43]. In contrast to reports of synaptically driven hyperexcitability in vivo, studies carried out using acute brain slices from AD mice demonstrate decreased basal hippocampal synaptic transmission, sometimes accompanied by increased paired-pulse ratio of evoked field potentials, indicative of decreased presynaptic glutamate release probability [29,44,45]. In addition, the membrane afterhyperpolarization mediated by activation of postsynaptic Ca^{2+} activated SK2 channels is increased in a 3xTg AD mouse model [46], leading to decreased postsynaptic membrane excitability and possible decreased synaptic transmission. It should also be noted that the decreased hippocampal synaptic transmission recently observed in a 5xTg AD mouse model was coupled with increased postsynaptic membrane excitability, due to an RyR2 mediated decrease in A-type K$^+$ current (I$_A$) [47], thus further illustrating the complexity of synaptic effects observed in AD mouse models. It is also noteworthy that the in vivo hyperexcitability studies mentioned above were conducted in animals anesthetized using the volatile inhalational anesthetic isoflurane. As isoflurane has been shown to result in increased cytosolic Ca^{2+} in hippocampal neurons [48], possibly due to IP_3 receptor activation [49], effects which are exaggerated in AD mice [50], isoflurane anesthesia could be a potential mediator of the Ca^{2+} hyperexcitability observed in vivo in AD mice.

While the last two decades have seen a large increase in the number of studies using AD mouse models, a more recent development has been in the use of human induced neurons (HiNs), which are neurons derived from tissue samples taken from patients, to study synaptic transmission [51–53]. In a recent study, an increased frequency of spontaneous excitatory postsynaptic current (EPSC's) was observed in AD derived HiNs, indicating an impulse-independent spontaneous increase in presynaptic glutamate release probability which is consistent with findings in human and animal studies [52]. Studies in patients with mild cognitive impairment have demonstrated hippocampal hyperactivity and decreased hippocampal volume [54,55], indicating a possible correlation between increased hippocampal activity and neurodegeneration.

More recently, proteomics has emerged as a method that allows for high throughput analysis of protein expression in small tissue samples [56], including post-mortem brain tissue from AD patients [57,58], and which has allowed for the study of changes in the interaction between presynaptic proteins [59], including SNAP25 and syntaxin [60]. Increased SNAP25 and syntaxin interaction results in reduced glutamatergic synaptic transmission [61,62] and decreased interaction between these proteins has been observed in the brains of AD patients, along with decreased levels of Complexin II [63], effects which would be expected to result in increased excitatory synaptic transmission [64,65]. SNAP25 has also been shown to negatively interact with presynaptic VGCCs to control presynaptic Ca^{2+} and affect neurotransmitter release [66,67], and the demonstrated therapeutic efficacy of putative AD medications such as levetiracetam, which targets presynaptic VGCCs [68,69], suggests that presynaptic Ca^{2+} channels could serve as a therapeutic target for AD.

1.2. Acetylcholine Signaling and α7nAChR Function in AD

nAChRs are essential for normal cognitive function [70,71], and this family of receptors includes the highly Ca^{2+} permeable homomeric α7nAChR isoform [72,73]. α7nAChRs are expressed throughout the septo-hippocampal circuit, both on medial septal nucleus/diagonal band cholinergic neurons [74], and also in the hippocampus. Cholinergic neurons in particular show significant degeneration in the course of AD [75,76], and this has resulted in development of medications to enhance cholinergic transmission, presumably through the activation of postsynaptic nAChRs. Hippocampal α7nAChRs are expressed presynaptically on mossy fiber terminals [77], and postsynaptically on CA1 interneurons [78,79], with activation of both resulting in Ca^{2+} influx [79–83]. Nicotine enhances hippocampal excitatory synaptic transmission via activation of α7nAChRs on mossy fiber terminals in the hippocampal CA3 region [80,81,84] and activation of CA1 α7nAChRs facilitates hippocampal LTP [85].

Aβ binds with high affinity to α7nAChRs [86,87], resulting in noncompetitive block of α7nAChR function, including at presynaptic α7nAChRs [88]. Cortical and hippocampal α7nAChR expression is reduced in AD patients [89,90] and AD mice [91] and Aβ binding to α7nAChRs results in the endocytosis of the Aβ α7nAChR complex with resulting accumulation within the lysosomal compartment [92]. Aβ binding to α7nAChRs results in Ca^{2+} influx, both in oocytes and presynaptic terminals in hippocampus [93,94]. Low micromolar concentrations of Aβ trigger glutamate release in the hippocampal dentate gyrus, CA3 and CA1 subfields via α7nAChRs [95] and picomolar concentrations of Aβ enhance hippocampal LTP via α7nAChRs [96]. In addition, α7nAChR activation rescues LTP deficits in hippocampal slices taken from Aβ infused rat brains, and Aβ treated hippocampal slices [97,98] and chronic treatment with an α7nAChR agonist restores cognition in AD mice [99]. Cells treated with the acetylcholinesterase inhibitor donepezil, used clinically in the treatment of AD, also showed reduced glutamate NMDAR mediated Ca^{2+} influx, an effect which was blocked by an α7nAChR antagonist [100]. Galantamine, also an acetylcholinesterase inhibitor, has been shown to positively modulate human α7nAChRs expressed in xenopus oocytes, thus allowing for a dual effect of increased synaptic acetylcholine and α7nAChR potentiation [101].

Although a role for α7nAChRs in the etiology of AD has not been established, many animal studies have demonstrated cognitive enhancing effects of compounds targeting α7nAChRs [102], including the α7nAChR partial agonist EVP-6124, which has also been shown to enhance cognition in patients with mild to moderate AD [103]. EVP-6124 and the α7nAChR positive allosteric modulator AVL-3288 have been shown to be well tolerated in patients [104,105], but some concerns exist about effects of potentiation of α7nAChR mediated Ca^{2+} effects in AD. In addition to α7nAChR mediated Ca^{2+} influx, activation of α7nAChRs triggers CICR via ryanodine sensitive Ca^{2+} stores [106], including at presynaptic α7nAChRs on hippocampal mossy fiber terminals [107], which are known to have strong RyR expression [108]. As RyR mediated CICR may be increased in AD, positive allosteric modulation of α7nAChRs may facilitate pre- and postsynaptic Ca^{2+} overload via already increased RyR function [109], possibly exacerbating AD related synaptic deficits. Based on these studies, the use of α7nAChR compounds in the treatment of cognitive impairment and AD looks promising, but caution should be exercised regarding the use of drugs which result in overt α7nAChR potentiation.

1.3. Potential Therapies for the Treatment of Synaptic Ca^{2+} Dysregulation in AD

As of now, there are only two FDA-approved classes of drugs used in the symptomatic treatment of AD: the noncompetitive NMDA antagonist memantine, and the acetylcholinesterase inhibitors, donepezil, galantamine and rivastigmine, with both classes of drugs having a synaptic site of action. Although both memantine and donepezil have been shown to be moderately effective in the treatment of AD symptoms, there is an urgent need for disease-modifying approaches, which currently requires the identification of novel compounds and receptor targets at the pre- or postsynaptic level. While a number of studies have identified promising small molecules targeting NMDARs [19,51], α7nAChRs [85,99], RyRs [110] and SERCA [111], few have made it past the preclinical stage of testing. In addition, the smoking cessation medication varenicline, which is an agonist at the α7nAChR [112], has been tested as a treatment for AD, but without any observed beneficial effects in patients [113]. Despite its failure, the clinical trial for varenicline illustrates the use of existing FDA approved medications as a strategy in the treatment of AD, bypassing many of the arduous and expensive aspects of drug development. The RyR modulator dantrolene (Ryanodex) is an FDA approved medication that has been shown to be effective in reversing many of the synaptic and cognitive effects seen in mouse models of AD [8,114–116], and has good CNS penetration when given orally or by nasal administration [117]. In addition, the clinically used L-type VGCC inhibitor isradipine has been shown to be neuroprotective in an AD mouse model [22], as has the beta-blocker carvedilol [47], however results from a recent large clinical study suggested no benefit of the VGCC antagonist nilvadipine as a treatment for AD [118]. Although the failure of large scale clinical trials for VGCC inhibitors as a treatment for AD has resulted in diminished enthusiasm for their use, the antiepileptic drug levetiracetam, which inhibits presynaptic VGCCs [68], has been shown to be beneficial in AD patients [69] and clinical trials for its use in the treatment of AD are ongoing [119]. The relative success of levetiracetam, along with the well documented failure of clinical trials targeting amyloid, strengthens the case for the use of synaptically targeted drugs in the treatment of AD and argues for the testing of FDA-approved medications as an important therapeutic strategy in the treatment of AD.

1.4. ER Ca^{2+} Channels in Synaptic Dysregulation in AD

Cytosolic Ca^{2+} levels are tightly regulated and maintained at low nM concentrations, despite a much higher extracellular Ca^{2+} concentration, and similarly elevated Ca^{2+} levels within intracellular organelles such as the ER. The ER is located throughout the cell, including pre- and postsynaptically, at synaptic terminals and dendritic spines respectively. RyRs, as well as having a role in gating ER Ca^{2+}, are sensitive to changes in cytosolic Ca^{2+} through the process of CICR, with increases in postsynaptic Ca^{2+} resulting in RyR activation and release of ER Ca^{2+} into the cytosol. CICR has the effect of amplifying postsynaptic Ca^{2+} generated from influx via Ca^{2+} permeable receptors/ion channels such as NMDA receptors (NMDAR) and voltage gated Ca^{2+} channels (VGCCs), and RyR mediated CICR is upregulated in neurons in 3xTg AD mice [9]. Although RyR mediated amplification of postsynaptic Ca^{2+} allows for a large rapid increase in cytosolic Ca^{2+}, this increased Ca^{2+} is usually rapidly removed from the cytosol, against a concentration gradient, by Ca^{2+} ATPases including the SERCA pump, which is also located on the ER membrane. The ER membrane also expresses IP_3 receptors, although these are not thought to be synaptically expressed [120].

Mutations in the presenilin 1 (PS1) gene are linked to familial AD, an early onset form of the disease, and these mutations have specific functional implications for Ca^{2+} regulation. Although PS1 is a part of the γ-secretase complex which cleaves amyloid precursor protein (APP), it is also expressed on the ER membrane where it regulates RyR and IP_3R channel properties [121–124], and may serve as a Ca^{2+} leak channel [125,126]. Mutations or altered expression of PS1 also affect the expression and sensitivity of neighboring RyRs [124]. RyRs play an important role in Ca^{2+} regulation, and RyR dysfunction is implicated in the Ca^{2+} dysregulation observed in AD [1]. RyR expression and RyR mediated Ca^{2+} responses are increased in the soma and dendritic spines of hippocampal and cortical pyramidal neurons of AD mice expressing PS1 mutations (Figures 1 and 2) [9,30,127], effects which are normalized by acute or chronic treatment with dantrolene, a negative allosteric RyR modulator [50,114]. In particular, the RyR2 isoform, which is overexpressed in the hippocampus of AD mice [30,114], plays an important role in maintenance of synaptic function [128] and shortening of the RyR2 mean channel open time reverses the synaptic dysfunction and Ca^{2+} dyshomeostasis observed in an AD mouse model [47]. Human-induced neurons (HiN) derived from fibroblasts from AD patients expressing the PS1 mutation also display increased RyR expression and evoked RyR Ca^{2+} release [53], and RyR expression is also increased in post-mortem brains of AD patients, and patients with mild cognitive impairment [129,130]. Postsynaptic Ca^{2+} responses to high frequency stimulation (HFS) are increased in hippocampal and cortical neurons from several AD mouse models [8,9,131], effects which are mediated by RyR activation (Figure 1) [9]. Presynaptic RyR function is also increased in 3xTg AD mice and RyR activation by caffeine decreases the paired-pulse ratio of evoked CA1 field potentials to a greater extent in AD mice, as well as restoring normal frequency of spontaneously released vesicles, indicating increased facilitation of glutamate release by presynaptic RyRs [30]. Further indications of pathogenic synaptic effects resulting from altered RyR-Ca^{2+} signaling is the restoration of reduced presynaptic vesicle stores observed in AD mice back to normal levels upon treatment with Ryanodex [24].

In addition to effects on basal synaptic transmission, changes in RyR-Ca^{2+} signaling may also have implications for synaptic plasticity and LTP, which are impaired in AD mice [8,45]. High frequency stimulation (HFS) of hippocampal CA3-CA1 Schaffer collaterals, which generates LTP, initially results in a period of short-term, presynaptically mediated plasticity known as post tetanic potentiation (PTP), which results from an accumulation of presynaptic Ca^{2+} and is accompanied by an increased release probability of glutamate. This form of short term plasticity is necessary for the synaptic tagging processes involved in LTP [132], however in 3xTg AD mice, PTP is reduced, and this diminished short-term plasticity is followed by decreased LTP [8]. Chronic treatment with the RyR modulator dantrolene has been shown to restore PTP and LTP to control levels seen in non-AD mice, and this effect was accompanied by a restoration of presynaptic vesicles in the active zone, illustrating a role for aberrant presynaptic RyR-Ca^{2+} signaling in the impaired short and long-term synaptic plasticity observed in AD mice [8]. Presenilin deletion decreases Ca^{2+} effects of RyR activation, due to decreased RyR expression [121] and selective deletion of presynaptic presenilin decreases the release probability of glutamate, and LTP, an effect that is mimicked and occluded by RyR inhibition [133]. Thus, it would seem alterations in PS1 expression/function result in diminished LTP, either due to decreased or increased presynaptic RyR function, emphasizing the importance of RyR stabilization in maintenance of normal synaptic function.

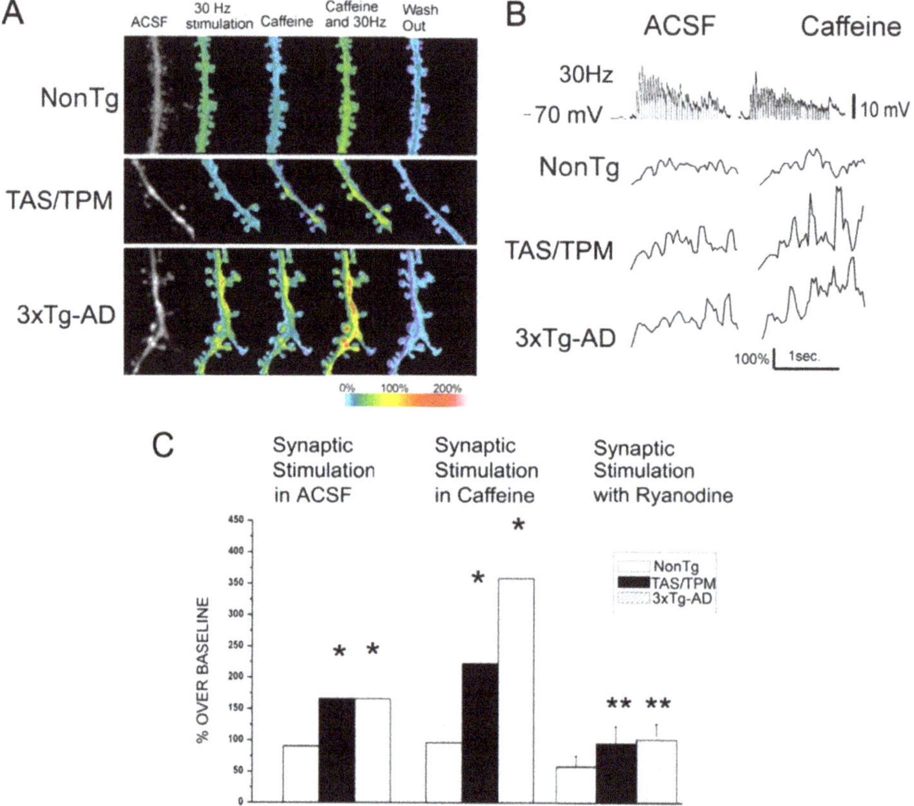

Figure 1. Synergistic Ca^{2+} interactions between RyR and glutamatergic synaptic transmission in AD
mouse cortical neurons. (**A**) Pseudocolored images of relative Ca^{2+} changes in representative NonTg
(top), TAS/TPM (a double transgenic AD mouse model) (middle), and 3xTg-AD (a triple transgenic AD
mouse model) (bottom) neurons in the following conditions (from left to right): baseline 30 Hz synaptic
stimulation (1.5 s), caffeine alone (10 mm), 30 Hz synaptic stimulation plus caffeine, and washout.
(**B**) Representative Ca^{2+} response traces after 30 Hz synaptic stimulation (voltage trace shown in top)
shown as percentage over baseline, in control aCSF (left panels) and in 10 mM caffeine (right panels)
for NonTg (top), TAS/TPM (middle), and 3xTg-AD (bottom) neurons. (**C**) Bar graphs show averaged
(mean ± SE) Ca^{2+} responses integrated over a 1.5 s time period of 30 Hz synaptic stimulation in
control ACSF (left grouping), synaptic stimulation plus caffeine (center), and synaptic stimulation with
ryanodine in the pipette (right grouping) for the NonTg, TAS/TPM, and 3xTg-AD neurons. Statistically
significant differences are indicated by asterisks (one-way ANOVA, $p < 0.05$). * Significantly different
from NonTg within treatment group; ** significantly different from synaptic stimulation in aCSF within
transgenic strain (modified from [9]).

Low resting cytosolic Ca^{2+} is maintained in part due to the actions of Ca^{2+}-ATPases, which rapidly
remove Ca^{2+} from the cell cytosol, against a concentration gradient. One of the major cellular
Ca^{2+}-ATPases is the SERCA pump, located on the ER membrane. SERCA function is facilitated by
presenilin, and knockdown of the genes for presenilin 1 and presenilin 2 results in elevated cytosolic
Ca^{2+}, due to decreased SERCA mediated clearance of cytosolic Ca^{2+} [134]. Overexpression of the
SERCA2b isoform typically found in neurons, and which physically interacts with the PS1 and PS2
Ca^{2+} channels on the ER membrane, results in increased Aβ [134]. Further evidence for presenilin's
role in SERCA function comes from a study showing that cells expressing a PS1 mutation show

an exaggerated cytosolic Ca^{2+} response to SERCA inhibition by thapsigargin, indicating increased SERCA function [135]. SERCA inhibition also mimics the effects of selective deletion of presynaptic presenilin, on synaptic transmission and LTP [133], and inhibition of presynaptic SERCA function by cyclopiazonic acid diminished the Ca^{2+} hyperactivity observed in cortical neurons of AD mice in vivo [43]. In addition to SERCA, STIM Ca^{2+} sensors and Orai Ca^{2+} channels facilitate ER Ca^{2+} filling through the process of store-operated-Ca^{2+} entry (SOCE), a process which is deficient in PS1 mutant expressing neurons [136,137]. As ER Ca^{2+} release is elevated in AD neurons (Figure 2), in parallel with diminished SOCE activity, this opens up the possibility of an increased role for SERCA in this maladaptive pathology.

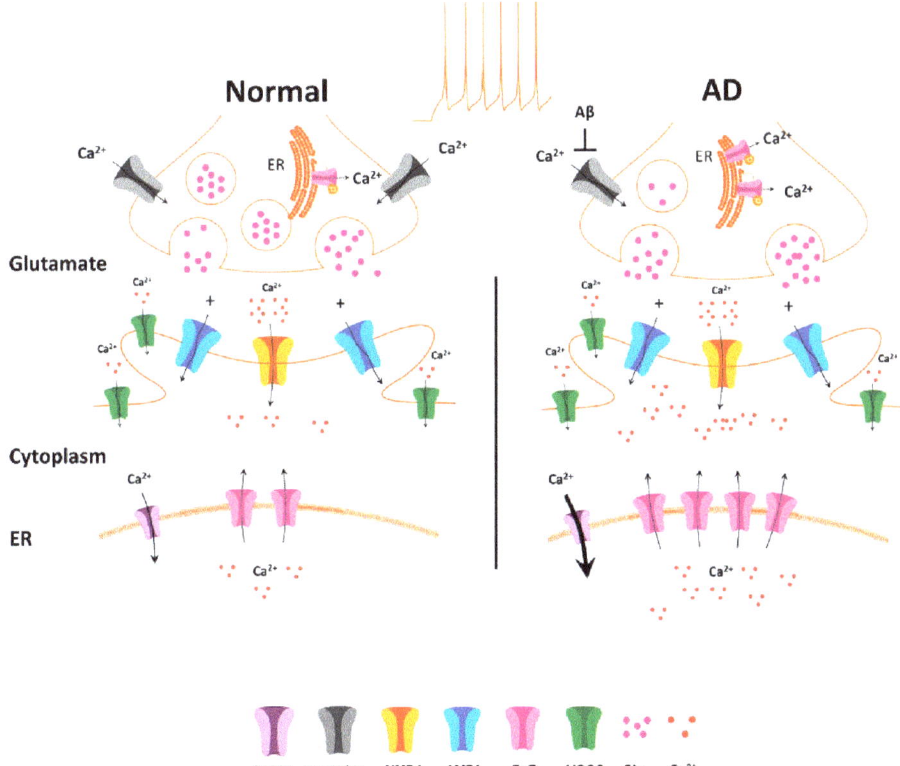

Figure 2. Schematic of synaptic Ca^{2+} dysregulation in AD. Under normal circumstances (left), impulse mediated increases in presynaptic Ca^{2+} result in neurotransmitter release via a ready releasable vesicular pool. In addition, activation of presynaptic α7nAChRs triggers further Ca^{2+} influx, thus facilitating impulse mediated release, and presynaptic effects may be further increased via presynaptic RyR mediated Ca^{2+}-induced-Ca^{2+} release (CICR). In mouse models of AD, although intrinsic cell excitability is not increased, presynaptic RyR mediated CICR may be increased, resulting in increased release probability of glutamate and depletion of vesicle stores. Aβ binding to presynaptic α7nAChRs may result in occlusion of the binding site, with decreased function and eventual decreased presynaptic α7nAChR expression due to endocytosis. In AD, increased ER Ca^{2+} stores, along with increased RyR expression results in increased postsynaptic CICR, which may facilitate stimulus-evoked postsynaptic Ca^{2+} increases.

The characteristic features of AD, including maladaptive protein accumulation, increased free radicals and metabolic disruptions, are concurrent with aberrant intracellular Ca^{2+} signaling and contributes to the activation of the ER stress response in cells [138]. In an attempt to restore ER homeostasis, the ER triggers the unfolded protein response (UPR) by increasing the expression of transcription factors (ATP6c, XBPIs, and ATF4) which provides tolerance to cellular stress [139]. If the UPR is incompetent in decreasing stress, the ER triggers cell death by apoptosis [140,141] or autophagy [142]. Several animal [143–145] and human studies [138,146] report that AD mutations cause alterations in the UPR, thus in AD, aberrant ER-Ca^{2+} release disrupts the neuron's compensatory mechanisms to restore cellular homeostasis and increases the vulnerability of neurons to stress and death.

2. Ca^{2+} Mishandling Impairs Cellular Organelle Functions in AD

In addition to synaptic signaling deficits, Ca^{2+} dyshomeostasis also has profound effects on the function of cell organelles, including the mitochondria and lysosomes, both of which play an important role in maintaining cellular and synaptic function. Like the ER, mitochondria and lysosomes act as intracellular Ca^{2+} stores, and dyshomeostasis of mitochondrial and lysosomal Ca^{2+} is emerging as a potential new source of cell dysfunction in AD, with profound implications for cellular and synaptic health.

2.1. Ca^{2+} Dysregulation Disrupts Mitochondrial Bioenergetics in AD

The mitochondria's ability to buffer intracellular Ca^{2+} signaling is critical for neuronal signal transductions, ATP synthesis, and coordination with other organelles in physiological and pathological conditions. Mitochondrial dysfunction is a well-established characteristic of AD manifesting as increased free radical production and rate of oxidative damage, decreased ATP/ADP ratio and impaired bioenergetics [147–150]. Numerous differentially expressed mitochondria regulatory genes (133 in total) have been identified in the AD cohort and found that genes coding for mitochondrial oxidative phosphorylation were downregulated in both early and late AD brain specimens–specifically, NADH ubiquinone oxidoreductase subunits and complex I components which transfer electrons to the respiratory chain [151,152]. Proteomic and protein expression studies also confirmed dysregulated mitochondrial oxidative phosphorylation complexes [153] and defective enzymatic activity in the citric acid cycle and electron transport chain (ETC) [154,155].

The main function of the mitochondria is the production of ATP. Unavoidably, the by-products of electron transport in aerobic respiration are reactive oxygen species (ROS), due to electron leaks at complex I and III. Ca^{2+} overload, as is the case in AD, hinders glucose metabolism by disrupting components of the ETC such as, mitochondria complex I and II, tricarboxylic acid cycle (TCA), pyruvate dehydrogenase complex (PDHC), α-ketoglutarate dehydrogenase complex (KGDHC), malate dehydrogenase (MDH), and increasing ROS production while decreasing ATP production [156]. Redox proteomics studies identify increased oxidatively modified proteins, specifically antioxidant enzymes such as glutathione-S-transferase Mu, peroxiredoxin 6, multidrug-resistant protein 1 or 3, and GSH, in various brain regions of MCI and AD patients [157]. Additionally, enzymes involved in respiration were oxidized, specifically ATP synthase, aconitase, and creatine kinase [157]. This suggests that increased oxidative stress, as a consequence of mitochondrial Ca^{2+} overload, contributed to mitochondrial dysfunction and impaired energy metabolism in AD.

The dynamic function of the mitochondria requires crosstalk between other major organelles, such as the ER (Figure 3). Mitochondria's physical coupling to the ER is crucial for efficient Ca^{2+} transfer and cellular homeostasis. Mitochondrial Ca^{2+} uptake controls the rate of energy production, regulates intracellular Ca^{2+} signaling, and mediates cell death. Ca^{2+} transfer between these organelles is facilitated via the mitochondrial-associated membrane proteins (MAM). Numerous molecular proteins have been identified to support this physical interaction. Of interest, the glucose-regulated protein 75 (GRP75) is linked to IP_3R and facilitates Ca^{2+} into the mitochondrial intermembrane space. From there, voltage-dependent anion-selective channel protein 1 (VDAC1) on the outer mitochondrial membrane, and the mitochondrial Ca^{2+} uniporter (MCU) on the inner mitochondrial membrane, transfer the

Ca^{2+} to the mitochondrial matrix to stimulate the mitochondrial dehydrogenase and increase ETC activity and ATP synthesis. [158–163]. These MAMs play a crucial role in regulating mitochondrial Ca^{2+} uptake. In AD, many genes involved in mitochondrial Ca^{2+} transport are altered [164–166]. Of note, genes encoding mitochondrial Ca^{2+} influx, such as MCU, are downregulated whereas genes encoding mitochondrial Ca^{2+} efflux, such as NCLX, are upregulated, suggesting a compensatory mechanism to avoid excessive Ca^{2+} uptake. Additionally, studies report that, soluble Aβ aggregations increase cytosolic Ca^{2+}, leading to mitochondrial Ca^{2+} overload via MCU. Excessive Ca^{2+} taken up by mitochondria leads to caspase activation and neuronal cell death [167].

In addition to Ca^{2+}, VDAC1 supports transport of superoxide anions [168] and since IP$_3$R and RyRs have been shown to be redox-sensitive, ROS and Ca^{2+} may play a regulatory role in ER-mitochondria communication. During AD- associated Ca^{2+} overload, mitochondrial Ca^{2+} influx elevates oxidative stress and increases ROS production. In AD, accumulated ROS has profound effects on cellular functions by oxidizing several proteins, such as the redox-sensitive RyR and IP$_3$R channels. Thus, Ca^{2+} induced ROS increase and ROS-mediated Ca^{2+} increase creates a self-amplifying loop that furthers neurotoxicity and cellular dyshomeostasis, and neuronal death [169–172]. In AD, MAM proteins are shown to be associated with presenilin 1 and 2, suggesting *PSEN1/2* may alter mitochondrial Ca^{2+} transport. Additionally, increased contact sites between ER and mitochondria, via MAM proteins in AD result in elevation in ER-mitochondrial Ca^{2+} signaling and increased mitochondrial superoxide production [173–178].

Ca^{2+} is released from the mitochondria through the Na$^+$/Ca^{2+} exchanger (NCLX) and the permeability transition pore [179]. The two functional states of the PTP regulates the amount of Ca^{2+} released; where the low conductance state amplifies Ca^{2+} waves and the high conductance state releases a surge of Ca^{2+} and apoptotic signals such as cytochrome C [180,181]. The biochemical signatures underlying apoptosis, rather than necrosis, indicates a choreographed and organized shutdown of the neuron. In AD, with continuous, prolonged increase in mitochondrial Ca^{2+} concentration, Ca^{2+} released from the mitochondria signals for apoptosis and increases AD pathology [172]. However the role of NCLX in AD needs further exploration as recent work suggests that impairment in glucose metabolism might reverse NCLX activity [182,183]. Additionally, impairment in NCLX accelerated memory declined and increased amyloidosis and tau pathology [184]. Mitochondrial calcium homeostasis may also rely on the activity of the plasma membrane NCLX, whose expression has been seen in differential patterns of mitochondrial expression dependent on cell type, and disruptions in expression may contribute to AD pathology [185–187].

Mitochondrial morphology and distribution are also crucial for neuronal homeostasis and synaptic function. Mitochondria undergo fusion and fission in the cytoplasm, which is a process to maintain a healthy pool of mitochondria with proper distribution. These mechanisms are controlled by DLP$_1$ for fission and Mfn$_1$, Mfn$_2$, and OPA$_1$ for fusion [188]. In AD, importantly, Aβ-induced and/or oxidative stress induced Ca^{2+} signaling led to increased DLP$_1$ activation, resulting in excessive mitochondrial translocation and fission [189,190]. Recent studies reported that mitochondrial fragmentation, along with extensive oxidative stress and neuroinflammation, lead to neuronal loss in the cortex and hippocampus [191,192]. Excessive mitochondrial fragmentation as a result of improper Ca^{2+} handling and increased oxidative stress disrupts mitochondrial function, advancing AD pathology. Disrupted DLP$_1$ and Mfn$_2$ function is also responsible for reduced mitochondrial distribution. In AD, mitochondria are less abundant in neuronal processes of susceptible pyramidal neurons [193,194]. Increased tau phosphorylation negatively regulates mitochondrial movement in neurons. Tau phosphorylated at the AT8 sites inhibited mitochondrial movement in neurite processes of PC12 cells and mouse cortical neurons due to impaired microtubule spacing [195,196].

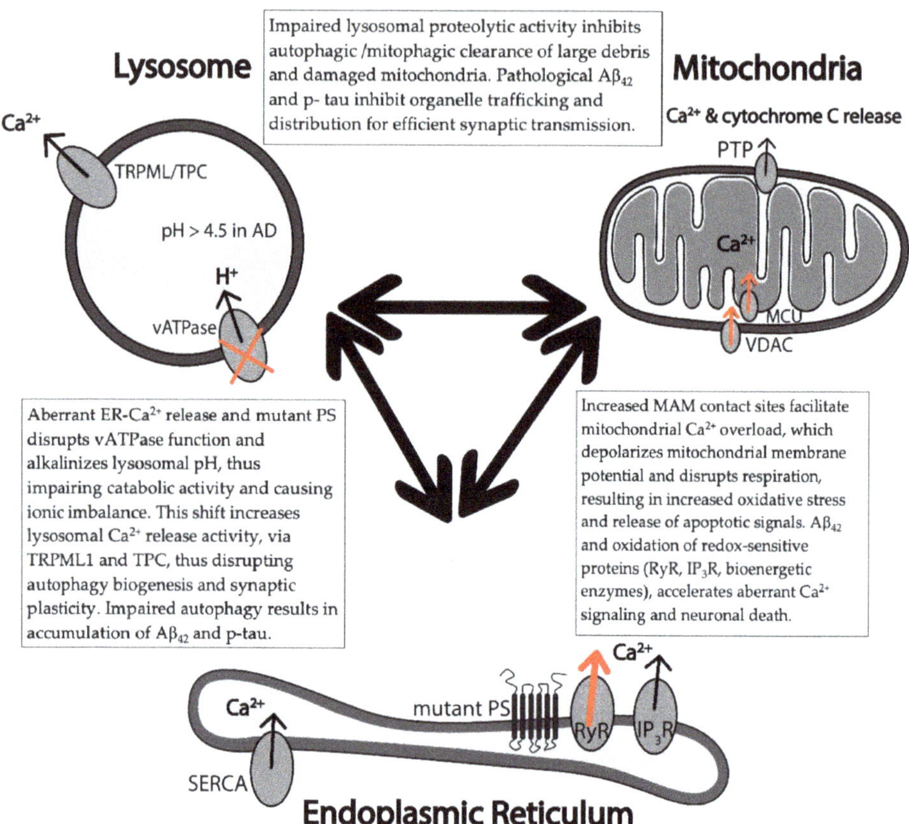

Figure 3. Aberrant Ca^{2+} disrupts inter-organelle functional relationships in early AD pathology. Schematic of feed-forward cascades among various neuronal Ca^{2+} handling organelles: Endoplasmic reticulutm (ER), lysosome, and mitochondria, in early AD pathology. Excess ER Ca^{2+} release through RyR and IP$_3$R cause mitochondrial Ca^{2+} overload that disrupts mitochondrial bioenergetics, resulting in increased oxidative stress and apoptosis. Additionally, ER Ca^{2+} disrupts lysosome-mediated clearance of maladaptive protein deposits and damaged organelle, such as Aβ, p-tau, and mitochondria, respectively. Aberrant intracellular Ca^{2+} signaling disrupts lysosomal Ca^{2+} release via TRPML and dysregulates autophagosome biosynthesis and impairs synaptic plasticity. Presenilin (PS) mutations disrupt vATPase trafficking resulting in alkaline lysosomes, thereby disrupting lysosomal ionic balance and lysosomal Ca^{2+} store. This alkaline environment impairs proteolysis and impairs autophagic clearance.

Damaged mitochondria are cleared through mitophagy, the selective degradation of mitochondria by autophagy following organelle damage or extreme cellular stress [197]. Mitophagy initiation involves the recruitment of PINK1 and PARKIN to the outer mitochondrial membrane, which tags the damaged mitochondria for degradation [198,199]. Additionally, mitophagy involves VDAC1 and MAM sensors, implicating the need for proper inter-organelle Ca^{2+} signaling and colocalization of the two organelles. Intracellular Ca^{2+} signaling relieves the inhibitory mammalian target of rapamycin (mTOR) block, thus activating mitophagy and initiating autophagosome biogenesis (ATG 32, 8, and 11) [200–202]. In AD patients, disruptions in mitophagy have been seen in the presence of Aβ, APP, and mutant PS1 expression. Aberrant inter-organelle Ca^{2+} signaling, as seen in AD, may disrupt degradation of damaged organelle via inactivation of lysosomal proteolysis and increase accumulation of cellular debris [198,199,203–205].

2.2. Ca²⁺ Dysregulation Impairs Lysosome-Autophagosome Mediated Protein Degradation

Lysosomal Ca^{2+} stores are responsible for regulating autophagy—a catabolic pathway utilizing the enzymatic activity of lysosomes to degrade and recycle large, bulky cellular debris, aggregated proteins, and damaged organelles. There are three types of autophagy: macroautophagy, microautophagy, and chaperone-mediated autophagy, however for simplicity, this review will focus on macroautophagy and be referred to as "autophagy". More detailed descriptions of the aforementioned can be found in these reviews [163,206]. Autophagy is a systematic, dynamic degradation process regulated by the fusion of cargo vesicles, autophagosomes, to degradative compartments, lysosomes, with active hydrolases and proteases. While other cells rely on division to dilute cellular debris, neurons are specialized, post-mitotic cells that require efficient basal autophagy regulation to prevent accumulation of misfolded proteins and damaged organelles. Autophagy depends on the close proximity and communication between lysosomes and ER [207,208], therefore disruptions in inter-organelle Ca^{2+} signaling hinders clearance of pathological protein deposits.

Transcriptomic profiles from the collective ongoing studies known as Rush Memory and Aging Project (ROSMAP), reveal clusters of genes that are associated with pathological protein handling. Specifically, higher expression of SORL1 and ABCA7 transcripts are associated with tau tangle pathology, while elevated BIN1 transcripts are associated with beta amyloid in AD brains [209]. PLXNB1 abundance is associated with increased amyloid load and higher paired helical filaments (PHF) tau tangle density. Notably, BIN1, ABCA7, and SORL1 have functions in endocytic transport, APP metabolism and lysosome recycling, and thus are ideally positioned to serve a role in AD proteinopathy [210]. Altered expression of protein handling genes is linked to blunted endosomal trafficking, diminished degradative potential of lysosomes, and reduced autophagy-mediated clearance [210].

The key, critical feature of lysosomes is the acidic lumen (pH ~4.5) necessary for protein degradation and autophagosome digestion. The acidic pH is maintained by an active vacuolar-ATPase H^+ pump (vATPase) driving the influx of H^+ into the lysosome [211–213]. Genetic evidence linking endosomal H^+ exchangers with AD suggest that proton leak pathways may regulate pathological $A\beta$ generation and contribute to disease etiology [214]. In the ROSMAP AD population, there is downregulation of vATPase subunit (V1) genes, as well as the transcription factor "EB" (TFEB), a master regulator of lysosomal biogenesis that is associated with regulated autophagy [151]. Mutations in PS prevents the glycosylation, downstream maturation, and trafficking of the vATPase to the lysosome, resulting in an alkaline lysosomal lumen [215,216]. The alkaline environment inactivates protease activity, such as cathepsin B, which halts degradation of APP metabolites and dysregulates biogenesis of lysosomes and autophagosomes [217]. Additionally, cathepsins may also play a role in lysosomal trafficking along neuronal axons and dendrites, which is essential in mediating proper disposal of cellular debris. Studies showed that disrupting lysosomal proteolysis by inhibiting cathepsins or suppressing lysosomal acidification slowed axonal transport and caused selective accumulation within dystrophic neurites, a key feature of AD [218–220]. These are abnormally swollen regions of axons and dendrites filled mainly with autophagosomes and lysosomes, which implies improper transport of degradative organelles. Aberrant Ca^{2+} signaling, as seen in AD, can hinder lysosomal acidification and impair proteolytic enzymes in lysosomes, further AD proteinopathy and impair lysososmal trafficking, resulting in neuritic dystrophy.

Additionally, an increase in the lysosomal pH disrupts the homeostatic mechanisms to maintain the lysosomal membrane potential. The alkaline lumen depolarizes the lysosomal membrane via activation of lysosomal voltage-activated Na^+ channels (lysoNa$_V$). The Na^+ efflux potentiates the influx of protons by the vATPase to restore lysosomal acidity. However, lysoNa$_V$ are also Ca^{2+} permeable, therefore aberrant Ca^{2+} increase and changes in the Ca^{2+} concentration gradient, as seen in early pathology of AD, can hinder lysosomal acidity by reducing the driving force of H^+ influx to restore lysosomal function [213,221,222].

This shift to a more alkaline lysosomal lumen causes hyperactivity of the lysosomal Ca^{2+} efflux channels, lysosomal transient receptor potential Ca^{2+} channel mucolipin subfamily member (TRPML1) and two-pore channel (TPC) [205,215,216,223–225] (although debated [226–228]). Lysosomal Ca^{2+} efflux through TRPML1, activates a calcineurin-dependent pathway that, via TFEB, enhances the transcription of genes involved in autophagy and lysosomal expression, such as LC3-II, ATG9B, UVRAG, WIPI, SQSTM1, MAPLC3B, GLA, GNS, HEXA, MCOLN1, TMEM55B, and ATP6V1H [229–231].

This lysosomal-mediated Ca^{2+} release is also responsible for fusion of autophagosomes to lysosomes. In a manner similar to neuronal vesicular fusion, lysosomal Ca^{2+} efflux channels such as, P/Q type VGCC, facilitate fusion between lysosomal tethering proteins, such as synaptogamin 7, SNAP29, and SNARE VAMP 7/8, to autophagosomal tethering proteins such as syntaxin 17 and possible SNARE proteins [232–235]. Aberrant Ca^{2+} concentration, as in AD, can influence fusion of autophagosomes to malfunctioned lysosome, resulting in accumulation of cargo vesicles with maladaptive proteins, furthering neurotoxicity.

Recent studies have shown that lysosomal functions go beyond their primary role as the degradative compartment within a neuron. Lysosomal Ca^{2+} stores are also involved in maintaining synaptic transmission. When mGluR1 is activated, NAADP-evoked lysosomal Ca^{2+} release from lysosomal Ca^{2+} channels, presumably through TPC, is amplified into Ca^{2+} waves via RyR activation [236]. This signal inactivates SK channels and prevents local hyperpolarization, which allows for greater Ca^{2+} entry through GluN receptors and facilitates the induction of LTP [237]. When VGCC is activated, NAADP-evoked lysosomal Ca^{2+} release facilitates fusion to the plasma membrane and release of cathepsin B. Cathepsin B then regulates structural plasticity and dendritic spine formation [238,239], however in pathological conditions, protease activity is inhibited and therefore synaptic dysfunctions occur. Recent work has shown that blocking endogenous cathepsin inhibitors, such as cystatin B, decreases Aβ accumulation, autophagic-lysosomal pathology, and cognitive improvement in AD mice [240].

The systemic destruction of collective organelles leads to the neuron's demise. The close proximity of the ER to the lysosome can enhance AD pathology (Figure 3) and PS mutations may alter the RyR-lysosomes trigger zone. In healthy neurons, RyR-mediated Ca^{2+} amplification suppresses autophagic flux [207] and induces LTP [237–239] however, increased Ca^{2+} signaling, as seen in neurodegenerative diseases, can alter this trigger zone and therefore disrupt autophagic clearance and synaptic plasticity.

Funding: This work was supported by NIH RF1 AG065628, R56 AG055497, and RF1 AG047237 (GES).

Acknowledgments: The authors thank Nikki Barrington for editorial assistance in the preparation of this review.

Conflicts of Interest: The authors declare no conflict of interest.

References

1. Stutzmann, G.E. The pathogenesis of Alzheimers disease is it a lifelong "calciumopathy"? *Neuroscientist* **2007**, *13*, 546–559. [CrossRef] [PubMed]
2. Selkoe, D.J. Alzheimer's disease is a synaptic failure. *Science* **2002**, *298*, 789–791. [CrossRef] [PubMed]
3. DeKosky, S.T.; Scheff, S.W. Synapse loss in frontal cortex biopsies in Alzheimer's disease: Correlation with cognitive severity. *Ann. Neurol.* **1990**, *27*, 457–464. [CrossRef] [PubMed]
4. Terry, R.D.; Masliah, E.; Salmon, D.P.; Butters, N.; DeTeresa, R.; Hill, R.; Hansen, L.A.; Katzman, R. Physical basis of cognitive alterations in Alzheimer's disease: Synapse loss is the major correlate of cognitive impairment. *Ann. Neurol.* **1991**, *30*, 572–580. [CrossRef] [PubMed]
5. Hsia, A.Y.; Masliah, E.; McConlogue, L.; Yu, G.Q.; Tatsuno, G.; Hu, K.; Kholodenko, D.; Malenka, R.C.; Nicoll, R.A.; Mucke, L. Plaque-independent disruption of neural circuits in Alzheimer's disease mouse models. *Proc. Natl. Acad. Sci. USA* **1999**, *96*, 3228–3233. [CrossRef] [PubMed]
6. Busche, M.A.; Chen, X.; Henning, H.A.; Reichwald, J.; Staufenbiel, M.; Sakmann, B.; Konnerth, A. Critical role of soluble amyloid-β for early hippocampal hyperactivity in a mouse model of Alzheimer's disease. *Proc. Natl. Acad. Sci. USA* **2012**, *109*, 8740–8745. [CrossRef]

7. Busche, M.A.; Eichhoff, G.; Adelsberger, H.; Abramowski, D.; Wiederhold, K.-H.; Haass, C.; Staufenbiel, M.; Konnerth, A.; Garaschuk, O. Clusters of hyperactive neurons near amyloid plaques in a mouse model of Alzheimer's disease. *Science* **2008**, *321*, 1686–1689. [CrossRef]

8. Chakroborty, S.; Hill, E.S.; Christian, D.T.; Helfrich, R.; Riley, S.; Schneider, C.; Kapecki, N.; Mustaly-Kalimi, S.; Seiler, F.A.; Peterson, D.A.; et al. Reduced presynaptic vesicle stores mediate cellular and network plasticity defects in an early-stage mouse model of Alzheimer's disease. *Mol. Neurodegener.* **2019**, *14*, 7. [CrossRef]

9. Goussakov, I.; Miller, M.B.; Stutzmann, G.E. NMDA-mediated Ca^{2+} influx drives aberrant ryanodine receptor activation in dendrites of young Alzheimer's disease mice. *J. Neurosci.* **2010**, *30*, 12128–12137. [CrossRef]

10. Murchison, D.; McDermott, A.N.; LaSarge, C.L.; Peebles, K.A.; Bizon, J.L.; Griffith, W.H. Enhanced Calcium Buffering in F344 Rat Cholinergic Basal Forebrain Neurons Is Associated with Age-Related Cognitive Impairment. *J. Neurophysiol.* **2009**, *102*, 2194–2207. [CrossRef]

11. Moyer, J.R.; Furtak, S.C.; McGann, J.P.; Brown, T.H. Aging-related changes in calcium binding proteins in rat perirhinal cortex. *Neurobiol. Aging* **2011**, *32*, 1693–1706. [CrossRef] [PubMed]

12. Oh, M.M.; Oliveira, F.A.; Waters, J.; Disterhoft, J.F. Altered Calcium Metabolism in Aging CA1 Hippocampal Pyramidal Neurons. *J. Neurosci.* **2013**, *33*, 7905–7911. [CrossRef] [PubMed]

13. Kuchibhotla, K.V.; Goldman, S.T.; Lattarulo, C.R.; Wu, H.-Y.; Hyman, B.T.; Bacskai, B.J. Abeta plaques lead to aberrant regulation of calcium homeostasis in vivo resulting in structural and functional disruption of neuronal networks. *Neuron* **2008**, *59*, 214–225. [CrossRef] [PubMed]

14. Lopez, J.R.; Lyckman, A.; Oddo, S.; Laferla, F.M.; Querfurth, H.W.; Shtifman, A. Increased intraneuronal resting [Ca^{2+}] in adult Alzheimer's disease mice. *J. Neurochem.* **2008**, *105*, 262–271. [CrossRef]

15. Uryash, A.; Flores, V.; Adams, J.A.; Allen, P.D.; Lopez, J.R. Memory and Learning Deficits Are Associated With Ca^{2+} Dyshomeostasis in Normal Aging. *Front. Aging Neurosci.* **2020**, *12*, 224. [CrossRef]

16. Thibault, O.; Hadley, R.; Landfield, P.W. Elevated postsynaptic [Ca^{2+}]$_i$ and L-type calcium channel activity in aged hippocampal neurons: Relationship to impaired synaptic plasticity. *J. Neurosci.* **2001**, *21*, 9744–9756. [CrossRef]

17. Bliss, T.V.; Collingridge, G.L. A synaptic model of memory: Long-term potentiation in the hippocampus. *Nature* **1993**, *361*, 31–39. [CrossRef]

18. Talantova, M.; Sanz-Blasco, S.; Zhang, X.; Xia, P.; Akhtar, M.W.; Okamoto, S.; Dziewczapolski, G.; Nakamura, T.; Cao, G.; Pratt, A.E.; et al. Aβ induces astrocytic glutamate release, extrasynaptic NMDA receptor activation, and synaptic loss. *Proc. Natl. Acad. Sci. USA* **2013**, *110*, E2518–E2527. [CrossRef]

19. Yan, J.; Bengtson, C.P.; Buchthal, B.; Hagenston, A.M.; Bading, H. Coupling of NMDA receptors and TRPM4 guides discovery of unconventional neuroprotectants. *Science* **2020**, *370*. [CrossRef]

20. Winblad, B.; Jones, R.W.; Wirth, Y.; Stöffler, A.; Möbius, H.J. Memantine in moderate to severe Alzheimer's disease: A meta-analysis of randomised clinical trials. *Dement. Geriatr. Cogn. Disord.* **2007**, *24*, 20–27. [CrossRef]

21. Gauthier, S.; Loft, H.; Cummings, J. Improvement in behavioural symptoms in patients with moderate to severe Alzheimer's disease by memantine: A pooled data analysis. *Int. J. Geriatr. Psychiatry* **2008**, *23*, 537–545. [CrossRef] [PubMed]

22. Copenhaver, P.F.; Anekonda, T.S.; Musashe, D.; Robinson, K.M.; Ramaker, J.M.; Swanson, T.L.; Wadsworth, T.L.; Kretzschmar, D.; Woltjer, R.L.; Quinn, J.F. A translational continuum of model systems for evaluating treatment strategies in Alzheimer's disease: Isradipine as a candidate drug. *Dis. Model. Mech.* **2011**, *4*, 634–648. [CrossRef] [PubMed]

23. Scheff, S.W.; Price, D.A.; Schmitt, F.A.; Mufson, E.J. Hippocampal synaptic loss in early Alzheimer's disease and mild cognitive impairment. *Neurobiol. Aging* **2006**, *27*, 1372–1384. [CrossRef] [PubMed]

24. Jacobsen, J.S.; Wu, C.-C.; Redwine, J.M.; Comery, T.A.; Arias, R.; Bowlby, M.; Martone, R.; Morrison, J.H.; Pangalos, M.N.; Reinhart, P.H.; et al. Early-onset behavioral and synaptic deficits in a mouse model of Alzheimer's disease. *Proc. Natl. Acad. Sci. USA* **2006**, *103*, 5161–5166. [CrossRef] [PubMed]

25. Spires, T.L.; Meyer-Luehmann, M.; Stern, E.A.; McLean, P.J.; Skoch, J.; Nguyen, P.T.; Bacskai, B.J.; Hyman, B.T. Dendritic spine abnormalities in amyloid precursor protein transgenic mice demonstrated by gene transfer and intravital multiphoton microscopy. *J. Neurosci.* **2005**, *25*, 7278–7287. [CrossRef]

26. Lanz, T.A.; Carter, D.B.; Merchant, K.M. Dendritic spine loss in the hippocampus of young PDAPP and Tg2576 mice and its prevention by the ApoE2 genotype. *Neurobiol. Dis.* **2003**, *13*, 246–253. [CrossRef]

27. Fitzjohn, S.M.; Morton, R.A.; Kuenzi, F.; Rosahl, T.W.; Shearman, M.; Lewis, H.; Smith, D.; Reynolds, D.S.; Davies, C.H.; Collingridge, G.L.; et al. Age-related impairment of synaptic transmission but normal long-term potentiation in transgenic mice that overexpress the human APP695SWE mutant form of amyloid precursor protein. *J. Neurosci.* **2001**, *21*, 4691–4698. [CrossRef]
28. Chapman, P.F.; White, G.L.; Jones, M.W.; Cooper-Blacketer, D.; Marshall, V.J.; Irizarry, M.; Younkin, L.; Good, M.A.; Bliss, T.V.; Hyman, B.T.; et al. Impaired synaptic plasticity and learning in aged amyloid precursor protein transgenic mice. *Nat. Neurosci.* **1999**, *2*, 271–276. [CrossRef]
29. Larson, J.; Lynch, G.; Games, D.; Seubert, P. Alterations in synaptic transmission and long-term potentiation in hippocampal slices from young and aged PDAPP mice. *Brain Res.* **1999**, *840*, 23–35. [CrossRef]
30. Chakroborty, S.; Goussakov, I.; Miller, M.B.; Stutzmann, G.E. Deviant ryanodine receptor-mediated calcium release resets synaptic homeostasis in presymptomatic 3xTg-AD mice. *J. Neurosci.* **2009**, *29*, 9458–9470. [CrossRef]
31. Zhang, H.; Wu, L.; Pchitskaya, E.; Zakharova, O.; Saito, T.; Saido, T.; Bezprozvanny, I. Neuronal Store-Operated Calcium Entry and Mushroom Spine Loss in Amyloid Precursor Protein Knock-In Mouse Model of Alzheimer's Disease. *J. Neurosci.* **2015**, *35*, 13275–13286. [CrossRef] [PubMed]
32. Lue, L.F.; Kuo, Y.M.; Roher, A.E.; Brachova, L.; Shen, Y.; Sue, L.; Beach, T.; Kurth, J.H.; Rydel, R.E.; Rogers, J. Soluble amyloid beta peptide concentration as a predictor of synaptic change in Alzheimer's disease. *Am. J. Pathol.* **1999**, *155*, 853–862. [CrossRef]
33. Arbel-Ornath, M.; Hudry, E.; Boivin, J.R.; Hashimoto, T.; Takeda, S.; Kuchibhotla, K.V.; Hou, S.; Lattarulo, C.R.; Belcher, A.M.; Shakerdge, N.; et al. Soluble oligomeric amyloid-β induces calcium dyshomeostasis that precedes synapse loss in the living mouse brain. *Mol. Neurodegener.* **2017**, *12*, 27. [CrossRef] [PubMed]
34. Calabrese, B.; Shaked, G.M.; Tabarean, I.V.; Braga, J.; Koo, E.H.; Halpain, S. Rapid, concurrent alterations in pre- and postsynaptic structure induced by naturally-secreted amyloid-β protein. *Mol. Cell. Neurosci.* **2007**, *35*, 183–193. [CrossRef] [PubMed]
35. Shankar, G.M.; Bloodgood, B.L.; Townsend, M.; Walsh, D.M.; Selkoe, D.J.; Sabatini, B.L. Natural oligomers of the Alzheimer amyloid-beta protein induce reversible synapse loss by modulating an NMDA-type glutamate receptor-dependent signaling pathway. *J. Neurosci.* **2007**, *27*, 2866–2875. [CrossRef]
36. Esteras, N.; Kundel, F.; Amodeo, G.F.; Pavlov, E.V.; Klenerman, D.; Abramov, A.Y. Insoluble tau aggregates induce neuronal death through modification of membrane ion conductance, activation of voltage-gated calcium channels and NADPH oxidase. *FEBS J.* **2020**. [CrossRef]
37. Zhou, L.; McInnes, J.; Wierda, K.; Holt, M.; Herrmann, A.G.; Jackson, R.J.; Wang, Y.-C.; Swerts, J.; Beyens, J.; Miskiewicz, K.; et al. Tau association with synaptic vesicles causes presynaptic dysfunction. *Nat. Commun.* **2017**, *8*, 15295. [CrossRef]
38. Gómez-Ramos, A.; Díaz-Hernández, M.; Cuadros, R.; Hernández, F.; Avila, J. Extracellular tau is toxic to neuronal cells. *FEBS Lett.* **2006**, *580*, 4842–4850. [CrossRef]
39. Yin, Y.; Gao, D.; Wang, Y.; Wang, Z.-H.; Wang, X.; Ye, J.; Wu, D.; Fang, L.; Pi, G.; Yang, Y.; et al. Tau accumulation induces synaptic impairment and memory deficit by calcineurin-mediated inactivation of nuclear CaMKIV/CREB signaling. *Proc. Natl. Acad. Sci. USA* **2016**, *113*, E3773–E3781. [CrossRef]
40. Keskin, A.D.; Kekuš, M.; Adelsberger, H.; Neumann, U.; Shimshek, D.R.; Song, B.; Zott, B.; Peng, T.; Förstl, H.; Staufenbiel, M.; et al. BACE inhibition-dependent repair of Alzheimer's pathophysiology. *Proc. Natl. Acad. Sci. USA* **2017**, *114*, 8631–8636. [CrossRef]
41. Koffie, R.M.; Meyer-Luehmann, M.; Hashimoto, T.; Adams, K.W.; Mielke, M.L.; Garcia-Alloza, M.; Micheva, K.D.; Smith, S.J.; Kim, M.L.; Lee, V.M.; et al. Oligomeric amyloid beta associates with postsynaptic densities and correlates with excitatory synapse loss near senile plaques. *Proc. Natl. Acad. Sci. USA* **2009**, *106*, 4012–4017. [CrossRef] [PubMed]
42. Briggs, C.A.; Schneider, C.; Richardson, J.C.; Stutzmann, G.E. β amyloid peptide plaques fail to alter evoked neuronal calcium signals in APP/PS1 Alzheimer's disease mice. *Neurobiol. Aging* **2013**, *34*, 1632–1643. [CrossRef] [PubMed]
43. Lerdkrai, C.; Asavapanumas, N.; Brawek, B.; Kovalchuk, Y.; Mojtahedi, N.; del Moral, M.O.; Garaschuk, O. Intracellular Ca^{2+} stores control in vivo neuronal hyperactivity in a mouse model of Alzheimer's disease. *Proc. Natl. Acad. Sci. USA* **2018**, *115*, E1279–E1288. [CrossRef] [PubMed]

44. He, Y.; Wei, M.; Wu, Y.; Qin, H.; Li, W.; Ma, X.; Cheng, J.; Ren, J.; Shen, Y.; Chen, Z.; et al. Amyloid β oligomers suppress excitatory transmitter release via presynaptic depletion of phosphatidylinositol-4,5-bisphosphate. *Nat. Commun.* **2019**, *10*, 1193. [CrossRef] [PubMed]

45. Oddo, S.; Caccamo, A.; Shepherd, J.D.; Murphy, M.P.; Golde, T.E.; Kayed, R.; Metherate, R.; Mattson, M.P.; Akbari, Y.; LaFerla, F.M. Triple-transgenic model of Alzheimer's disease with plaques and tangles: Intracellular Abeta and synaptic dysfunction. *Neuron* **2003**, *39*, 409–421. [CrossRef]

46. Chakroborty, S.; Kim, J.; Schneider, C.; Jacobson, C.; Molgó, J.; Stutzmann, G.E. Early presynaptic and postsynaptic calcium signaling abnormalities mask underlying synaptic depression in presymptomatic Alzheimer's disease mice. *J. Neurosci.* **2012**, *32*, 8341–8353. [CrossRef] [PubMed]

47. Yao, J.; Sun, B.; Institoris, A.; Zhan, X.; Guo, W.; Song, Z.; Liu, Y.; Hiess, F.; Boyce, A.K.J.; Ni, M.; et al. Limiting RyR2 Open Time Prevents Alzheimer's Disease-Related Neuronal Hyperactivity and Memory Loss but Not β-Amyloid Accumulation. *Cell Rep.* **2020**, *32*, 108169. [CrossRef] [PubMed]

48. Kindler, C.H.; Eilers, H.; Donohoe, P.; Ozer, S.; Bickler, P.E. Volatile anesthetics increase intracellular calcium in cerebrocortical and hippocampal neurons. *Anesthesiology* **1999**, *90*, 1137–1145. [CrossRef]

49. Wei, H.; Liang, G.; Yang, H.; Wang, Q.; Hawkins, B.; Madesh, M.; Wang, S.; Eckenhoff, R.G. The Common Inhalational Anesthetic Isoflurane Induces Apoptosis via Activation of Inositol 1,4,5-Trisphosphate Receptors. *Anesthesiol. J. Am. Soc. Anesthesiol.* **2008**, *108*, 251–260. [CrossRef]

50. Stutzmann, G.E.; Smith, I.; Caccamo, A.; Oddo, S.; Laferla, F.M.; Parker, I. Enhanced ryanodine receptor recruitment contributes to Ca^{2+} disruptions in young, adult, and aged Alzheimer's disease mice. *J. Neurosci.* **2006**, *26*, 5180–5189. [CrossRef]

51. Ghatak, S.; Dolatabadi, N.; Gao, R.; Wu, Y.; Scott, H.; Trudler, D.; Sultan, A.; Ambasudhan, R.; Nakamura, T.; Masliah, E.; et al. NitroSynapsin ameliorates hypersynchronous neural network activity in Alzheimer hiPSC models. *Mol. Psychiatry* **2020**. [CrossRef] [PubMed]

52. Ghatak, S.; Dolatabadi, N.; Trudler, D.; Zhang, X.; Wu, Y.; Mohata, M.; Ambasudhan, R.; Talantova, M.; Lipton, S.A. Mechanisms of hyperexcitability in Alzheimer's disease hiPSC-derived neurons and cerebral organoids vs. isogenic controls. *eLife* **2019**, *8*. [CrossRef] [PubMed]

53. Schrank, S.; McDaid, J.; Briggs, C.A.; Mustaly-Kalimi, S.; Brinks, D.; Houcek, A.; Singer, O.; Bottero, V.; Marr, R.A.; Stutzmann, G.E. Human-Induced Neurons from Presenilin 1 Mutant Patients Model Aspects of Alzheimer's Disease Pathology. *Int. J. Mol. Sci.* **2020**, *21*, 1030. [CrossRef] [PubMed]

54. Huijbers, W.; Mormino, E.C.; Schultz, A.P.; Wigman, S.; Ward, A.M.; Larvie, M.; Amariglio, R.E.; Marshall, G.A.; Rentz, D.M.; Johnson, K.A.; et al. Amyloid-β deposition in mild cognitive impairment is associated with increased hippocampal activity, atrophy and clinical progression. *Brain J. Neurol.* **2015**, *138*, 1023–1035. [CrossRef] [PubMed]

55. Hämäläinen, A.; Pihlajamäki, M.; Tanila, H.; Hänninen, T.; Niskanen, E.; Tervo, S.; Karjalainen, P.A.; Vanninen, R.L.; Soininen, H. Increased fMRI responses during encoding in mild cognitive impairment. *Neurobiol. Aging* **2007**, *28*, 1889–1903. [CrossRef] [PubMed]

56. Drummond, E.; Wisniewski, T. Using Proteomics to Understand Alzheimer's Disease Pathogenesis. In *Alzheimer's Disease*; Wisniewski, T., Ed.; Codon Publications: Brisbane, Australia, 2019; ISBN 978-0-646-80968-7.

57. Honer, W.G.; Ramos-Miguel, A.; Alamri, J.; Sawada, K.; Barr, A.M.; Schneider, J.A.; Bennett, D.A. The synaptic pathology of cognitive life. *Dialogues Clin. Neurosci.* **2019**, *21*, 271–279. [CrossRef] [PubMed]

58. Bennett, D.A.; Buchman, A.S.; Boyle, P.A.; Barnes, L.L.; Wilson, R.S.; Schneider, J.A. Religious Orders Study and Rush Memory and Aging Project. *J. Alzheimers Dis. JAD* **2018**, *64*, S161–S189. [CrossRef]

59. Dieterich, D.C.; Kreutz, M.R. Proteomics of the Synapse—A Quantitative Approach to Neuronal Plasticity. *Mol. Cell. Proteomics MCP* **2016**, *15*, 368–381. [CrossRef]

60. Honer, W.G.; Barr, A.M.; Sawada, K.; Thornton, A.E.; Morris, M.C.; Leurgans, S.E.; Schneider, J.A.; Bennett, D.A. Cognitive reserve, presynaptic proteins and dementia in the elderly. *Transl. Psychiatry* **2012**, *2*, e114. [CrossRef]

61. Jeans, A.F.; Oliver, P.L.; Johnson, R.; Capogna, M.; Vikman, J.; Molnár, Z.; Babbs, A.; Partridge, C.J.; Salehi, A.; Bengtsson, M.; et al. A dominant mutation in Snap25 causes impaired vesicle trafficking, sensorimotor gating, and ataxia in the blind-drunk mouse. *Proc. Natl. Acad. Sci. USA* **2007**, *104*, 2431–2436. [CrossRef]

62. Toft-Bertelsen, T.L.; Ziomkiewicz, I.; Houy, S.; Pinheiro, P.S.; Sørensen, J.B. Regulation of Ca^{2+} channels by SNAP-25 via recruitment of syntaxin-1 from plasma membrane clusters. *Mol. Biol. Cell* **2016**, *27*, 3329–3341. [CrossRef] [PubMed]

63. Ramos-Miguel, A.; Sawada, K.; Jones, A.A.; Thornton, A.E.; Barr, A.M.; Leurgans, S.E.; Schneider, J.A.; Bennett, D.A.; Honer, W.G. Presynaptic proteins complexin-I and complexin-II differentially influence cognitive function in early and late stages of Alzheimer's disease. *Acta Neuropathol.* **2017**, *133*, 395–407. [CrossRef] [PubMed]

64. Verderio, C.; Pozzi, D.; Pravettoni, E.; Inverardi, F.; Schenk, U.; Coco, S.; Proux-Gillardeaux, V.; Galli, T.; Rossetto, O.; Frassoni, C.; et al. SNAP-25 Modulation of Calcium Dynamics Underlies Differences in GABAergic and Glutamatergic Responsiveness to Depolarization. *Neuron* **2004**, *41*, 599–610. [CrossRef]

65. Malsam, J.; Bärfuss, S.; Trimbuch, T.; Zarebidaki, F.; Sonnen, A.F.-P.; Wild, K.; Scheutzow, A.; Rohland, L.; Mayer, M.P.; Sinning, I.; et al. Complexin Suppresses Spontaneous Exocytosis by Capturing the Membrane-Proximal Regions of VAMP2 and SNAP25. *Cell Rep.* **2020**, *32*, 107926. [CrossRef] [PubMed]

66. Rettig, J.; Sheng, Z.H.; Kim, D.K.; Hodson, C.D.; Snutch, T.P.; Catterall, W.A. Isoform-specific interaction of the alpha1A subunits of brain Ca^{2+} channels with the presynaptic proteins syntaxin and SNAP-25. *Proc. Natl. Acad. Sci. USA* **1996**, *93*, 7363–7368. [CrossRef] [PubMed]

67. Antonucci, F.; Corradini, I.; Morini, R.; Fossati, G.; Menna, E.; Pozzi, D.; Pacioni, S.; Verderio, C.; Bacci, A.; Matteoli, M. Reduced SNAP-25 alters short-term plasticity at developing glutamatergic synapses. *EMBO Rep.* **2013**, *14*, 645–651. [CrossRef] [PubMed]

68. Vogl, C.; Mochida, S.; Wolff, C.; Whalley, B.J.; Stephens, G.J. The synaptic vesicle glycoprotein 2A ligand levetiracetam inhibits presynaptic Ca^{2+} channels through an intracellular pathway. *Mol. Pharmacol.* **2012**, *82*, 199–208. [CrossRef]

69. Cumbo, E.; Ligori, L.D. Levetiracetam, lamotrigine, and phenobarbital in patients with epileptic seizures and Alzheimer's disease. *Epilepsy Behav. EB* **2010**, *17*, 461–466. [CrossRef]

70. Newhouse, P.A.; Potter, A.; Corwin, J.; Lenox, R. Acute nicotinic blockade produces cognitive impairment in normal humans. *Psychopharmacology* **1992**, *108*, 480–484. [CrossRef]

71. Kadir, A.; Almkvist, O.; Wall, A.; Långström, B.; Nordberg, A. PET imaging of cortical 11C-nicotine binding correlates with the cognitive function of attention in Alzheimer's disease. *Psychopharmacology* **2006**, *188*, 509–520. [CrossRef]

72. Fucile, S. Ca^{2+} permeability of nicotinic acetylcholine receptors. *Cell Calcium* **2004**, *35*, 1–8. [CrossRef] [PubMed]

73. Séguéla, P.; Wadiche, J.; Dineley-Miller, K.; Dani, J.A.; Patrick, J.W. Molecular cloning, functional properties, and distribution of rat brain alpha 7: A nicotinic cation channel highly permeable to calcium. *J. Neurosci.* **1993**, *13*, 596–604. [CrossRef] [PubMed]

74. Thinschmidt, J.S.; Frazier, C.J.; King, M.A.; Meyer, E.M.; Papke, R.L. Medial septal/diagonal band cells express multiple functional nicotinic receptor subtypes that are correlated with firing frequency. *Neurosci. Lett.* **2005**, *389*, 163–168. [CrossRef]

75. Davies, P.; Maloney, A.J. Selective loss of central cholinergic neurons in Alzheimer's disease. *Lancet Lond. Engl.* **1976**, *2*, 1403. [CrossRef]

76. Mufson, E.J.; Counts, S.E.; Perez, S.E.; Ginsberg, S.D. Cholinergic system during the progression of Alzheimer's disease: Therapeutic implications. *Expert Rev. Neurother.* **2008**, *8*, 1703–1718. [CrossRef] [PubMed]

77. Grybko, M.; Sharma, G.; Vijayaraghavan, S. Functional distribution of nicotinic receptors in CA3 region of the hippocampus. *J. Mol. Neurosci. MN* **2010**, *40*, 114–120. [CrossRef]

78. Frazier, C.J.; Rollins, Y.D.; Breese, C.R.; Leonard, S.; Freedman, R.; Dunwiddie, T.V. Acetylcholine activates an alpha-bungarotoxin-sensitive nicotinic current in rat hippocampal interneurons, but not pyramidal cells. *J. Neurosci.* **1998**, *18*, 1187–1195. [CrossRef]

79. Khiroug, L.; Giniatullin, R.; Klein, R.C.; Fayuk, D.; Yakel, J.L. Functional mapping and Ca^{2+} regulation of nicotinic acetylcholine receptor channels in rat hippocampal CA1 neurons. *J. Neurosci.* **2003**, *23*, 9024–9031. [CrossRef]

80. Gray, R.; Rajan, A.S.; Radcliffe, K.A.; Yakehiro, M.; Dani, J.A. Hippocampal synaptic transmission enhanced by low concentrations of nicotine. *Nature* **1996**, *383*, 713–716. [CrossRef]

81. Radcliffe, K.A.; Dani, J.A. Nicotinic stimulation produces multiple forms of increased glutamatergic synaptic transmission. *J. Neurosci.* **1998**, *18*, 7075–7083. [CrossRef]

82. Fayuk, D.; Yakel, J.L. Ca^{2+} permeability of nicotinic acetylcholine receptors in rat hippocampal CA1 interneurones. *J. Physiol.* **2005**, *566*, 759–768. [CrossRef] [PubMed]

83. Cheng, Q.; Yakel, J.L. Presynaptic α7 nicotinic acetylcholine receptors enhance hippocampal mossy fiber glutamatergic transmission via PKA activation. *J. Neurosci.* **2014**, *34*, 124–133. [CrossRef] [PubMed]

84. Sharma, G.; Grybko, M.; Vijayaraghavan, S. Action potential-independent and nicotinic receptor-mediated concerted release of multiple quanta at hippocampal CA3-mossy fiber synapses. *J. Neurosci.* **2008**, *28*, 2563–2575. [CrossRef] [PubMed]

85. Townsend, M.; Whyment, A.; Walczak, J.-S.; Jeggo, R.; van den Top, M.; Flood, D.G.; Leventhal, L.; Patzke, H.; Koenig, G. α7-nAChR agonist enhances neural plasticity in the hippocampus via a GABAergic circuit. *J. Neurophysiol.* **2016**, *116*, 2663–2675. [CrossRef] [PubMed]

86. Wang, H.Y.; Lee, D.H.; D'Andrea, M.R.; Peterson, P.A.; Shank, R.P.; Reitz, A.B. beta-Amyloid(1-42) binds to alpha7 nicotinic acetylcholine receptor with high affinity. Implications for Alzheimer's disease pathology. *J. Biol. Chem.* **2000**, *275*, 5626–5632. [CrossRef] [PubMed]

87. Wang, H.Y.; Lee, D.H.; Davis, C.B.; Shank, R.P. Amyloid peptide Abeta(1-42) binds selectively and with picomolar affinity to alpha7 nicotinic acetylcholine receptors. *J. Neurochem.* **2000**, *75*, 1155–1161. [CrossRef]

88. Liu, Q.; Kawai, H.; Berg, D.K. beta -Amyloid peptide blocks the response of alpha 7-containing nicotinic receptors on hippocampal neurons. *Proc. Natl. Acad. Sci. USA* **2001**, *98*, 4734–4739. [CrossRef]

89. Burghaus, L.; Schütz, U.; Krempel, U.; de Vos, R.A.; Jansen Steur, E.N.; Wevers, A.; Lindstrom, J.; Schröder, H. Quantitative assessment of nicotinic acetylcholine receptor proteins in the cerebral cortex of Alzheimer patients. *Brain Res. Mol. Brain Res.* **2000**, *76*, 385–388. [CrossRef]

90. Guan, Z.-Z.; Zhang, X.; Ravid, R.; Nordberg, A. Decreased Protein Levels of Nicotinic Receptor Subunits in the Hippocampus and Temporal Cortex of Patients with Alzheimer's Disease. *J. Neurochem.* **2000**, *74*, 237–243. [CrossRef]

91. Oddo, S.; Caccamo, A.; Green, K.N.; Liang, K.; Tran, L.; Chen, Y.; Leslie, F.M.; LaFerla, F.M. Chronic nicotine administration exacerbates tau pathology in a transgenic model of Alzheimer's disease. *Proc. Natl. Acad. Sci. USA* **2005**, *102*, 3046–3051. [CrossRef]

92. D'Andrea, M.R.; Nagele, R.G. Targeting the alpha 7 nicotinic acetylcholine receptor to reduce amyloid accumulation in Alzheimer's disease pyramidal neurons. *Curr. Pharm. Des.* **2006**, *12*, 677–684. [CrossRef] [PubMed]

93. Dougherty, J.J.; Wu, J.; Nichols, R.A. Beta-amyloid regulation of presynaptic nicotinic receptors in rat hippocampus and neocortex. *J. Neurosci.* **2003**, *23*, 6740–6747. [CrossRef] [PubMed]

94. Dineley, K.T.; Bell, K.A.; Bui, D.; Sweatt, J.D. beta-Amyloid peptide activates alpha 7 nicotinic acetylcholine receptors expressed in Xenopus oocytes. *J. Biol. Chem.* **2002**, *277*, 25056–25061. [CrossRef]

95. Hascup, K.N.; Hascup, E.R. Soluble Amyloid-β42 Stimulates Glutamate Release through Activation of the α7 Nicotinic Acetylcholine Receptor. *J. Alzheimers Dis. JAD* **2016**, *53*, 337–347. [CrossRef] [PubMed]

96. Puzzo, D.; Privitera, L.; Leznik, E.; Fà, M.; Staniszewski, A.; Palmeri, A.; Arancio, O. Picomolar Amyloid-β Positively Modulates Synaptic Plasticity and Memory in Hippocampus. *J. Neurosci.* **2008**, *28*, 14537–14545. [CrossRef]

97. Kroker, K.S.; Moreth, J.; Kussmaul, L.; Rast, G.; Rosenbrock, H. Restoring long-term potentiation impaired by amyloid-beta oligomers: Comparison of an acetylcholinesterase inhibitior and selective neuronal nicotinic receptor agonists. *Brain Res. Bull.* **2013**, *96*, 28–38. [CrossRef]

98. Ondrejcak, T.; Wang, Q.; Kew, J.N.C.; Virley, D.J.; Upton, N.; Anwyl, R.; Rowan, M.J. Activation of α7 nicotinic acetylcholine receptors persistently enhances hippocampal synaptic transmission and prevents Aß-mediated inhibition of LTP in the rat hippocampus. *Eur. J. Pharmacol.* **2012**, *677*, 63–70. [CrossRef]

99. Medeiros, R.; Castello, N.A.; Cheng, D.; Kitazawa, M.; Baglietto-Vargas, D.; Green, K.N.; Esbenshade, T.A.; Bitner, R.S.; Decker, M.W.; LaFerla, F.M. α7 Nicotinic receptor agonist enhances cognition in aged 3xTg-AD mice with robust plaques and tangles. *Am. J. Pathol.* **2014**, *184*, 520–529. [CrossRef]

100. Shen, H.; Kihara, T.; Hongo, H.; Wu, X.; Kem, W.R.; Shimohama, S.; Akaike, A.; Niidome, T.; Sugimoto, H. Neuroprotection by donepezil against glutamate excitotoxicity involves stimulation of alpha7 nicotinic receptors and internalization of NMDA receptors. *Br. J. Pharmacol.* **2010**, *161*, 127–139. [CrossRef]

101. Texidó, L.; Ros, E.; Martín-Satué, M.; López, S.; Aleu, J.; Marsal, J.; Solsona, C. Effect of galantamine on the human alpha7 neuronal nicotinic acetylcholine receptor, the Torpedo nicotinic acetylcholine receptor and spontaneous cholinergic synaptic activity. *Br. J. Pharmacol.* **2005**, *145*, 672–678. [CrossRef]

102. Yang, T.; Xiao, T.; Sun, Q.; Wang, K. The current agonists and positive allosteric modulators of α7 nAChR for CNS indications in clinical trials. *Acta Pharm. Sin. B* **2017**, *7*, 611–622. [CrossRef] [PubMed]

103. Deardorff, W.J.; Shobassy, A.; Grossberg, G.T. Safety and clinical effects of EVP-6124 in subjects with Alzheimer's disease currently or previously receiving an acetylcholinesterase inhibitor medication. *Expert Rev. Neurother.* **2015**, *15*, 7–17. [CrossRef] [PubMed]

104. Barbier, A.J.; Hilhorst, M.; Van Vliet, A.; Snyder, P.; Palfreyman, M.G.; Gawryl, M.; Dgetluck, N.; Massaro, M.; Tiessen, R.; Timmerman, W.; et al. Pharmacodynamics, pharmacokinetics, safety, and tolerability of encenicline, a selective α7 nicotinic receptor partial agonist, in single ascending-dose and bioavailability studies. *Clin. Ther.* **2015**, *37*, 311–324. [CrossRef]

105. Gee, K.W.; Olincy, A.; Kanner, R.; Johnson, L.; Hogenkamp, D.; Harris, J.; Tran, M.; Edmonds, S.A.; Sauer, W.; Yoshimura, R.; et al. First in human trial of a type I positive allosteric modulator of alpha7-nicotinic acetylcholine receptors: Pharmacokinetics, safety, and evidence for neurocognitive effect of AVL-3288. *J. Psychopharmacol. Oxf. Engl.* **2017**, *31*, 434–441. [CrossRef]

106. Dajas-Bailador, F.A.; Mogg, A.J.; Wonnacott, S. Intracellular Ca^{2+} signals evoked by stimulation of nicotinic acetylcholine receptors in SH-SY5Y cells: Contribution of voltage-operated Ca^{2+} channels and Ca^{2+} stores. *J. Neurochem.* **2002**, *81*, 606–614. [CrossRef] [PubMed]

107. Sharma, G.; Vijayaraghavan, S. Modulation of presynaptic store calcium induces release of glutamate and postsynaptic firing. *Neuron* **2003**, *38*, 929–939. [CrossRef]

108. Padua, R.A.; Nagy, J.I.; Geiger, J.D. Subcellular localization of ryanodine receptors in rat brain. *Eur. J. Pharmacol.* **1996**, *298*, 185–189. [CrossRef]

109. Guerra-Álvarez, M.; Moreno-Ortega, A.J.; Navarro, E.; Fernández-Morales, J.C.; Egea, J.; López, M.G.; Cano-Abad, M.F. Positive allosteric modulation of alpha-7 nicotinic receptors promotes cell death by inducing Ca^{2+} release from the endoplasmic reticulum. *J. Neurochem.* **2015**, *133*, 309–319. [CrossRef]

110. Lacampagne, A.; Liu, X.; Reiken, S.; Bussiere, R.; Meli, A.C.; Lauritzen, I.; Teich, A.F.; Zalk, R.; Saint, N.; Arancio, O.; et al. Post-translational remodeling of ryanodine receptor induces calcium leak leading to Alzheimer's disease-like pathologies and cognitive deficits. *Acta Neuropathol.* **2017**, *134*, 749–767. [CrossRef]

111. Krajnak, K.; Dahl, R. A new target for Alzheimer's disease: A small molecule SERCA activator is neuroprotective in vitro and improves memory and cognition in APP/PS1 mice. *Bioorg. Med. Chem. Lett.* **2018**, *28*, 1591–1594. [CrossRef]

112. Mihalak, K.B.; Carroll, F.I.; Luetje, C.W. Varenicline is a partial agonist at alpha4beta2 and a full agonist at alpha7 neuronal nicotinic receptors. *Mol. Pharmacol.* **2006**, *70*, 801–805. [CrossRef] [PubMed]

113. Kim, S.Y.; Choi, S.H.; Rollema, H.; Schwam, E.M.; McRae, T.; Dubrava, S.; Jacobsen, J. Phase II crossover trial of varenicline in mild-to-moderate Alzheimer's disease. *Dement. Geriatr. Cogn. Disord.* **2014**, *37*, 232–245. [CrossRef]

114. Chakroborty, S.; Briggs, C.; Miller, M.B.; Goussakov, I.; Schneider, C.; Kim, J.; Wicks, J.; Richardson, J.C.; Conklin, V.; Cameransi, B.G.; et al. Stabilizing ER Ca^{2+} Channel Function as an Early Preventative Strategy for Alzheimer's Disease. *PLoS ONE* **2012**, *7*, e52056. [CrossRef] [PubMed]

115. Oulès, B.; Del Prete, D.; Greco, B.; Zhang, X.; Lauritzen, I.; Sevalle, J.; Moreno, S.; Paterlini-Bréchot, P.; Trebak, M.; Checler, F.; et al. Ryanodine Receptor Blockade Reduces Amyloid-β Load and Memory Impairments in Tg2576 Mouse Model of Alzheimer Disease. *J. Neurosci.* **2012**, *32*, 11820–11834. [CrossRef]

116. Peng, J.; Liang, G.; Inan, S.; Wu, Z.; Joseph, D.J.; Meng, Q.; Peng, Y.; Eckenhoff, M.F.; Wei, H. Dantrolene ameliorates cognitive decline and neuropathology in Alzheimer triple transgenic mice. *Neurosci. Lett.* **2012**, *516*, 274–279. [CrossRef]

117. Wang, J.; Shi, Y.; Yu, S.; Wang, Y.; Meng, Q.; Liang, G.; Eckenhoff, M.F.; Wei, H. Intranasal administration of dantrolene increased brain concentration and duration. *PLoS ONE* **2020**, *15*, e0229156. [CrossRef]

118. Lawlor, B.; Segurado, R.; Kennelly, S.; Olde Rikkert, M.G.M.; Howard, R.; Pasquier, F.; Börjesson-Hanson, A.; Tsolaki, M.; Lucca, U.; Molloy, D.W.; et al. Nilvadipine in mild to moderate Alzheimer disease: A randomised controlled trial. *PLoS Med.* **2018**, *15*, e1002660. [CrossRef]

119. Study of AGB101 in Mild Cognitive Impairment Due to Alzheimer's Disease—Full Text View—ClinicalTrials.gov. Available online: https://clinicaltrials.gov/ct2/show/NCT03486938 (accessed on 15 August 2020).

120. Sharp, A.H.; McPherson, P.S.; Dawson, T.M.; Aoki, C.; Campbell, K.P.; Snyder, S.H. Differential immunohistochemical localization of inositol 1,4,5-trisphosphate- and ryanodine-sensitive Ca^{2+} release channels in rat brain. *J. Neurosci.* **1993**, *13*, 3051–3063. [CrossRef]
121. Wu, B.; Yamaguchi, H.; Lai, F.A.; Shen, J. Presenilins regulate calcium homeostasis and presynaptic function via ryanodine receptors in hippocampal neurons. *Proc. Natl. Acad. Sci. USA* **2013**, *110*, 15091–15096. [CrossRef]
122. Payne, A.J.; Gerdes, B.C.; Naumchuk, Y.; McCalley, A.E.; Kaja, S.; Koulen, P. Presenilins regulate the cellular activity of ryanodine receptors differentially through isotype-specific N-terminal cysteines. *Exp. Neurol.* **2013**, *250*, 143–150. [CrossRef]
123. Cheung, K.-H.; Shineman, D.; Müller, M.; Cárdenas, C.; Mei, L.; Yang, J.; Tomita, T.; Iwatsubo, T.; Lee, V.M.-Y.; Foskett, J.K. Mechanism of Ca^{2+} disruption in Alzheimer's disease by presenilin regulation of InsP3 receptor channel gating. *Neuron* **2008**, *58*, 871–883. [CrossRef] [PubMed]
124. Chan, S.L.; Mayne, M.; Holden, C.P.; Geiger, J.D.; Mattson, M.P. Presenilin-1 mutations increase levels of ryanodine receptors and calcium release in PC12 cells and cortical neurons. *J. Biol. Chem.* **2000**, *275*, 18195–18200. [CrossRef] [PubMed]
125. Tu, H.; Nelson, O.; Bezprozvanny, A.; Wang, Z.; Lee, S.-F.; Hao, Y.-H.; Serneels, L.; De Strooper, B.; Yu, G.; Bezprozvanny, I. Presenilins form ER Ca^{2+} leak channels, a function disrupted by familial Alzheimer's disease-linked mutations. *Cell* **2006**, *126*, 981–993. [CrossRef] [PubMed]
126. Nelson, O.; Tu, H.; Lei, T.; Bentahir, M.; de Strooper, B.; Bezprozvanny, I. Familial Alzheimer disease-linked mutations specifically disrupt Ca^{2+} leak function of presenilin 1. *J. Clin. Investig.* **2007**, *117*, 1230–1239. [CrossRef] [PubMed]
127. Aloni, E.; Oni-Biton, E.; Tsoory, M.; Moallem, D.H.; Segal, M. Synaptopodin Deficiency Ameliorates Symptoms in the 3xTg Mouse Model of Alzheimer's Disease. *J. Neurosci.* **2019**, *39*, 3983–3992. [CrossRef] [PubMed]
128. Bertan, F.; Wischhof, L.; Sosulina, L.; Mittag, M.; Dalügge, D.; Fornarelli, A.; Gardoni, F.; Marcello, E.; Di Luca, M.; Fuhrmann, M.; et al. Loss of Ryanodine Receptor 2 impairs neuronal activity-dependent remodeling of dendritic spines and triggers compensatory neuronal hyperexcitability. *Cell Death Differ.* **2020**, 1–20. [CrossRef]
129. Kelliher, M.; Fastbom, J.; Cowburn, R.F.; Bonkale, W.; Ohm, T.G.; Ravid, R.; Sorrentino, V.; O'Neill, C. Alterations in the ryanodine receptor calcium release channel correlate with Alzheimer's disease neurofibrillary and beta-amyloid pathologies. *Neuroscience* **1999**, *92*, 499–513. [CrossRef]
130. Bruno, A.; Huang, J.; Bennett, D.A.; Marr, R.; Hastings, M.L.; Stutzmann, G.E. Altered Ryanodine Receptor Expression in Mild Cognitive Impairment and Alzheimer's Disease. *Neurobiol. Aging* **2012**, *33*, 1001.e1–1001.e6. [CrossRef]
131. Chakroborty, S.; Kim, J.; Schneider, C.; West, A.R.; Stutzmann, G.E. Nitric Oxide Signaling Is Recruited as a Compensatory Mechanism for Sustaining Synaptic Plasticity in Alzheimer's Disease Mice. *J. Neurosci.* **2015**, *35*, 6893–6902. [CrossRef]
132. Redondo, R.L.; Morris, R.G.M. Making memories last: The synaptic tagging and capture hypothesis. *Nat. Rev. Neurosci.* **2011**, *12*, 17–30. [CrossRef]
133. Zhang, C.; Wu, B.; Beglopoulos, V.; Wines-Samuelson, M.; Zhang, D.; Dragatsis, I.; Südhof, T.C.; Shen, J. Presenilins are essential for regulating neurotransmitter release. *Nature* **2009**, *460*, 632–636. [CrossRef] [PubMed]
134. Green, K.N.; Demuro, A.; Akbari, Y.; Hitt, B.D.; Smith, I.F.; Parker, I.; LaFerla, F.M. SERCA pump activity is physiologically regulated by presenilin and regulates amyloid beta production. *J. Cell Biol.* **2008**, *181*, 1107–1116. [CrossRef] [PubMed]
135. Guo, Q.; Sopher, B.L.; Furukawa, K.; Pham, D.G.; Robinson, N.; Martin, G.M.; Mattson, M.P. Alzheimer's presenilin mutation sensitizes neural cells to apoptosis induced by trophic factor withdrawal and amyloid beta-peptide: Involvement of calcium and oxyradicals. *J. Neurosci.* **1997**, *17*, 4212–4222. [CrossRef] [PubMed]
136. Leissring, M.A.; Akbari, Y.; Fanger, C.M.; Cahalan, M.D.; Mattson, M.P.; LaFerla, F.M. Capacitative Calcium Entry Deficits and Elevated Luminal Calcium Content in Mutant Presenilin-1 Knockin Mice. *J. Cell Biol.* **2000**, *149*, 793–798. [CrossRef]
137. Tong, B.C.-K.; Lee, C.S.-K.; Cheng, W.-H.; Lai, K.-O.; Foskett, J.K.; Cheung, K.-H. Familial Alzheimer's disease–associated presenilin 1 mutants promote γ-secretase cleavage of STIM1 to impair store-operated Ca^{2+} entry. *Sci. Signal.* **2016**, *9*, ra89. [CrossRef]

138. Hoozemans, J.J.M.; van Haastert, E.S.; Nijholt, D.A.T.; Rozemuller, A.J.M.; Eikelenboom, P.; Scheper, W. The Unfolded Protein Response Is Activated in Pretangle Neurons in Alzheimer's Disease Hippocampus. *Am. J. Pathol.* **2009**, *174*, 1241–1251. [CrossRef]

139. Uddin, M.S.; Tewari, D.; Sharma, G.; Kabir, M.T.; Barreto, G.E.; Bin-Jumah, M.N.; Perveen, A.; Abdel-Daim, M.M.; Ashraf, G.M. Molecular Mechanisms of ER Stress and UPR in the Pathogenesis of Alzheimer's Disease. *Mol. Neurobiol.* **2020**, *57*, 2902–2919. [CrossRef]

140. Szegezdi, E.; Logue, S.E.; Gorman, A.M.; Samali, A. Mediators of endoplasmic reticulum stress-induced apoptosis. *EMBO Rep.* **2006**, *7*, 880–885. [CrossRef]

141. Gorman, A.M.; Healy, S.J.M.; Jäger, R.; Samali, A. Stress management at the ER: Regulators of ER stress-induced apoptosis. *Pharmacol. Ther.* **2012**, *134*, 306–316. [CrossRef]

142. Ogata, M.; Hino, S.; Saito, A.; Morikawa, K.; Kondo, S.; Kanemoto, S.; Murakami, T.; Taniguchi, M.; Tanii, I.; Yoshinaga, K.; et al. Autophagy Is Activated for Cell Survival after Endoplasmic ReticulumStress. *Mol. Cell. Biol.* **2006**, *26*, 9220–9231. [CrossRef]

143. Cissé, M.; Duplan, E.; Lorivel, T.; Dunys, J.; Bauer, C.; Meckler, X.; Gerakis, Y.; Lauritzen, I.; Checler, F. The transcription factor XBP1s restores hippocampal synaptic plasticity and memory by control of the Kalirin-7 pathway in Alzheimer model. *Mol. Psychiatry* **2017**, *22*, 1562–1575. [CrossRef]

144. Ma, T.; Trinh, M.A.; Wexler, A.J.; Bourbon, C.; Gatti, E.; Pierre, P.; Cavener, D.R.; Klann, E. Suppression of eIF2α kinases alleviates Alzheimer's disease–related plasticity and memory deficits. *Nat. Neurosci.* **2013**, *16*, 1299–1305. [CrossRef] [PubMed]

145. Soejima, N.; Ohyagi, Y.; Nakamura, N.; Himeno, E.; Iinuma, K.M.; Sakae, N.; Yamasaki, R.; Tabira, T.; Murakami, K.; Irie, K.; et al. Intracellular accumulation of toxic turn amyloid-β is associated with endoplasmic reticulum stress in Alzheimer's disease. *Curr. Alzheimer Res.* **2013**, *10*, 11–20. [PubMed]

146. Katayama, T.; Imaizumi, K.; Sato, N.; Miyoshi, K.; Kudo, T.; Hitomi, J.; Morihara, T.; Yoneda, T.; Gomi, F.; Mori, Y.; et al. Presenilin-1 mutations downregulate the signalling pathway of the unfolded-protein response. *Nat. Cell Biol.* **1999**, *1*, 479–485. [CrossRef] [PubMed]

147. Hidalgo, C.; Arias-Cavieres, A. Calcium, Reactive Oxygen Species, and Synaptic Plasticity. *Physiology* **2016**, *31*, 201–215. [CrossRef] [PubMed]

148. Mostafavi, S.; Gaiteri, C.; Sullivan, S.E.; White, C.C.; Tasaki, S.; Xu, J.; Taga, M.; Klein, H.-U.; Patrick, E.; Komashko, V.; et al. A molecular network of the aging human brain provides insights into the pathology and cognitive decline of Alzheimer's disease. *Nat. Neurosci.* **2018**, *21*, 811–819. [CrossRef] [PubMed]

149. Tramutola, A.; Lanzillotta, C.; Perluigi, M.; Butterfield, D.A. Oxidative stress, protein modification and Alzheimer disease. *Brain Res. Bull.* **2017**, *133*, 88–96. [CrossRef]

150. Zhang, H.; Menzies, K.J.; Auwerx, J. The role of mitochondria in stem cell fate and aging. *Development* **2018**, *145*. [CrossRef]

151. Canchi, S.; Raao, B.; Masliah, D.; Rosenthal, S.B.; Sasik, R.; Fisch, K.M.; De Jager, P.L.; Bennett, D.A.; Rissman, R.A. Integrating Gene and Protein Expression Reveals Perturbed Functional Networks in Alzheimer's Disease. *Cell Rep.* **2019**, *28*, 1103–1116.e4. [CrossRef]

152. Manczak, M.; Park, B.S.; Jung, Y.; Reddy, P.H. Differential expression of oxidative phosphorylation genes in patients with Alzheimer's disease. *NeuroMolecular Med.* **2004**, *5*, 147–162. [CrossRef]

153. Adav, S.S.; Park, J.E.; Sze, S.K. Quantitative profiling brain proteomes revealed mitochondrial dysfunction in Alzheimer's disease. *Mol. Brain* **2019**, *12*. [CrossRef] [PubMed]

154. Bubber, P.; Haroutunian, V.; Fisch, G.; Blass, J.P.; Gibson, G.E. Mitochondrial abnormalities in Alzheimer brain: Mechanistic implications. *Ann. Neurol.* **2005**, *57*, 695–703. [CrossRef] [PubMed]

155. Parker, W.D.; Parks, J.; Filley, C.M.; Kleinschmidt-DeMasters, B.K. Electron transport chain defects in Alzheimer's disease brain. *Neurology* **1994**, *44*, 1090–1096. [CrossRef] [PubMed]

156. Gibson, G.E.; Thakkar, A. Interactions of Mitochondria/Metabolism and Calcium Regulation in Alzheimer's Disease: A Calcinist Point of View. *Neurochem. Res.* **2017**, *42*, 1636–1648. [CrossRef]

157. Swomley, A.M.; Butterfield, D.A. Oxidative stress in Alzheimer disease and mild cognitive impairment: Evidence from human data provided by redox proteomics. *Arch. Toxicol.* **2015**, *89*, 1669–1680. [CrossRef] [PubMed]

158. Duchen, M.R. Contributions of mitochondria to animal physiology: From homeostatic sensor to calcium signalling and cell death. *J. Physiol.* **1999**, *516*, 1–17. [CrossRef]

159. Hajnóczky, G.; Hager, R.; Thomas, A.P. Mitochondria suppress local feedback activation of inositol 1,4, 5-trisphosphate receptors by Ca^{2+}. *J. Biol. Chem.* **1999**, *274*, 14157–14162. [CrossRef]
160. Robb-Gaspers, L.D.; Burnett, P.; Rutter, G.A.; Denton, R.M.; Rizzuto, R.; Thomas, A.P. Integrating cytosolic calcium signals into mitochondrial metabolic responses. *EMBO J.* **1998**, *17*, 4987–5000. [CrossRef]
161. Rowland, A.A.; Voeltz, G.K. Endoplasmic reticulum-mitochondria contacts: Function of the junction. *Nat. Rev. Mol. Cell Biol.* **2012**, *13*, 607–615. [CrossRef]
162. Schon, E.A.; Area-Gomez, E. Mitochondria-associated ER membranes in Alzheimer disease. *Mol. Cell. Neurosci.* **2013**, *55*, 26–36. [CrossRef]
163. Szabadkai, G.; Bianchi, K.; Várnai, P.; De Stefani, D.; Wieckowski, M.R.; Cavagna, D.; Nagy, A.I.; Balla, T.; Rizzuto, R. Chaperone-mediated coupling of endoplasmic reticulum and mitochondrial Ca^{2+} channels. *J. Cell Biol.* **2006**, *175*, 901–911. [CrossRef]
164. Wang, M.; Beckmann, N.D.; Roussos, P.; Wang, E.; Zhou, X.; Wang, Q.; Ming, C.; Neff, R.; Ma, W.; Fullard, J.F.; et al. The Mount Sinai cohort of large-scale genomic, transcriptomic and proteomic data in Alzheimer's disease. *Sci. Data* **2018**, *5*, 180185. [CrossRef] [PubMed]
165. De Jager, P.L.; Ma, Y.; McCabe, C.; Xu, J.; Vardarajan, B.N.; Felsky, D.; Klein, H.-U.; White, C.C.; Peters, M.A.; Lodgson, B.; et al. A multi-omic atlas of the human frontal cortex for aging and Alzheimer's disease research. *Sci. Data* **2018**, *5*, 180142. [CrossRef] [PubMed]
166. Hodes, R.J.; Buckholtz, N. Accelerating Medicines Partnership: Alzheimer's Disease (AMP-AD) Knowledge Portal Aids Alzheimer's Drug Discovery through Open Data Sharing. *Expert Opin. Ther. Targets* **2016**, *20*, 389–391. [CrossRef] [PubMed]
167. Calvo-Rodriguez, M.; Hou, S.S.; Snyder, A.C.; Kharitonova, E.K.; Russ, A.N.; Das, S.; Fan, Z.; Muzikansky, A.; Garcia-Alloza, M.; Serrano-Pozo, A.; et al. Increased mitochondrial calcium levels associated with neuronal death in a mouse model of Alzheimer's disease. *Nat. Commun.* **2020**, *11*, 2146. [CrossRef] [PubMed]
168. Han, D.; Antunes, F.; Canali, R.; Rettori, D.; Cadenas, E. Voltage-dependent anion channels control the release of the superoxide anion from mitochondria to cytosol. *J. Biol. Chem.* **2003**, *278*, 5557–5563. [CrossRef]
169. Paula-Lima, A.C.; Adasme, T.; SanMartín, C.; Sebollela, A.; Hetz, C.; Carrasco, M.A.; Ferreira, S.T.; Hidalgo, C. Amyloid β-Peptide Oligomers Stimulate RyR-Mediated Ca^{2+} Release Inducing Mitochondrial Fragmentation in Hippocampal Neurons and Prevent RyR-Mediated Dendritic Spine Remodeling Produced by BDNF. *Antioxid. Redox Signal.* **2010**, *14*, 1209–1223. [CrossRef]
170. SanMartín, C.D.; Veloso, P.; Adasme, T.; Lobos, P.; Bruna, B.; Galaz, J.; García, A.; Hartel, S.; Hidalgo, C.; Paula-Lima, A.C. RyR2-Mediated Ca^{2+} Release and Mitochondrial ROS Generation Partake in the Synaptic Dysfunction Caused by Amyloid β Peptide Oligomers. *Front. Mol. Neurosci.* **2017**, *10*. [CrossRef]
171. Smaili, S.S.; Russell, J.T. Permeability transition pore regulates both mitochondrial membrane potential and agonist-evoked Ca^{2+} signals in oligodendrocyte progenitors. *Cell Calcium* **1999**, *26*, 121–130. [CrossRef]
172. Wang, W.; Zhao, F.; Ma, X.; Perry, G.; Zhu, X. Mitochondria dysfunction in the pathogenesis of Alzheimer's disease: Recent advances. *Mol. Neurodegener.* **2020**, *15*, 30. [CrossRef]
173. Area-Gomez, E.; del Carmen Lara Castillo, M.; Tambini, M.D.; Guardia-Laguarta, C.; de Groof, A.J.; Madra, M.; Ikenouchi, J.; Umeda, M.; Bird, T.D.; Sturley, S.L.; et al. Upregulated function of mitochondria-associated ER membranes in Alzheimer disease. *EMBO J.* **2012**, *31*, 4106–4123. [CrossRef] [PubMed]
174. Area-Gomez, E.; de Groof, A.; Bonilla, E.; Montesinos, J.; Tanji, K.; Boldogh, I.; Pon, L.; Schon, E.A. A key role for MAM in mediating mitochondrial dysfunction in Alzheimer disease. *Cell Death Dis.* **2018**, *9*, 1–10. [CrossRef] [PubMed]
175. Del Prete, D.; Suski, J.M.; Oulès, B.; Debayle, D.; Gay, A.S.; Lacas-Gervais, S.; Bussiere, R.; Bauer, C.; Pinton, P.; Paterlini-Bréchot, P.; et al. Localization and Processing of the Amyloid-β Protein Precursor in Mitochondria-Associated Membranes. *J. Alzheimers Dis. JAD* **2017**, *55*, 1549–1570. [CrossRef] [PubMed]
176. Hedskog, L.; Pinho, C.M.; Filadi, R.; Rönnbäck, A.; Hertwig, L.; Wiehager, B.; Larssen, P.; Gellhaar, S.; Sandebring, A.; Westerlund, M.; et al. Modulation of the endoplasmic reticulum-mitochondria interface in Alzheimer's disease and related models. *Proc. Natl. Acad. Sci. USA* **2013**, *110*, 7916–7921. [CrossRef]
177. Sarasija, S.; Laboy, J.T.; Ashkavand, Z.; Bonner, J.; Tang, Y.; Norman, K.R. Presenilin mutations deregulate mitochondrial Ca^{2+} homeostasis and metabolic activity causing neurodegeneration in Caenorhabditis elegans. *eLife* **2018**, *7*, e33052. [CrossRef]

178. Völgyi, K.; Badics, K.; Sialana, F.J.; Gulyássy, P.; Udvari, E.B.; Kis, V.; Drahos, L.; Lubec, G.; Kékesi, K.A.; Juhász, G. Early Presymptomatic Changes in the Proteome of Mitochondria-Associated Membrane in the APP/PS1 Mouse Model of Alzheimer's Disease. *Mol. Neurobiol.* **2018**, *55*, 7839–7857. [CrossRef]

179. Luongo, T.S.; Lambert, J.P.; Gross, P.; Nwokedi, M.; Lombardi, A.A.; Shanmughapriya, S.; Carpenter, A.C.; Kolmetzky, D.; Gao, E.; van Berlo, J.H.; et al. The Mitochondrial Na$^+$/Ca^{2+} Exchanger Is Essential for Ca^{2+} Homeostasis and Viability. Available online: https://pubmed.ncbi.nlm.nih.gov/28445457/ (accessed on 26 October 2020).

180. Kroemer, G.; Galluzzi, L.; Brenner, C. Mitochondrial membrane permeabilization in cell death. *Physiol. Rev.* **2007**, *87*, 99–163. [CrossRef]

181. Smaili, S.S.; Pereira, G.J.S.; Costa, M.M.; Rocha, K.K.; Rodrigues, L.; do Carmo, L.G.; Hirata, H.; Hsu, Y.-T. The Role of Calcium Stores in Apoptosis and Autophagy. *Curr. Mol. Med.* **2013**, *13*, 252–265. [CrossRef]

182. Magi, S.; Piccirillo, S.; Maiolino, M.; Lariccia, V.; Amoroso, S. NCX1 and EAAC1 transporters are involved in the protective action of glutamate in an in vitro Alzheimer's disease-like model. *Cell Calcium* **2020**, *91*, 102268. [CrossRef]

183. Zilberter, Y.; Zilberter, M. The vicious circle of hypometabolism in neurodegenerative diseases: Ways and mechanisms of metabolic correction. *J. Neurosci. Res.* **2017**, *95*, 2217–2235. [CrossRef]

184. Jadiya, P.; Kolmetzky, D.W.; Tomar, D.; Di Meco, A.; Lombardi, A.A.; Lambert, J.P.; Luongo, T.S.; Ludtmann, M.H.; Praticò, D.; Elrod, J.W. Impaired mitochondrial calcium efflux contributes to disease progression in models of Alzheimer's disease. *Nat. Commun.* **2019**, *10*, 3885. [CrossRef] [PubMed]

185. Piccirillo, S.; Magi, S.; Castaldo, P.; Preziuso, A.; Lariccia, V.; Amoroso, S. NCX and EAAT transporters in ischemia: At the crossroad between glutamate metabolism and cell survival. *Cell Calcium* **2020**, *86*, 102160. [CrossRef] [PubMed]

186. Magi, S.; Piccirillo, S.; Preziuso, A.; Amoroso, S.; Lariccia, V. Mitochondrial localization of NCXs: Balancing calcium and energy homeostasis. *Cell Calcium* **2020**, *86*, 102162. [CrossRef] [PubMed]

187. Gobbi, P.; Castaldo, P.; Minelli, A.; Salucci, S.; Magi, S.; Corcione, E.; Amoroso, S. Mitochondrial localization of Na$^+$/Ca^{2+} exchangers NCX1-3 in neurons and astrocytes of adult rat brain in situ. *Pharmacol. Res.* **2007**, *56*, 556–565. [CrossRef] [PubMed]

188. Mishra, P.; Chan, D.C. Mitochondrial dynamics and inheritance during cell division, development and disease. *Nat. Rev. Mol. Cell Biol.* **2014**, *15*, 634–646. [CrossRef]

189. Kim, B.; Park, J.; Chang, K.-T.; Lee, D.-S. Peroxiredoxin 5 prevents amyloid-beta oligomer-induced neuronal cell death by inhibiting ERK-Drp1-mediated mitochondrial fragmentation. *Free Radic. Biol. Med.* **2016**, *90*, 184–194. [CrossRef]

190. Kim, D.I.; Lee, K.H.; Gabr, A.A.; Choi, G.E.; Kim, J.S.; Ko, S.H.; Han, H.J. Aβ-Induced Drp1 phosphorylation through Akt activation promotes excessive mitochondrial fission leading to neuronal apoptosis. *Biochim. Biophys. Acta* **2016**, *1863*, 2820–2834. [CrossRef]

191. Han, S.; Nandy, P.; Austria, Q.; Siedlak, S.L.; Torres, S.; Fujioka, H.; Wang, W.; Zhu, X. Mfn2 Ablation in the Adult Mouse Hippocampus and Cortex Causes Neuronal Death. *Cells* **2020**, *9*, 116. [CrossRef]

192. Jiang, S.; Nandy, P.; Wang, W.; Ma, X.; Hsia, J.; Wang, C.; Wang, Z.; Niu, M.; Siedlak, S.L.; Torres, S.; et al. Mfn2 ablation causes an oxidative stress response and eventual neuronal death in the hippocampus and cortex. *Mol. Neurodegener.* **2018**, *13*, 5. [CrossRef]

193. Pickett, E.K.; Rose, J.; McCrory, C.; McKenzie, C.-A.; King, D.; Smith, C.; Gillingwater, T.H.; Henstridge, C.M.; Spires-Jones, T.L. Region-specific depletion of synaptic mitochondria in the brains of patients with Alzheimer's disease. *Acta Neuropathol. (Berl.)* **2018**, *136*, 747–757. [CrossRef]

194. Sheng, Z.-H.; Cai, Q. Mitochondrial transport in neurons: Impact on synaptic homeostasis and neurodegeneration. *Nat. Rev. Neurosci.* **2012**, *13*, 77–93. [CrossRef] [PubMed]

195. Ebneth, A.; Godemann, R.; Stamer, K.; Illenberger, S.; Trinczek, B.; Mandelkow, E. Overexpression of tau protein inhibits kinesin-dependent trafficking of vesicles, mitochondria, and endoplasmic reticulum: Implications for Alzheimer's disease. *J. Cell Biol.* **1998**, *143*, 777–794. [CrossRef] [PubMed]

196. Shahpasand, K.; Uemura, I.; Saito, T.; Asano, T.; Hata, K.; Shibata, K.; Toyoshima, Y.; Hasegawa, M.; Hisanaga, S. Regulation of Mitochondrial Transport and Inter-Microtubule Spacing by Tau Phosphorylation at the Sites Hyperphosphorylated in Alzheimer's Disease. *J. Neurosci.* **2012**, *32*, 2430–2441. [CrossRef]

197. Lemasters, J.J. Selective Mitochondrial Autophagy, or Mitophagy, as a Targeted Defense Against Oxidative Stress, Mitochondrial Dysfunction, and Aging. *Rejuvenation Res.* **2005**, *8*, 3–5. [CrossRef] [PubMed]

198. Kerr, J.S.; Adriaanse, B.A.; Greig, N.H.; Mattson, M.P.; Cader, M.Z.; Bohr, V.A.; Fang, E.F. Mitophagy and Alzheimer's Disease: Cellular and Molecular Mechanisms. *Trends Neurosci.* **2017**, *40*, 151–166. [CrossRef] [PubMed]

199. Martín-Maestro, P.; Gargini, R.; Perry, G.; Avila, J.; García-Escudero, V. PARK2 enhancement is able to compensate mitophagy alterations found in sporadic Alzheimer's disease. *Hum. Mol. Genet.* **2016**, *25*, 792–806. [CrossRef]

200. Cárdenas, C.; Foskett, J.K. Mitochondrial Ca^{2+} signals in autophagy. *Cell Calcium* **2012**, *52*, 44–51. [CrossRef]

201. Ding, W.X.; Yin, X.M. Mitophagy: Mechanisms, pathophysiological roles, and analysis. *Biol. Chem.* **2012**, *393*, 547–564. [CrossRef]

202. Okamoto, K.; Kondo-Okamoto, N.; Ohsumi, Y. Mitochondria-anchored receptor Atg32 mediates degradation of mitochondria via selective autophagy. *Dev. Cell* **2009**, *17*, 87–97. [CrossRef]

203. Martín-Maestro, P.; Gargini, R.; A Sproul, A.; García, E.; Antón, L.C.; Noggle, S.; Arancio, O.; Avila, J.; García-Escudero, V. Mitophagy Failure in Fibroblasts and iPSC-Derived Neurons of Alzheimer's Disease-Associated Presenilin 1 Mutation. *Front. Mol. Neurosci.* **2017**, *10*, 291. [CrossRef]

204. Nguyen, T.N.; Padman, B.S.; Lazarou, M. Deciphering the Molecular Signals of PINK1/Parkin Mitophagy. *Trends Cell Biol.* **2016**, *26*, 733–744. [CrossRef] [PubMed]

205. Mustaly, S.; Littlefield, A.; Stutzmann, G.E. Calcium Signaling Deficits in Glia and Autophagic Pathways Contributing to Neurodegenerative Disease. *Antioxid. Redox Signal.* **2018**, *29*, 1158–1175. [CrossRef] [PubMed]

206. Li, W.; Li, J.; Bao, J. Microautophagy: Lesser-known self-eating. *Cell. Mol. Life Sci.* **2012**, *69*, 1125–1136. [CrossRef]

207. Vervliet, T.; Pintelon, I.; Welkenhuyzen, K.; Bootman, M.D.; Bannai, H.; Mikoshiba, K.; Martinet, W.; Nadif Kasri, N.; Parys, J.B.; Bultynck, G. Basal ryanodine receptor activity suppresses autophagic flux. *Biochem. Pharmacol.* **2017**, *132*, 133–142. [CrossRef] [PubMed]

208. Wu, Y.; Whiteus, C.; Xu, C.S.; Hayworth, K.J.; Weinberg, R.J.; Hess, H.F.; De Camilli, P. Contacts between the endoplasmic reticulum and other membranes in neurons. *Proc. Natl. Acad. Sci. USA* **2017**, *114*, E4859–E4867. [CrossRef] [PubMed]

209. Yu, L.; Chibnik, L.B.; Srivastava, G.P.; Pochet, N.; Yang, J.; Xu, J.; Kozubek, J.; Obholzer, N.; Leurgans, S.E.; Schneider, J.A.; et al. Association of Brain DNA methylation in SORL1, ABCA7, HLA-DRB5, SLC24A4, and BIN1 with pathological diagnosis of Alzheimer disease. *JAMA Neurol.* **2015**, *72*, 15–24. [CrossRef]

210. Van Acker, Z.P.; Bretou, M.; Annaert, W. Endo-lysosomal dysregulations and late-onset Alzheimer's disease: Impact of genetic risk factors. *Mol. Neurodegener.* **2019**, *14*, 20. [CrossRef]

211. De Duve, C.; Wattiaux, R. Functions of Lysosomes. *Annu. Rev. Physiol.* **1966**, *28*, 435–492. [CrossRef]

212. Nakamura, N.; Matsuura, A.; Wada, Y.; Ohsumi, Y. Acidification of vacuoles is required for autophagic degradation in the yeast, Saccharomyces cerevisiae. *J. Biochem.* **1997**, *121*, 338–344. [CrossRef]

213. Xu, H.; Ren, D. Lysosomal Physiology. *Annu. Rev. Physiol.* **2015**, *77*, 57–80. [CrossRef]

214. Prasad, H.; Rao, R. The Na$^+$/H$^+$ Exchanger NHE6 Modulates Endosomal pH to Control Processing of Amyloid Precursor Protein in a Cell Culture Model of Alzheimer Disease. *J. Biol. Chem.* **2015**, *290*, 5311–5327. [CrossRef] [PubMed]

215. Lee, J.H.; Yu, W.H.; Kumar, A.; Lee, S.; Mohan, P.S.; Peterhoff, C.M.; Wolfe, D.M.; Martinez-Vicente, M.; Massey, A.C.; Sovak, G.; et al. Lysosomal proteolysis and autophagy require presenilin 1 and are disrupted by Alzheimer-related PS1 mutations. *Cell* **2010**, *141*, 1146–1158. [CrossRef] [PubMed]

216. Lee, J.H.; McBrayer, M.K.; Wolfe, D.M.; Haslett, L.J.; Kumar, A.; Sato, Y.; Lie, P.P.Y.; Mohan, P.; Coffey, E.E.; Kompella, U.; et al. Presenilin 1 Maintains Lysosomal Ca^{2+} Homeostasis via TRPML1 by Regulating vATPase-Mediated Lysosome Acidification. *Cell Rep.* **2015**, *12*, 1430–1444. [CrossRef] [PubMed]

217. Man, S.M.; Kanneganti, T.-D. Regulation of lysosomal dynamics and autophagy by CTSB/cathepsin B. *Autophagy* **2016**, *12*, 2504–2505. [CrossRef] [PubMed]

218. Lee, S.; Sato, Y.; Nixon, R.A. Lysosomal Proteolysis Inhibition Selectively Disrupts Axonal Transport of Degradative Organelles and Causes an Alzheimer's-Like Axonal Dystrophy. *J. Neurosci.* **2011**, *31*, 7817–7830. [CrossRef] [PubMed]

219. Lee, S.; Sato, Y.; Nixon, R.A. Primary lysosomal dysfunction causes cargo-specific deficits of axonal transport leading to Alzheimer-like neuritic dystrophy. *Autophagy* **2011**, *7*, 1562–1563. [CrossRef] [PubMed]

220. Torres, M.; Jimenez, S.; Sanchez-Varo, R.; Navarro, V.; Trujillo-Estrada, L.; Sanchez-Mejias, E.; Carmona, I.; Davila, J.C.; Vizuete, M.; Gutierrez, A.; et al. Defective lysosomal proteolysis and axonal transport are early pathogenic events that worsen with age leading to increased APP metabolism and synaptic Abeta in transgenic APP/PS1 hippocampus. *Mol. Neurodegener.* **2012**, *7*, 59. [CrossRef]

221. Calcraft, P.J.; Ruas, M.; Pan, Z.; Cheng, X.; Arredouani, A.; Hao, X.; Tang, J.; Rietdorf, K.; Teboul, L.; Chuang, K.-T.; et al. NAADP mobilizes calcium from acidic organelles through two-pore channels. *Nature* **2009**, *459*, 596–600. [CrossRef]

222. Cang, C.; Bekele, B.; Ren, D. The voltage-gated sodium channel TPC1 confers endolysosomal excitability. *Nat. Chem. Biol.* **2014**, *10*, 463–469. [CrossRef]

223. Carmona-Gutierrez, D.; Hughes, A.L.; Madeo, F.; Ruckenstuhl, C. The crucial impact of lysosomes in aging and longevity. *Ageing Res. Rev.* **2016**, *32*, 2–12. [CrossRef]

224. Colacurcio, D.J.; Nixon, R.A. Disorders of lysosomal acidificstion-The emerging role of v-ATPase in aging and neurodegenerative disease. *Ageing Res. Rev.* **2016**, *32*, 75–88. [CrossRef] [PubMed]

225. Neely Kayala, K.M.; Dickinson, G.D.; Minassian, A.; Walls, K.C.; Green, K.N.; Laferla, F.M. Presenilin-null cells have altered two-pore calcium channel expression and lysosomal calcium: Implications for lysosomal function. *Brain Res.* **2012**, *1489*, 8–16. [CrossRef] [PubMed]

226. Coen, K.; Flannagan, R.S.; Baron, S.; Carraro-Lacroix, L.R.; Wang, D.; Vermeire, W.; Michiels, C.; Munck, S.; Baert, V.; Sugita, S.; et al. Lysosomal calcium homeostasis defects, not proton pump defects, cause endo-lysosomal dysfunction in PSEN-deficient cells. *J. Cell Biol.* **2012**, *198*, 23–35. [CrossRef] [PubMed]

227. Mauvezin, C.; Nagy, P.; Juhász, G.; Neufeld, T.P. Autophagosome–lysosome fusion is independent of V-ATPase-mediated acidification. *Nat. Commun.* **2015**, *6*, 7007. [CrossRef]

228. Zhang, X.; Garbett, K.; Veeraraghavalu, K.; Wilburn, B.; Gilmore, R.; Mirnics, K.; Sisodia, S.S. A Role for Presenilins in Autophagy Revisited: Normal Acidification of Lysosomes in Cells Lacking PSEN1 and PSEN2. *J. Neurosci.* **2012**, *32*, 8633–8648. [CrossRef]

229. Medina, D.L.; Di Paola, S.; Peluso, I.; Armani, A.; De Stefani, D.; Venditti, R.; Montefusco, S.; Scotto-Rosato, A.; Prezioso, C.; Forrester, A.; et al. Lysosomal calcium signalling regulates autophagy through calcineurin and TFEB. *Nat. Cell Biol.* **2015**, *17*, 288–299. [CrossRef]

230. Sardiello, M.; Palmieri, M.; di Ronza, A.; Medina, D.L.; Valenza, M.; Gennarino, V.A.; Malta, C.D.; Donaudy, F.; Embrione, V.; Polishchuk, R.S.; et al. A Gene Network Regulating Lysosomal Biogenesis and Function. *Science* **2009**, *325*, 473–477. [CrossRef]

231. Settembre, C.; Zoncu, R.; Medina, D.L.; Vetrini, F.; Erdin, S.; Erdin, S.; Huynh, T.; Ferron, M.; Karsenty, G.; Vellard, M.C.; et al. A lysosome-to-nucleus signalling mechanism senses and regulates the lysosome via mTOR and TFEB. *EMBO J.* **2012**, *31*, 1095–1108. [CrossRef]

232. Glavan, G.; Schliebs, R.; Živin, M. Synaptotagmins in neurodegeneration. *Anat. Rec.* **2009**, *292*, 1849–1862. [CrossRef]

233. Jun, K.; Piedras-Rentería, E.S.; Smith, S.M.; Wheeler, D.B.; Lee, S.B.; Lee, T.G.; Chin, H.; Adams, M.E.; Scheller, R.H.; Tsien, R.W.; et al. Ablation of P/Q-type Ca^{2+} channel currents, altered synaptic transmission, and progressive ataxia in mice lacking the α1A-subunit. *Proc. Natl. Acad. Sci. USA* **1999**, *96*, 15245–15250. [CrossRef]

234. Saftig, P.; Klumperman, J. Lysosome biogenesis and lysosomal membrane proteins: Trafficking meets function. *Nat. Rev. Mol. Cell Biol.* **2009**, *10*, 623–635. [CrossRef] [PubMed]

235. Tian, X.; Gala, U.; Zhang, Y.; Shang, W.; Jaiswal, S.N.; di Ronza, A.; Jaiswal, M.; Yamamoto, S.; Sandoval, H.; Duraine, L.; et al. A voltage-gated calcium channel regulates lysosomal fusion with endosomes and autophagosomes and is required for neuronal homeostasis. *PLoS Biol.* **2015**, *13*. [CrossRef] [PubMed]

236. Kinnear, N.P.; Wyatt, C.N.; Clark, J.H.; Calcraft, P.J.; Fleischer, S.; Jeyakumar, L.H.; Nixon, G.F.; Evans, A.M. Lysosomes co-localize with ryanodine receptor subtype 3 to form a trigger zone for calcium signalling by NAADP in rat pulmonary arterial smooth muscle. *Cell Calcium* **2008**, *44*, 190–201. [CrossRef] [PubMed]

237. Foster, W.J.; Taylor, H.B.C.; Padamsey, Z.; Jeans, A.F.; Galione, A.; Emptage, N.J. Hippocampal mGluR1-dependent long-term potentiation requires NAADP-mediated acidic store Ca^{2+} signaling. *Sci. Signal.* **2018**, *11*. [CrossRef]

238. Padamsey, Z.; McGuinness, L.; Bardo, S.J.; Reinhart, M.; Tong, R.; Hedegaard, A.; Hart, M.L.; Emptage, N.J. Activity-Dependent Exocytosis of Lysosomes Regulates the Structural Plasticity of Dendritic Spines. *Neuron* **2017**, *93*, 132–146. [CrossRef]

239. Padamsey, Z.; McGuinness, L.; Emptage, N.J. Inhibition of lysosomal Ca^{2+} signalling disrupts dendritic spine structure and impairs wound healing in neurons. *Commun. Integr. Biol.* **2017**, *10*, e1344802. [CrossRef]

240. Yang, D.S.; Stavrides, P.; Mohan, P.S.; Kaushik, S.; Kumar, A.; Ohno, M.; Schmidt, S.D.; Wesson, D.; Bandyopadhyay, U.; Jiang, Y.; et al. Reversal of autophagy dysfunction in the TgCRND8 mouse model of Alzheimer's disease ameliorates amyloid pathologies and memory deficits. *Brain* **2011**, *134*, 258–277. [CrossRef]

Publisher's Note: MDPI stays neutral with regard to jurisdictional claims in published maps and institutional affiliations.

Review

Mitochondrial Calcium Deregulation in the Mechanism of Beta-Amyloid and Tau Pathology

Noemi Esteras * and Andrey Y. Abramov *

Department of Clinical and Movement Neurosciences, UCL Queen Square Institute of Neurology, Queen Square, London WC1N 3BG, UK
* Correspondence: n.gallego@ucl.ac.uk (N.E.); a.abramov@ucl.ac.uk (A.Y.A.)

Received: 31 August 2020; Accepted: 19 September 2020; Published: 21 September 2020

Abstract: Aggregation and deposition of β-amyloid and/or tau protein are the key neuropathological features in neurodegenerative disorders such as Alzheimer's disease (AD) and other tauopathies including frontotemporal dementia (FTD). The interaction between oxidative stress, mitochondrial dysfunction and the impairment of calcium ions (Ca^{2+}) homeostasis induced by misfolded tau and β-amyloid plays an important role in the progressive neuronal loss occurring in specific areas of the brain. In addition to the control of bioenergetics and ROS production, mitochondria are fine regulators of the cytosolic Ca^{2+} homeostasis that induce vital signalling mechanisms in excitable cells such as neurons. Impairment in the mitochondrial Ca^{2+} uptake through the mitochondrial Ca^{2+} uniporter (MCU) or release through the Na^+/Ca^{2+} exchanger may lead to mitochondrial Ca^{2+} overload and opening of the permeability transition pore inducing neuronal death. Recent evidence suggests an important role for these mechanisms as the underlying causes for neuronal death in β-amyloid and tau pathology. The present review will focus on the mechanisms that lead to cytosolic and especially mitochondrial Ca^{2+} disturbances occurring in AD and tau-induced FTD, and propose possible therapeutic interventions for these disorders.

Keywords: calcium; mitochondria; tau; β-amyloid; MCU; NCLX; VGCCs; glutamate; mPTP

1. Introduction

Neurodegenerative disorders, characterised by progressive neuronal loss in specific areas of the brain, nowadays represent one of the biggest medical and social challenges: very few therapeutic strategies are available to slow down the course of these diseases. Aggregation and deposition of misfolded proteins are histopathological hallmarks in these conditions. Among them, β-amyloid plaques found in Alzheimer's disease (AD) and tau aggregates present in AD, frontotemporal dementia (FTD) and up to other 20 diseases collectively termed tauopathies are one of the most studied [1]. Many actors seem to play an essential role in the pathogenesis of these disorders. The interplay between oxidative stress, mitochondrial dysfunction and calcium ions (Ca^{2+}) impairment has been shown to mediate neuronal dysfunction and death in patients' cells and cellular and animal models of β-amyloid and tau pathology. The present review will focus on the Ca^{2+} signalling impairment, with a special emphasis on the mitochondrial Ca^{2+} dysbalance occurring in AD and tau-induced FTD.

2. Calcium Homeostasis in Neurons

Ca^{2+} signalling is a key mechanism in critical events for cell life, from gene transcription or cell growth, to cell-specific mechanisms, such as muscle contraction or egg fertilisation. In neurons, Ca^{2+} is involved in most aspects of neuronal function: differentiation and migration, synaptic transmission and plasticity, vesicle release, cell death and survival or neuronal–glial communication [2,3]. Indeed, impairment of Ca^{2+} homeostasis has been widely studied and reported to be crucial in the development of neurodegenerative disorders, such as AD, Parkinson's disease, amyotrophic lateral sclerosis or Friedrich Ataxia [4–7].

Ca^{2+} act as second messengers that transmit external signals to its intracellular targets. Ca^{2+} signals are generated by a fine regulation between Ca^{2+} influx and removal, which induces transient fluctuations in the cytosolic $[Ca^{2+}]$. The different kinetics, frequency, amplitudes or spatial locations of these transients entails a signalling mechanism able to exert specific impacts in their downstream effectors [8]. Due to its implications in cellular function, cytosolic Ca^{2+} levels must be therefore tightly regulated, and this becomes essential in excitable cells such as neurons. While a deficient Ca^{2+} signalling might perturb synaptic transmission [9], sustained elevated levels of cytosolic Ca^{2+} are detrimental for neurons: Ca^{2+} overload promotes cell death through different mechanisms, such as necrosis, apoptosis or the more recent ferroptosis, all in which mitochondria play also an essential role [10,11].

In neurons, free cytosolic Ca^{2+} levels are kept at ~100 nM in resting conditions, while the extracellular concentration reaches the millimolar range, defining a substantial concentration gradient of 10^4. The majority of Ca^{2+} influx from the extracellular site in neurons occurs through different ion channels located in the plasma membrane, either voltage- or ligand-operated, upon specific stimulation (Figure 1). In the first case, depolarisation of the neurons leads to the opening of different voltage-gated Ca^{2+} channels (VGCCs) [12], while in the second, the opening of the ionotropic glutamate receptors triggered by the binding of the excitatory neurotransmitter glutamate is the most important example [13]. Regulated Ca^{2+} release to the cytosol can also occur from intracellular Ca^{2+} stores such as the highly dynamic endoplasmic reticulum (ER) [14]. In this case, agonist binding to Ryanodine (RyRs) or inositol 1,4,5-triphosphate (IP3) receptors leads to a release of Ca^{2+} that plays an important role in many neuronal functions [15]. Depletion of the ER induces the activation of the Store-Operated Calcium Entry (SOCE) in order to replenish the organelle. In neurons, STIM proteins located in the ER sense the decrease in $[Ca^{2+}]$, accumulate close to the ER-plasma membrane junctions, and interact with the Store-Operated Calcium Channels (SOCCs) in the plasma membrane, allowing the entrance of Ca^{2+}. Orai channels were identified as components of the SOCCs [16], while transient receptor potential channels (TRPC) also play a relevant role [17].

Either way, the duration and spread of the Ca^{2+} signals is controlled by several clearance mechanisms, which dissipate the massive increase in the cytosolic $[Ca^{2+}]$ and restore it to its basal levels to maintain Ca^{2+} homeostasis. These mechanisms include the efflux of Ca^{2+} by transporters through the plasma membrane, uptake by organelles such as the ER and the mitochondria and binding to Ca^{2+}-buffering proteins.

The main transporters implicated in the efflux of Ca^{2+} out of the neurons are the high-affinity, low capacity plasma membrane Ca^{2+}-ATPase (PMCA), which hydrolyses ATP to pump Ca^{2+} against gradient, and the low affinity, high capacity Na^+/Ca^{2+} exchanger (NCX), which is abundant in neurons and uses the electrochemical gradient of Na^+ to extrude Ca^{2+} [18]. NCX is reversible, and under specific circumstances of Na^+ and Ca^{2+} gradient and membrane potential can work in the opposite direction, extruding Na^+ and letting Ca^{2+} in [19]. The two main intracellular stores that also collaborate in the uptake of cytosolic Ca^{2+} are the mitochondria (which will be discussed in detail later) and the ER, which uses the Sarco-Endoplasmic Reticulum Ca^{2+}-ATPase (SERCA) to pump Ca^{2+} into the ER lumen, at the expense of ATP hydrolysis. Finally, several cytosolic Ca^{2+}-binding proteins also cooperate in the Ca^{2+} homeostasis by binding and buffering free cytosolic Ca^{2+}. The most important belong to the EF-hand family, and include parvalbumin, calbindin D-28k and calretinin, which are expressed in different areas of the brain [20]. Other members of the family, such as calmodulin, S100 proteins or neuronal Ca^{2+} sensors (NCS), are also Ca^{2+}-binding proteins that act as Ca^{2+} sensors which transduce the signal to downstream effectors. The latter involve a complex network of signalling cascades that ultimately have specific cellular effects: Ca^{2+}-calmodulin kinase II, which regulates long-term potentiation, learning and memory; protein kinase A, which modulates neuronal excitability;

or calpains, a family of proteases that cleave amyloid precursor protein APP or tau protein are just a few examples [21,22].

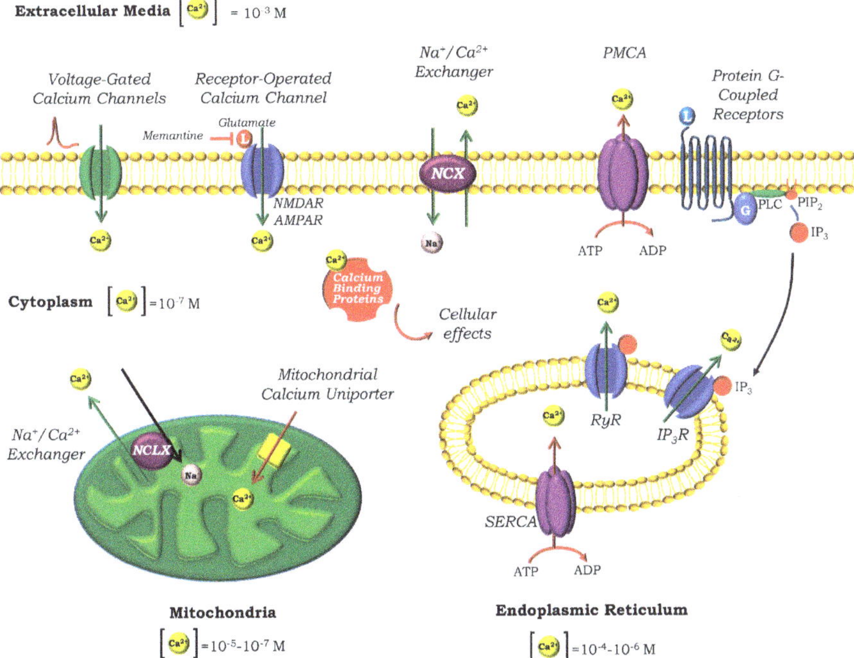

Figure 1. Calcium homeostasis in neurons. Ca^{2+} signals are shaped by a fine regulation between cytosolic Ca^{2+} influx and efflux. The main sources for Ca^{2+} influx are the extracellular media and intracellular stores such as the endoplasmic reticulum (ER). Depolarisation of the neurons leads to the opening of the voltage-gated calcium channels (VGCCs) in the plasma membrane, while ligand binding triggers the opening of the receptor-operated calcium channels (ROCs). AMPA and especially NMDA receptors, both activated by glutamate, are the most important ROCs in the neurons. AD-approved drug memantine is an inhibitor of the NMDARs. Ca^{2+} can also be released to the cytosol from the ER, after activation of the Ryanodine or inositol 1,4,5-triphosphate (IP_3) receptors. Binding of a ligand (such as glutamate) to a G-protein-coupled receptor in the plasma membrane (such as specific metabotropic glutamate receptors) activates phospholipase C (PLC), leading to the cleavage of the membrane phospholipid phosphatidylinositol 4,5-bisphosphate (PIP_2), resulting in the release of the soluble second messenger IP3, that diffuses through the cell and binds its receptor. Cytosolic Ca^{2+} binds specific Ca^{2+} binding proteins, which transduce the signal to its final effectors. Excess cytosolic Ca^{2+} is removed from the cytosol by different mechanisms: (i) efflux through the plasma membrane by the Na^+/Ca^{2+} exchanger NCX and the plasma membrane Ca^{2+}-ATPase (PMCA), (ii) uptake to the ER by the sarco/endoplasmic reticulum Ca^{2+}-ATPase (SERCA), (iii) uptake to the mitochondria by the mitochondrial Ca^{2+} uniporter MCU and (iv) buffering by Ca^{2+} binding proteins. Mitochondrial Ca^{2+} homeostasis is maintained by the efflux through the mitochondrial Na^+/Ca^{2+} exchanger NCLX.

3. Mitochondria and Ca^{2+} Homeostasis

Mitochondria play a fundamental role in the rapid buffering and shaping of the cytosolic Ca^{2+} transients. These are mobile organelles, which can be strategically recruited in close proximity to microdomains with a high cytosolic [Ca^{2+}], such as the synapses, acting as highly localised Ca^{2+} buffers able to shape the local Ca^{2+} signals and regulate neuronal activity [23].

With additional essential roles, such as ROS production or triggering of apoptosis, the best-known mitochondrial function is controlling the bioenergetics of the cell. Substrate oxidation in the Krebs' cycle occurs in the matrix and provides the electron transport chain (ETC) in the inner mitochondrial membrane with NADH and $FADH_2$. Electrons transfer from these donors to its final acceptor O_2 through the ETC is coupled with the translocation of protons to the intermembrane space. This creates an electrochemical gradient whose major component is the membrane potential ($\Delta\Psi m$), which fuels ATP production in the ATP synthase. Importantly, mitochondrial Ca^{2+} uptake activates dehydrogenases at the ETC activating mitochondrial respiration and ATP production [24,25].

In addition to its bioenergetics purpose, $\Delta\Psi m$ is also used by the mitochondria to uptake Ca^{2+} into their matrix through the high capacity, low-affinity mitochondrial Ca^{2+} uniporter (MCU) located in the inner membrane. The molecular composition of the uniporter has been recently elucidated, and involves a protein complex consisting on MCU—the pore-forming component—and several regulatory units (MICU1, MICU2, MCUb, MCUR1 and EMRE) (reviewed in [26]). The role of MICU3, highly expressed in the brain, in enhancing MCU Ca^{2+} uptake has been recently described [27]. Current investigations are focused in understanding the regulation of MCU and the specific role of all these proteins in Ca^{2+} uptake [28,29]. First experiments with MCU knock-out animal models surprisingly revealed that these mice displayed only a mild muscular phenotype [30] and a relatively normal heart function [31]. However, MCU KO in a different genetic background [32] or MICU1 KO appeared to be lethal [33]. Other studies focused in brain function show that silencing of MCU during development induces memory impairment in Drosophila [34], and experiments in brain MCU-KO mitochondria revealed that MCU deletion did not completely block mitochondrial Ca^{2+} uptake, suggesting additional uptake pathways [35]. Indeed, it is still a matter of debate if the complex regulation of MCU can result in alternative uptake modes or if other MCU-independent mechanisms coexist and mediate Ca^{2+} uptake in the mitochondria [36,37].

The mitochondrial Na^+/Ca^{2+} exchanger NCLX, located in the inner mitochondrial membrane [38], was also molecularly identified not long ago [39] as being responsible for mitochondrial Ca^{2+} efflux in excitable cells. NCLX is a low affinity, high capacity transporter that uses the electrochemical gradient of Na^+ to extrude Ca^{2+} from mitochondria. Like other Na^+/Ca^{2+} exchangers, it is related to the plasma membrane NCX, but in addition to Na^+ is able to exchange Li^+ for Ca^{2+}. Considering that the rate of Ca^{2+} efflux is much slower than MCU-mediated influx, NCLX appears to mediate the rate limiting step in mitochondrial Ca^{2+} homeostasis [40]. Indeed, in contrast to MCU, NCLX deletion is associated with more severe phenotypes in vivo: conditional cardiac NCLX deletion in mice leads to myocardial dysfunction and fulminant heart failure [41]. Inhibition of NCLX in Parkinson's disease related mutation PINK1 cells leads to mitochondrial Ca^{2+} overload and cell death [42,43]. Regulation of the exchanger activity can occur via different mechanisms such as $[Ca^{2+}]$, via direct and indirect mechanisms like calpain-induced degradation, pH, PKC or PKA, as recently reviewed in [44]. In addition, a role for plasma membrane NCX1-3 in mediating mitochondrial Ca^{2+} efflux in brain cells has also been proposed [45–47].

Ca^{2+} uptake into the mitochondrial matrix stimulates mitochondrial bioenergetics. Several matrix dehydrogenaseses and metabolite carriers are activated by Ca^{2+}, increasing mitochondrial respiration and ATP production [48,49]. This suggests a physiological role for mitochondrial Ca^{2+} in the adaptation of the cell to the energy demands imposed by Ca^{2+} signalling. However, as with cytosolic $[Ca^{2+}]$, mitochondrial Ca^{2+} content must be tightly regulated. Mitochondrial Ca^{2+} overload, especially under conditions of oxidative stress, triggers the opening of the mitochondrial permeability transition pore (mPTP), a high-conductance mitochondrial channel whose composition and structure are still under debate. While firmly closed under physiological conditions, after mPTP opening, the mitochondrial inner membrane becomes unselectively permeable to small solutes, leading to $\Delta\Psi m$ collapse and eventually mitochondrial swelling and necrotic and apoptotic cell death. mPTP opening has been implicated as the mechanism of cell death in many human diseases, thus representing a major therapeutic target [50]. It should be noted that ROS are one of the most important triggers for mPTP in combination with Ca^{2+} overload [51]. Misfolded proteins, including β-amyloid and tau, are able to induce ROS production in enzymes (including ETC of mitochondria and NADPH oxidase) or produce free radicals in combination with heavy metals, and trigger mPTP opening [52,53].

4. Calcium Homeostasis Impairment in AD and Tauopathies

As mentioned before, Ca^{2+} homeostasis impairment has been linked to many different diseases, including neurodegenerative conditions.

AD is the most common neurodegenerative disorder and the principal cause of dementia. It affects millions of people worldwide, with numbers expecting to multiply in the next years, conveying a critical social and medical challenge. Clinically, AD is characterised by progressive memory loss and cognitive and behavioural impairment [54]. Neuronal and synaptic loss in specific areas of the hippocampus and neocortex, together with the presence of extracellular β-amyloid plaques and intracellular neurofibrillary tangles (NFTs) containing tau aggregates comprise the main neuropathological hallmarks of the disease [55]. Together with them, oxidative stress, mitochondrial dysfunction and altered Ca^{2+} homeostasis have emerged as important actors and been long studied in the last decades to try to unravel the interplay of all these factors in the pathogenesis of the disease [4,56,57]. Although the majority of the cases are sporadic, a small percentage of them are rare familiar cases with an early onset, linked to mutations in the amyloid precursor protein *APP* gene and presenilins 1 and 2 *PSEN1* and *PSEN2* genes, components of the γ-secretase complex involved in the amyloidogenic cleavage of APP that leads to β-amyloid formation [58]. This evidence suggests a critical role for β-amyloid in AD pathogenesis. Indeed, the Amyloid cascade hypothesis, formulated in the early 1990s by Hardy and Higgins [59], proposes that the deposition of β-amyloid is the causative agent of AD pathology, leading to NFTs, neuronal loss and dementia. However, β-amyloid deposits correlate weakly with neuronal death, while spreading of tau pathology through the brain and the number of NFTs are strongly associated with the progression of AD [60]. Tau is a soluble protein that plays a critical role in the stabilisation of the microtubules, but under pathological circumstances self-aggregates into paired-helical fragments (PHF) whose aggregation finally leads to NFTs. Importantly, abnormal tau hyperphosphorylation impacts its pathogenic role and aggregation capacity and indeed deposited tau is highly phosphorylated. In addition, tau isoform imbalance is sufficient to cause neurodegeneration [61]. Interestingly, mutations in the *MAPT* gene encoding tau protein are not linked to AD, but to frontotemporal dementia (FTD) and other tauopathies, a term that comprehends a wider range of neurodegenerative disorders in which deposits of tau are found in the brain [1]. FTD is characterised by the progressive neurodegeneration of frontal and temporal lobes of the brain, and comprises different molecular and clinical entities affecting behaviour, function and language of the patients, which are usually younger than those with AD [62,63]. Research in FTD has gained increasing attention in the recent years, but as in AD, there is still a lot to learn to be able to prevent or cure these disorders.

The important role of Ca^{2+} dysfunction in AD was first proposed by Khachaturian 25 years ago [64]. Growing body of evidence has been published since then, highlighting the multiple molecular mechanisms that can contribute to the Ca^{2+} homeostasis impairment in AD. The role of tau, and especially β-amyloid, has been extensively studied in different animal and cellular models.

4.1. Cytosolic Ca^{2+} Disturbances in AD and Tauopathies

β-amyloid was first shown to form Ca^{2+}-permeable pores in artificial membranes [65] that lead to dysregulated Ca^{2+} entry in the cytoplasm of brain cells [66,67]. Although less studied, we, and others, have shown that tau is also able to form ion channels under specific conditions [68,69]. Importantly, structure and aggregation stage determined the ability of both proteins to form pores.

Alteration of the glutamatergic signalling, involved in synaptic plasticity, learning and memory, also plays an important role in the Ca^{2+} imbalance and synaptic dysfunction in AD [70,71]. Glutamate is the major excitatory neurotransmitter in the brain and activates a family of metabotropic (G-coupled proteins) and ionotropic (ion channels) receptors. Among the latter, AMPA, and especially NMDA receptors, have attracted much attention due to its role in mediating glutamate excitotoxicity. Excitotoxicity is defined as the neuronal death induced by cellular overload of Ca^{2+} due to excessive stimulation of the glutamate receptors, caused, for example, by an excess of extracellular glutamate. It is involved in the mechanism of cell death in acute (stroke) and chronic neurodegenerative disorders and such has attracted great attention in the pathogenesis of AD [72]. Research has shown that β-amyloid oligomers can directly activate NMDA receptors [73], and specifically those containing the NR2B subunit [74,75]. Receptors expressing this subunit are preferentially localised in the extrasynaptic area and mediate excitotoxicity [76]. It was proposed that modulating the balance between synaptic NR2A and extrasynaptic NR2B may improve behaviour ability in β-amyloid treated mice [77]. Tau involvement in excitotoxicity has been also described in AD [78–80] and FTD [81]. For a review of the role of glutamate receptors in AD, see in [82]. Importantly, the non-competitive NMDA receptor antagonist memantine is one of the few drugs approved for use in AD.

The rest of the approved drugs for AD are cholinesterase inhibitors that aim to prevent acetylcholine degradation. Indeed, the cholinergic pathway has been long implicated in AD pathogenesis, and it was proposed that the loss of cholinergic neurotransmission leads to cognitive impairment [83]. Importantly, tau has been shown to play a role in the loss of cholinergic neurons through interaction with muscarinic receptors [84], and some of the toxic effects of β-amyloid are mediated by its interaction with nicotinic acetylcholine receptors. Interestingly, acetylcholine and antibodies against acetylcholine receptors protect neurons against β-amyloid-induced cell death but have no effect on the β-amyloid-induced Ca^{2+} deregulation [85].

Tau and β-amyloid-induced Ca^{2+} dysfunction through VGCCs have been described in different models of AD and tau-induced FTD [86,87]. We have recently shown that in vitro aggregated tau fibrils with the P301S mutation linked to FTD are able to incorporate into membranes and modify their ionic currents, as seen by BLM experiments [69]. When applied to primary neuronal cultures, this leads to the opening of neuronal VGCCs, inducing characteristic Ca^{2+} transients in these cells. Increased cytosolic $[Ca^{2+}]$ is able to activate NADPH oxidase, enhancing ROS production in neurons and leading to cell death. Ca^{2+} signals and increased ROS production were observed after the acute application of tau aggregates, suggesting a mechanism by which extracellular tau fibrils can incorporate into the membranes and lead to neuronal dysfunction in the neighbouring neurons [69]. Importantly, we show that tau-induced Ca^{2+} transients and NADPH-driven ROS production were prevented by nifedipine and verapamil, Ca^{2+} channels blockers commonly used in clinic for hypertension. Clinical trials with these compounds in patients with dementia show heterogeneous results, with many demonstrating no positive effect for Ca^{2+} blockers in reducing the rate of cognitive decline in AD patients [88]. However, the severity of the disease at the beginning of the treatment seems to influence the outcome [89]. Indeed, several studies have shown that hypertense patients on treatment with this group of drugs could have a reduced risk of dementia [90–92], suggesting a potential use of these medications for the prevention of the disease that needs to be further confirmed.

ER Ca^{2+} dysregulation also plays a role in AD. Many authors have shown an increased Ca^{2+} release from the ER both through RyRs [93,94] and IP3Rs [95,96] by different mechanisms. In addition, impairment of the STIM-mediated SOCE has been described in familiar models of the disease [97,98] and recent studies point at the role of tau in ER stress through TRPC and SOCE upregulation [99].

Ca^{2+} efflux through the plasma membrane by PMCA can be inhibited by β-amyloid and tau, as shown by Mata et al., whose findings are summarised in their review [100]. Both proteins appear to bind the transporter, with tau inhibitory effect occurring in the nanomolar range. In addition, the possible oxidation of PMCA induced by β-amyloid and tau might lead to a decrease in the ATPase activity [101]. β-amyloid is also able to interact with the plasma membrane NCX and reduce its activity [102]. Differing results have been published regarding the protein levels of NCX isotypes in the different areas of brains of patients [103]. However, studies in AD brains and neuronal cultures pointed the specific altered cleavage of NCX3 (and not NCX1) mediated by the Ca^{2+}-dependent protease calpain [104]. Interestingly, this feature appeared only in AD brains and not in brains from tauopathies like FTD, suggesting a specific role for β-amyloid and not tau. In addition, both NCX and PMCA can be downregulated in response to oxidative stress [105].

Calpain overactivation has been consistently reported in AD and tauopathy brains [106], and β-amyloid [107] and tauopathy models [108,109]. Other Ca^{2+}-dependent molecules, such as calmodulin and its binding proteins have a prominent role in AD [110] and have been suggested as potential biomarkers of the disease [111].

4.2. Mitochondrial Ca^{2+} Disturbances in AD and Tauopathies

Neurons, as excitable cells, are continuously firing action potentials and employing a vast Ca^{2+} signalling that comes at the expense of an increased metabolic demand to maintain Ca^{2+} homeostasis (for example through Ca^{2+}-ATPases) and re-establish electrochemical gradients. In this context, mitochondria play a fundamental role in maintaining the metabolic needs. This could represent a challenge in neurodegenerative disorders, in which mitochondrial dysfunction has been extensively described together with oxidative stress and Ca^{2+} impairment, all of which are implicated in the pathogenesis of the disease [53,57,112].

In addition, mitochondria themselves are direct sites of action of β-amyloid and tau. β-amyloid has been shown to be imported via TOM [113] and directly produced in this organelle [114], while a fraction of intracellular tau has been found to locate within the inner mitochondrial space [115]. Indeed, mitochondrial accumulation of tau in synaptosomes from AD brains appeared to correlate with synaptic loss [116].

Tau and beta-amyloid dysfunction have been widely linked to altered cytosolic Ca^{2+} homeostasis through the different mechanisms explained before. These scenarios compromise mitochondria in two ways: challenging mitochondrial Ca^{2+} buffering capacity, which might become overloaded, and, in addition, the cellular bioenergetics of cells in which mitochondrial function could be already impaired. Mitochondrial Ca^{2+} uptake by MCU is driven by the $\Delta\Psi m$, and therefore mitochondrial depolarisation might compromise the uptake of Ca^{2+} and its physiological role in bioenergetics, and in addition expose the cytosol to higher $[Ca^{2+}]$. On the other hand, mitochondrial depolarisation and bioenergetics dysfunction can be triggered by Ca^{2+} and prevented by the inhibition of mitochondrial Ca^{2+} uptake as previously shown in works by Abramov and Duchen [117–119].

Several reports highlight the role of the mitochondrial Ca^{2+} uptake in neuronal death induced by glutamate excitotoxicity [120,121]. Qiu et al. showed that MCU overexpression exacerbated excitotoxic cell death, while MCU silencing prevented NMDA-induced mitochondrial Ca^{2+} uptake protecting neurons from excitotoxic cell death [122]. Our group has recently described the protective role of a novel compound, TG-2112x, which is able to partially inhibit mitochondrial Ca^{2+} uptake without affecting $\Delta\Psi m$ or bioenergetics, and protects neurons against glutamate excitotoxicity [123]. These results from Angelova et al. suggest this compound as a new therapeutic opportunity in diseases such as AD in which excitotoxicity play a detrimental role.

Other authors have proposed that the induction of a mild mitochondrial uncoupling with different agents such as non-steroidal anti-inflammatory drugs (NSAIDs) could also reduce mitochondrial Ca^{2+} uptake and prevent overload induced by β-amyloid, protecting neurons against cell death in AD [124]. Results from trials have been however conflicting, probably due to the narrow effective dose window for this strategy, as high doses might lead to opposite effects and collapse the $\Delta\Psi m$ [125].

Recent in vivo imaging by Calvo-Rodriguez et al. in the APP/PS1 transgenic mouse model of AD has shown β-amyloid dependent mitochondrial Ca^{2+} overload in a subset of neurons in the brain of these mice, which preceded neuronal death and could be prevented by MCU inhibition [126]. Interestingly, neuronal death did not occur in neighbour cells with lower mitochondrial Ca^{2+} levels highlighting one more time the deleterious effect of mitochondrial Ca^{2+} overload. This work also evaluated available microarray and RNA-Sequencing datasheets to analyse the expression of mitochondrial Ca^{2+}-related genes in patients with AD and found that all the genes involved in mitochondrial uptake were downregulated, while *Slc8b1* gene encoding NCLX was significantly upregulated, suggesting a possible compensatory response to prevent mitochondrial Ca^{2+} overload [126]. However, other reports show contradictory results [127].

The mitochondria-associated ER membranes are subcompartments of the ER connected physically and biochemically to the mitochondria allowing the communication between both organelles and the transfer of Ca^{2+} from ER to mitochondria [128]. Many relevant functions for the pathogenesis of AD such as β-amyloid production appear to occur in these regions [129] and a higher degree of apposition between ER and mitochondria has been found in AD cells, brains and mouse models [130,131]. As seen in preselinin 2 (PS2) cellular and animal models of AD, the increased ER-mitochondria interactions enhance Ca^{2+} transfer, which might contribute to mitochondrial Ca^{2+} overload [132,133]. Some authors have shown that presenilins are able to form cation-permeable pores responsible for passive Ca^{2+} leak from the ER [134], thus contributing to the pathogenesis of the disease, although this hypothesis is under debate, with other authors showing opposite results [135].

mPTP opening induced by mitochondrial Ca^{2+} overload is one of the mechanisms of β-amyloid- and tau-induced mitochondrial dysfunction and cell death [6,8,136–138]. β-amyloid is able to interact with a key component of the pore, cyclophilin D, and potentiate mitochondrial dysfunction and mPTP formation [139]. Reduction in cyclophilin D expression, on the other hand, protects neurons and improves learning and memory in mouse models of AD [139]. In addition, treatment with the classical blocker of mPTP, cyclosporine A, or removal of polyphosphate, thought to be a component of the pore, are able to prevent β-amyloid-induced mPTP opening and cell death [119,137].

Impairment of mitochondrial Ca^{2+} efflux has not been explored in the pathogenesis of AD until very recently. Jadiya et al. have shown in different mouse and animal models of the disease that AD progression is associated with the loss of NCLX expression and functionality [127]. Importantly, genetic rescue of NCLX expression in neurons completely restored the cognitive decline and the cellular pathology in the AD mice.

Recent work from our group has shown for the first time the tau-induced altered mitochondrial Ca^{2+} efflux through NCLX in neurons [138]. In this study, we show that K18 tau, a fragment of the protein comprising the four repeat (4R) region of the protein, led to cytosolic Ca^{2+} oscillations in primary neurons after 24 h incubation. These oscillations were followed by mitochondrial Ca^{2+} uptake, and induced a gradual increase in basal cytosolic and mitochondrial Ca^{2+}, suggesting an impaired Ca^{2+} handling induced by tau [138]. Stimulation of a physiological Ca^{2+} signal with glutamate (in neurons) or ATP (in astrocytes) incubated with tau further evidenced a slower cytosolic and mitochondrial Ca^{2+} efflux in both cell types. Experiments in permeabilised cells confirmed that the impairment was mediated by NCLX, as showed by the altered Na^+ and Ca^{2+} currents. More importantly, tau-induced NCLX impairment led to a faster mitochondrial depolarisation when exposing the neurons to (pathological) repetitive Ca^{2+} stimulations, suggesting an increased vulnerability to Ca^{2+}-induced cell death [138]. iPSC-derived neurons from patients carrying the FTD-related 10+16 mutation in *MAPT* were also more vulnerable to physiological and pathological Ca^{2+} stimulation, and presented an increased susceptibility to mPTP opening [138].

10+16 *MAPT* mutation also impairs neuronal excitability [140] and bioenergetics of iPSC-derived neurons [141]. In contrast to other neurodegeneration models [142,143], 10+16 *MAPT* neurons display an increased mitochondrial membrane potential [141]. As a result, mitochondrial ROS production in the neurons is enhanced, leading to oxidative stress and neuronal death, all of which are prevented treating the cells with mitochondrial antioxidants.Oxidative stress, in combination with mitochondrial Ca^{2+} overload, are the triggers for mPTP opening. Preliminary data shows that mitochondrial antioxidants are able to protect the 10+16 neurons and reduce their susceptibility to mPTP opening (Figure 2). Confirming previous results [138], cellular Ca^{2+} overload induced by the ionophore ferutinin triggered mPTP opening and led to apoptosis in iPSC-neurons, which occurred significantly earlier in the FTD patients than in controls (Figure 2). Treatment with the mitochondrial antioxidant MitoTEMPO (MT, 1 h, 100 nM) significantly delayed the mPTP opening in the patients' neurons to times similar to control, thus counteracting their increased vulnerability. This, together with previous results [141], highlights the potential role of mitochondrial antioxidants in the prevention of neuronal death through different mechanisms that might include averting mitochondrial Ca^{2+} overload. Further investigations will be needed to prove this point and understand if this effect is merely due to the reduction of the already elevated mitochondrial ROS, or if mito ROS overproduction induced by tau might influence other aspects of cytosolic or mitochondrial Ca^{2+} homeostasis trough different mechanisms such as redox regulation. These results highlight the close interconnection between impaired bioenergetics, oxidative stress and Ca^{2+} signalling in tau pathology.

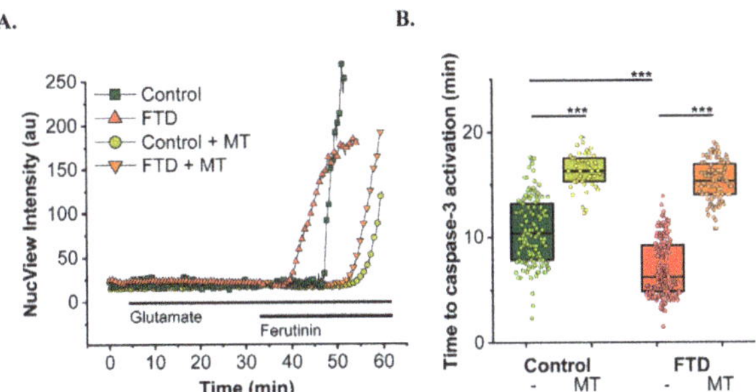

Figure 2. Mitochondrial antioxidants reduce FTD neurons vulnerability to mPTP opening. (**A**) Representative traces depict NucView intensity in iPSC-derived neurons from controls or FTD-related mutation 10+16 in *MAPT* treated or not with MitoTEMPO (MT) 100 nM and exposed to 50 μm glutamate or the electrogenic Ca^{2+} ionophore ferutinin [144,145]. Sudden increase in NucView fluorescence indicates caspase-3 activation. (**B**) Time to caspase-3 activation after Ca^{2+} overload with ferutinin. Box represents median and 25, 75 percentiles. $n = 126$ neurons analysed in control, FTD $n = 199$, control + MT, $n = 43$, FTD + MT, $n = 104$. *** $p < 0.001$, Mann–Whitney test. Method: iPSC-derived neurons were loaded with the non-fluorescent caspase-3 substrate NucView488 for 15 min. NucView is cleaved upon caspase-3 activation inducing a sudden increase in green fluorescence. Images were taken on a Zeiss 710 LSM confocal microscope with an integrated META detection system.

5. Conclusions

Ca^{2+}, and especially mitochondrial Ca^{2+} homeostasis, plays a key role in neurodegenerative disorders including AD and other tauopathies like FTD. As detailed in the present review, both β-amyloid and tau protein induce cytosolic and mitochondrial Ca^{2+} deregulation through different direct and indirect pathways that ultimately lead to neuronal dysfuntion and cell death. Importantly, isoform type, length, or aggregation stage, among other characteristics of these proteins have been shown to influence the pathogenic mechanism. Mitochondrial Ca^{2+} overload appears as a downstream key event in the process of neurodegeneration, and recent studies point at a direct role of these proteins in the impairment of mitochondrial Ca^{2+} influx (β-amyloid) and efflux (β-amyloid and tau). Specific targeting of the mechanisms leading to Ca^{2+} impairment, and especially mitochondria-targeted therapies emerge as potential treatments for these disorders.

Funding: This research received no external funding.

Acknowledgments: We would like to thank Dr. Selina Wray, who generated the iPSC lines from patients used in this manuscript.

Conflicts of Interest: The authors declare no conflict of interest.

References

1. Spillantini, M.G.; Goedert, M. Tau pathology and neurodegeneration. *Lancet Neurol.* **2013**, *12*, 609–622. [CrossRef]
2. Horigane, S.-I.; Ozawa, Y.; Yamada, H.; Takemoto-Kimura, S. Calcium signalling: A key regulator of neuronal migration. *J. Biochem.* **2019**, *165*, 401–409. [CrossRef]
3. Wojda, U.; Salinska, E.; Kuźnicki, J. Calcium ions in neuronal degeneration. *Iubmb Life* **2008**, *60*, 575–590. [CrossRef]
4. Alzheimer's Association Calcium Hypothesis Workgroup; Khachaturian, Z.S. Calcium Hypothesis of Alzheimer's disease and brain aging: A framework for integrating new evidence into a comprehensive theory of pathogenesis. *Alzheimer's Dement.* **2017**, *13*, 178–182. [CrossRef]
5. Britti, E.; Delaspre, F.; Tamarit, J.; Ros, J.; Ros, J. Mitochondrial calcium signalling and neurodegenerative diseases. *Neuronal Signal.* **2018**, *2*, NS20180061. [CrossRef]
6. Abeti, R.; Abramov, A.Y. Mitochondrial Ca2+ in neurodegenerative disorders. *Pharm. Res.* **2015**, *99*, 377–381. [CrossRef]
7. Ludtmann, M.H.R.; Abramov, A.Y. Mitochondrial calcium imbalance in Parkinson's disease. *Neurosci. Lett.* **2018**, *663*, 86–90. [CrossRef]
8. Semyanov, A.V. Spatiotemporal pattern of calcium activity in astrocytic network. *Cell Calcium* **2019**, *78*, 15–25. [CrossRef]
9. Shankar, G.M.; Bloodgood, B.L.; Townsend, M.; Walsh, D.M.; Selkoe, D.J.; Sabatini, B.L. Natural Oligomers of the Alzheimer Amyloid- Protein Induce Reversible Synapse Loss by Modulating an NMDA-Type Glutamate Receptor-Dependent Signaling Pathway. *J. Neurosci.* **2007**, *27*, 2866–2875. [CrossRef]
10. Orrenius, S.; Gogvadze, V.; Zhivotovsky, B. Calcium and mitochondria in the regulation of cell death. *Biochem. Biophys. Res. Commun.* **2015**, *460*, 72–81. [CrossRef]
11. Angelova, P.R.; Choi, M.L.; Berezhnov, A.V.; Horrocks, M.H.; Hughes, C.D.; De, S.; Rodrigues, M.; Yapom, R.; Little, D.; Dolt, K.S.; et al. Alpha synuclein aggregation drives ferroptosis: An interplay of iron, calcium and lipid peroxidation. *Cell Death Differ.* **2020**, *27*, 2781–2796. [CrossRef] [PubMed]
12. Simms, B.A.; Zamponi, G.W. Neuronal Voltage-Gated Calcium Channels: Structure, Function, and Dysfunction. *Neuron* **2014**, *82*, 24–45. [CrossRef] [PubMed]
13. Reiner, A.; Levitz, J. Glutamatergic Signaling in the Central Nervous System: Ionotropic and Metabotropic Receptors in Concert. *Neuron* **2018**, *98*, 1080–1098. [CrossRef] [PubMed]
14. Verkhratsky, A. Endoplasmic reticulum calcium signaling in nerve cells. *Boil. Res.* **2004**, *37*, 693–699. [CrossRef] [PubMed]
15. Karagas, N.E.; Venkatachalam, K. Roles for the Endoplasmic Reticulum in Regulation of Neuronal Calcium Homeostasis. *Cells* **2019**, *8*, 1232. [CrossRef]
16. Prakriya, M.; Feske, S.; Gwack, Y.; Srikanth, S.; Rao, A.; Hogan, P.G. Orai1 is an essential pore subunit of the CRAC channel. *Nat.* **2006**, *443*, 230–233. [CrossRef]
17. López, J.J.; Jardin, I.; Sánchez-Collado, J.; Salido, G.M.; Smani, T.; Rosado, J.A. TRPC Channels in the SOCE Scenario. *Cells* **2020**, *9*, 126. [CrossRef]
18. Brini, M.; Carafoli, E. The Plasma Membrane Ca2+ ATPase and the Plasma Membrane Sodium Calcium Exchanger Cooperate in the Regulation of Cell Calcium. *Cold Spring Harb. Perspect. Boil.* **2010**, *3*, a004168. [CrossRef]
19. Roome, C.J.; Power, E.M.; Empson, R. Transient reversal of the sodium/calcium exchanger boosts presynaptic calcium and synaptic transmission at a cerebellar synapse. *J. Neurophysiol.* **2013**, *109*, 1669–1680. [CrossRef]
20. Schwaller, B. Cytosolic Ca2+ Buffers. *Cold Spring Harb. Perspect. Boil.* **2010**, *2*, a004051. [CrossRef]
21. Burgoyne, R.D. Neuronal calcium sensor proteins: Generating diversity in neuronal Ca2+ signalling. *Nat. Rev. Neurosci.* **2007**, *8*, 182–193. [CrossRef]
22. Sharma, R.K.; Parameswaran, S. Calmodulin-binding proteins: A journey of 40 years. *Cell Calcium* **2018**, *75*, 89–100. [CrossRef] [PubMed]
23. Rizzuto, R.; De Stefani, D.; Raffaello, A.; Mammucari, C. Mitochondria as sensors and regulators of calcium signalling. *Nat. Rev. Mol. Cell Boil.* **2012**, *13*, 566–578. [CrossRef] [PubMed]
24. Abramov, A.Y.; Angelova, P.R. Cellular mechanisms of complex I-associated pathology. *Biochem. Soc. Trans.* **2019**, *47*, 1963–1969. [CrossRef]

25. McCormack, J.G.; Denton, R.M. The role of intramitochondrial Ca2+ in the regulation of oxidative phosphorylation in mammalian tissues. *Biochem. Soc. Trans.* **1993**, *21*, 793–799. [CrossRef]

26. Kamer, K.J.; Mootha, V.K. The molecular era of the mitochondrial calcium uniporter. *Nat. Rev. Mol. Cell Boil.* **2015**, *16*, 545–553. [CrossRef] [PubMed]

27. Patron, M.; Granatiero, V.; Espino, J.; Rizzuto, R.; De Stefani, D. MICU3 is a tissue-specific enhancer of mitochondrial calcium uptake. *Cell Death Differ.* **2018**, *26*, 179–195. [CrossRef]

28. Fan, M.; Zhang, J.; Tsai, C.-W.; Orlando, B.J.; Rodriguez, M.; Xu, Y.; Liao, M.; Tsai, M.-F.; Feng, L. Structure and mechanism of the mitochondrial Ca2+ uniporter holocomplex. *Nature* **2020**, *582*, 129–133. [CrossRef]

29. Wu, W.; Shen, Q.; Zhang, R.; Qiu, Z.; Wang, Y.; Zheng, J.; Jia, Z. The structure of the MICU 1- MICU 2 complex unveils the regulation of the mitochondrial calcium uniporter. *Embo J.* **2020**, *10*. [CrossRef]

30. Pan, X.; Liu, J.; Nguyen, T.; Liu, C.; Sun, J.; Teng, Y.; Fergusson, M.M.; Rovira, I.I.; Allen, M.; Springer, D.A.; et al. The physiological role of mitochondrial calcium revealed by mice lacking the mitochondrial calcium uniporter. *Nature* **2013**, *15*, 1464–1472. [CrossRef]

31. Holmstrom, K.; Pan, X.; Liu, J.C.; Menazza, S.; Liu, J.; Nguyen, T.T.; Pan, H.; Parks, R.J.; Anderson, S.A.; Noguchi, A.; et al. Assessment of cardiac function in mice lacking the mitochondrial calcium uniporter. *J. Mol. Cell. Cardiol.* **2015**, *85*, 178–182. [CrossRef] [PubMed]

32. Murphy, E.; Pan, X.; Nguyen, T.; Liu, J.; Holmstrom, K.; Finkel, T. Unresolved questions from the analysis of mice lacking MCU expression. *Biochem. Biophys. Res. Commun.* **2014**, *449*, 384–385. [CrossRef] [PubMed]

33. Antony, A.N.; Paillard, M.; Moffat, C.; Juskeviciute, E.; Correnti, J.; Bolon, B.; Rubin, E.; Csordás, G.; Seifert, E.L.; Hoek, J.B.; et al. MICU1 regulation of mitochondrial Ca2+ uptake dictates survival and tissue regeneration. *Nat. Commun.* **2016**, *7*, 10955. [CrossRef] [PubMed]

34. Drago, I.; Davis, R.L. Inhibiting the Mitochondrial Calcium Uniporter during Development Impairs Memory in Adult Drosophila. *Cell Rep.* **2016**, *16*, 2763–2776. [CrossRef]

35. Hamilton, J.; Brustovetsky, T.; Rysted, J.E.; Lin, Z.; Usachev, Y.M.; Brustovetsky, N. Deletion of mitochondrial calcium uniporter incompletely inhibits calcium uptake and induction of the permeability transition pore in brain mitochondria. *J. Boil. Chem.* **2018**, *293*, 15652–15663. [CrossRef]

36. Elustondo, P.A.; Nichols, M.; Robertson, G.S.; Pavlov, E.V. Mitochondrial Ca2+ uptake pathways. *J. Bioenerg. Biomembr.* **2016**, *49*, 113–119. [CrossRef]

37. De Stefani, D.; Patron, M.; Rizzuto, R. Structure and function of the mitochondrial calcium uniporter complex. *Biochim. Et Biophys. Acta (Bba)* **2015**, *1853*, 2006–2011. [CrossRef]

38. Carafoli, E.; Tiozzo, R.; Lugli, G.; Crovetti, F.; Kratzing, C. The release of calcium from heart mitochondria by sodium. *J. Mol. Cell. Cardiol.* **1974**, *6*, 361–371. [CrossRef]

39. Palty, R.; Silverman, W.F.; Hershfinkel, M.; Caporale, T.; Sensi, S.L.; Parnis, J.; Nolte, C.; Fishman, D.; Shoshan-Barmatz, V.; Herrmann, S.; et al. NCLX is an essential component of mitochondrial Na+/Ca2+ exchange. *Proc. Natl. Acad. Sci. USA* **2009**, *107*, 436–441. [CrossRef]

40. Drago, I.; De Stefani, D.; Rizzuto, R.; Pozzan, T. Mitochondrial Ca2+ uptake contributes to buffering cytoplasmic Ca2+ peaks in cardiomyocytes. *Proc. Natl. Acad. Sci. USA* **2012**, *109*, 12986–12991. [CrossRef]

41. Luongo, T.S.; Lambert, J.; Gross, P.; Nwokedi, M.; Lombardi, A.A.; Shanmughapriya, S.; Carpenter, A.C.; Kolmetzky, D.; Gao, E.; Van Berlo, J.H.; et al. The mitochondrial Na+/Ca2+ exchanger is essential for Ca2+ homeostasis and viability. *Nature* **2017**, *545*, 93–97. [CrossRef] [PubMed]

42. Gandhi, S.; Wood-Kaczmar, A.; Yao, Z.; Plun-Favreau, H.; Deas, E.; Klupsch, K.; Downward, J.; Latchman, D.S.; Tabrizi, S.J.; Wood, N.W.; et al. PINK1-Associated Parkinson's Disease Is Caused by Neuronal Vulnerability to Calcium-Induced Cell Death. *Mol. Cell* **2009**, *33*, 627–638. [CrossRef]

43. Kostić, M.; Ludtmann, M.H.R.; Bading, H.; Hershfinkel, M.; Steer, E.; Chu, C.T.; Abramov, A.Y.; Sekler, I. PKA Phosphorylation of NCLX Reverses Mitochondrial Calcium Overload and Depolarization, Promoting Survival of PINK1-Deficient Dopaminergic Neurons. *Cell Rep.* **2015**, *13*, 376–386. [CrossRef] [PubMed]

44. Kostic, M.; Sekler, I. Functional properties and mode of regulation of the mitochondrial Na+/Ca2+ exchanger, NCLX. *Semin. Cell Dev. Boil.* **2019**, *94*, 59–65. [CrossRef] [PubMed]

45. Wood-Kaczmar, A.; Deas, E.; Wood, N.W.; Abramov, A.Y. The Role of the Mitochondrial NCX in the Mechanism of Neurodegeneration in Parkinson's Disease. *Neurotransm. Interact. Cogn. Funct.* **2012**, *961*, 241–249. [CrossRef]

46. Gobbi, P.; Castaldo, P.; Minelli, A.; Salucci, S.; Magi, S.; Corcione, E.; Amoroso, S. Mitochondrial localization of Na+/Ca2+ exchangers NCX1–3 in neurons and astrocytes of adult rat brain in situ. *Pharm. Res.* **2007**, *56*, 556–565. [CrossRef]

47. Sisalli, M.J.; Secondo, A.; Esposito, A.; Valsecchi, V.; Savoia, C.; Di Renzo, G.F.; Annunziato, L.; Scorziello, A. Endoplasmic reticulum refilling and mitochondrial calcium extrusion promoted in neurons by NCX1 and NCX3 in ischemic preconditioning are determinant for neuroprotection. *Cell Death Differ.* **2014**, *21*, 1142–1149. [CrossRef]

48. Glancy, B.; Balaban, R.S. Role of Mitochondrial Ca2+in the Regulation of Cellular Energetics. *Biochemistry* **2012**, *51*, 2959–2973. [CrossRef]

49. Llorente-Folch, I.; Rueda, C.B.; Amigo, I.; Del Arco, A.; Saheki, T.; Pardo, B.; Satrústegui, J. Calcium-Regulation of Mitochondrial Respiration Maintains ATP Homeostasis and Requires ARALAR/AGC1-Malate Aspartate Shuttle in Intact Cortical Neurons. *J. Neurosci.* **2013**, *33*, 13957–13971. [CrossRef]

50. Briston, T.; Selwood, D.; Szabadkai, G.; Duchen, M.R. Mitochondrial Permeability Transition: A Molecular Lesion with Multiple Drug Targets. *Trends Pharm. Sci.* **2019**, *40*, 50–70. [CrossRef]

51. Angelova, P.R.; Abramov, A.Y. Role of mitochondrial ROS in the brain: From physiology to neurodegeneration. *FEBS Lett.* **2018**, *592*, 692–702. [CrossRef] [PubMed]

52. Abramov, A.Y.; Potapova, E.; Dremin, V.; Dunaev, A.V. Interaction of Oxidative Stress and Misfolded Proteins in the Mechanism of Neurodegeneration. *Life* **2020**, *10*, 101. [CrossRef] [PubMed]

53. Angelova, P.R.; Esteras, N.; Abramov, A.Y. Mitochondria and lipid peroxidation in the mechanism of neurodegeneration: Finding ways for prevention. *Med. Res. Rev.* **2020**. [CrossRef]

54. McKhann, G.M.; Knopman, D.S.; Chertkow, H.; Hyman, B.T.; Jack, C.R.; Kawas, C.H.; Klunk, W.E.; Koroshetz, W.J.; Manly, J.J.; Mayeux, R.; et al. The diagnosis of dementia due to Alzheimer's disease: Recommendations from the National Institute on Aging-Alzheimer's Association workgroups on diagnostic guidelines for Alzheimer's disease. *Alzheimer's Dement.* **2011**, *7*, 263–269. [CrossRef] [PubMed]

55. Ballard, C.; Gauthier, S.; Corbett, A.; Brayne, C.; Aarsland, D.; Jones, E. Alzheimer's disease. *Lancet* **2011**, *377*, 1019–1031. [CrossRef]

56. Popugaeva, E.; Pchitskaya, E.; Bezprozvanny, I. Dysregulation of neuronal calcium homeostasis in Alzheimer's disease - A therapeutic opportunity? *Biochem. Biophys. Res. Commun.* **2016**, *483*, 998–1004. [CrossRef]

57. Angelova, P.R.; Abramov, A.Y. Alpha-synuclein and beta-amyloid – different targets, same players: Calcium, free radicals and mitochondria in the mechanism of neurodegeneration. *Biochem. Biophys. Res. Commun.* **2017**, *483*, 1110–1115. [CrossRef]

58. Hardy, J. Amyloid, the presenilins and Alzheimer's disease. *Trends Neurosci.* **1997**, *20*, 154–159. [CrossRef]

59. Hardy, J.; Higgins, G.; Mayford, M.; Barzilai, A.; Keller, F.; Schacher, S.; Kandel, E. Alzheimer's disease: The amyloid cascade hypothesis. *Science* **1992**, *256*, 184–185. [CrossRef]

60. Arriagada, P.V.; Growdon, J.H.; Hedley-Whyte, E.T.; Hyman, B.T. Neurofibrillary tangles but not senile plaques parallel duration and severity of Alzheimer's disease. *Neurology* **1992**, *42*, 631. [CrossRef]

61. Rösler, T.W.; Marvian, A.T.; Brendel, M.; Nykänen, N.-P.; Höllerhage, M.; Schwarz, S.C.; Hopfner, F.; Koeglsperger, T.; Respondek, G.; Schweyer, K.; et al. Four-repeat tauopathies. *Prog. Neurobiol.* **2019**, *180*, 101644. [CrossRef] [PubMed]

62. Bang, J.; Spina, S.; Miller, B.L. Frontotemporal dementia. *Lancet* **2015**, *386*, 1672–1682. [CrossRef]

63. MacKenzie, I.R.A.; Neumann, M. Molecular neuropathology of frontotemporal dementia: Insights into disease mechanisms from postmortem studies. *J. Neurochem.* **2016**, *138*, 54–70. [CrossRef] [PubMed]

64. Khachaturian, Z.S. Calcium Hypothesis of Alzheimer's Disease and Brain Aginga. *Ann. N. Y. Acad. Sci.* **2006**, *747*, 1–11. [CrossRef]

65. Arispe, N.; Pollard, H.B.; Rojas, E. Giant multilevel cation channels formed by Alzheimer disease amyloid beta-protein [A beta P-(1-40)] in bilayer membranes. *Proc. Natl. Acad. Sci. USA* **1993**, *90*, 10573–10577. [CrossRef] [PubMed]

66. Abramov, A.Y.; Canevari, L.; Duchen, M.R. Changes in Intracellular Calcium and Glutathione in Astrocytes as the Primary Mechanism of Amyloid Neurotoxicity. *J. Neurosci.* **2003**, *23*, 5088–5095. [CrossRef]

67. Abramov, A.Y.; Canevari, L.; Duchen, M.R. Calcium signals induced by amyloid β peptide and their consequences in neurons and astrocytes in culture. *Biochim. Biophys. Acta (BBA)* **2004**, *1742*, 81–87. [CrossRef]

68. Patel, N.; Ramachandran, S.; Azimov, R.; Kagan, B.L.; Lal, R. Ion Channel Formation by Tau Protein: Implications for Alzheimer's Disease and Tauopathies. *Biochemistry* **2015**, *54*, 7320–7325. [CrossRef]

69. Esteras, N.; Kundel, F.; Amodeo, G.F.; Pavlov, E.V.; Klenerman, D.; Abramov, A.Y. Insoluble tau aggregates induce neuronal death through modification of membrane ion conductance, activation of voltage-gated calcium channels and NADPH oxidase. *FEBS J.* **2020**. [CrossRef]

70. Nalbantoglu, J.; Tirado-Santiago, G.; Lahsaïni, A.; Poirier, J.; Gonçalves, O.; Verge, G.; Momoli, F.; Welner, S.A.; Massicotte, G.; Julien, J.-P.; et al. Impaired learning and LTP in mice expressing the carboxy terminus of the Alzheimer amyloid precursor protein. *Nature* **1997**, *387*, 500–505. [CrossRef]

71. Snyder, E.M.; Nong, Y.; Almeida, C.G.; Paul, S.; Moran, T.; Choi, E.Y.; Nairn, A.C.; Salter, M.W.; Lombroso, P.J.; Gouras, G.K.; et al. Regulation of NMDA receptor trafficking by amyloid-β. *Nat. Neurosci.* **2005**, *8*, 1051–1058. [CrossRef] [PubMed]

72. Lewerenz, J.; Maher, P. Chronic Glutamate Toxicity in Neurodegenerative Diseases—What is the Evidence? *Front. Mol. Neurosci.* **2015**, *9*, 469. [CrossRef]

73. Texidó, L.; Martín-Satué, M.; Alberdi, E.; Solsona, C.; Matute, C. Amyloid β peptide oligomers directly activate NMDA receptors. *Cell Calcium* **2011**, *49*, 184–190. [CrossRef]

74. Ferreira, I.L.; Bajouco, L.; Mota, S.; Auberson, Y.; Oliveira, C.R.; Rego, A.C. Amyloid beta peptide 1–42 disturbs intracellular calcium homeostasis through activation of GluN2B-containing N-methyl-d-aspartate receptors in cortical cultures. *Cell Calcium* **2012**, *51*, 95–106. [CrossRef]

75. Rammes, G.; Seeser, F.; Mattusch, K.; Zhu, K.; Haas, L.; Kummer, M.; Heneka, M.; Herms, J.; Parsons, C.G. The NMDA receptor antagonist Radiprodil reverses the synaptotoxic effects of different amyloid-beta (Aβ) species on long-term potentiation (LTP). *Neuropharmacology* **2018**, *140*, 184–192. [CrossRef] [PubMed]

76. Liu, Y.; Wong, T.P.; Aarts, M.M.; Rooyakkers, A.; Liu, L.; Lai, T.W.; Wu, D.C.; Lu, J.; Tymianski, M.; Craig, A.M.; et al. NMDA Receptor Subunits Have Differential Roles in Mediating Excitotoxic Neuronal Death Both In Vitro and In Vivo. *J. Neurosci.* **2007**, *27*, 2846–2857. [CrossRef] [PubMed]

77. Huang, Y.; Shen, W.; Su, J.; Cheng, B.; Li, D.; Liu, G.; Zhou, W.; Zhang, Y.-X. Modulating the Balance of Synaptic and Extrasynaptic NMDA Receptors Shows Positive Effects against Amyloid-β-Induced Neurotoxicity. *J. Alzheimer's Dis.* **2017**, *57*, 885–897. [CrossRef]

78. Miyamoto, T.; Stein, L.R.; Thomas, R.; Djukic, B.; Taneja, P.; Knox, J.; Vossel, K.; Mucke, L. Phosphorylation of tau at Y18, but not tau-fyn binding, is required for tau to modulate NMDA receptor-dependent excitotoxicity in primary neuronal culture. *Mol. Neurodegener.* **2017**, *12*, 41. [CrossRef]

79. Monteiro-Fernandes, D.; Silva, J.; Soares-Cunha, C.; Dalla, C.; Kokras, N.; Arnaud, F.; Billiras, R.; Zhuravleva, V.; Waites, C.; Bretin, S.; et al. Allosteric modulation of AMPA receptors counteracts Tau-related excitotoxic synaptic signaling and memory deficits in stress- and Aβ-evoked hippocampal pathology. *Mol. Psychiatry* **2020**, 1–13. [CrossRef]

80. Ittner, L.M.; Ke, Y.D.; Delerue, F.; Bi, M.; Gladbach, A.; Van Eersel, J.; Wölfing, H.; Chieng, B.C.; Christie, M.J.; Napier, I.A.; et al. Dendritic Function of Tau Mediates Amyloid-β Toxicity in Alzheimer's Disease Mouse Models. *Cell* **2010**, *142*, 387–397. [CrossRef]

81. Decker, J.M.; Krüger, L.; Sydow, A.; Dennissen, F.; Siskova, Z.; Mandelkow, E.; Mandelkow, E. The Tau/A152T mutation, a risk factor for frontotemporal-spectrum disorders, leads to NR 2B receptor-mediated excitotoxicity. *EMBO Rep.* **2016**, *17*, 552–569. [CrossRef] [PubMed]

82. Wang, R.; Reddy, P.H. Role of Glutamate and NMDA Receptors in Alzheimer's Disease. *J. Alzheimer's Dis.* **2017**, *57*, 1041–1048. [CrossRef] [PubMed]

83. Bartus, R.; Dean, R.; Beer, B.; Lippa, A. The cholinergic hypothesis of geriatric memory dysfunction. *Science* **1982**, *217*, 408–414. [CrossRef] [PubMed]

84. Simon, D.; Hernandez, F.; Ávila, J. The Involvement of Cholinergic Neurons in the Spreading of Tau Pathology. *Front. Neurol.* **2013**, *4*, 74. [CrossRef] [PubMed]

85. Kamynina, A.V.; Holmstrom, K.; Koroev, D.O.; Volpina, O.M.; Abramov, A.Y. Acetylcholine and antibodies against the acetylcholine receptor protect neurons and astrocytes against beta-amyloid toxicity. *Int. J. Biochem. Cell Boil.* **2013**, *45*, 899–907. [CrossRef]

86. Ishii, M.; Hiller, A.J.; Pham, L.; McGuire, M.J.; Iadecola, C.; Wang, G. Amyloid-Beta Modulates Low-Threshold Activated Voltage-Gated L-Type Calcium Channels of Arcuate Neuropeptide Y Neurons Leading to Calcium Dysregulation and Hypothalamic Dysfunction. *J. Neurosci.* **2019**, *39*, 8816–8825. [CrossRef]

87. Furukawa, K.; Wang, Y.; Yao, P.J.; Fu, W.; Mattson, M.P.; Itoyama, Y.; Onodera, H.; D'Souza, I.; Poorkaj, P.H.; Bird, T.D.; et al. Alteration in calcium channel properties is responsible for the neurotoxic action of a familial frontotemporal dementia tau mutation. *J. Neurochem.* **2003**, *87*, 427–436. [CrossRef]

88. Lawlor, B.; Segurado, R.; Kennelly, S.; Rikkert, M.O.; Howard, R.J.; Pasquier, F.; Börjesson-Hanson, A.; Tsolaki, M.; Lucca, U.; Molloy, D.W.; et al. Nilvadipine in mild to moderate Alzheimer disease: A randomised controlled trial. *PLoS Med.* **2018**, *15*, e1002660. [CrossRef]

89. Abdullah, L.; Crawford, F.; Tsolaki, M.; Börjesson-Hanson, A.; Rikkert, M.O.; Pasquier, F.; Wallin, A.; Kennelly, S.; Ait-Ghezala, G.; Paris, D.; et al. The Influence of Baseline Alzheimer's Disease Severity on Cognitive Decline and CSF Biomarkers in the NILVAD Trial. *Front. Neurol.* **2020**, *11*, 149. [CrossRef]

90. Hwang, D.; Kim, S.; Choi, H.; Oh, I.-H.; Kim, B.S.; Choi, H.R.; Kim, S.Y.; Won, C. Calcium-Channel Blockers and Dementia Risk in Older Adults – National Health Insurance Service – Senior Cohort (2002–2013). *Circ. J.* **2016**, *80*, 2336–2342. [CrossRef]

91. Feldman, L.; Vinker, S.; Efrati, S.; Beberashvili, I.; Gorelik, O.; Wasser, W.; Shani, M. Amlodipine treatment of hypertension associates with a decreased dementia risk. *Clin. Exp. Hypertens.* **2016**, *38*, 545–549. [CrossRef] [PubMed]

92. Bohlken, J.; Jacob, L.; Kostev, K. The Relationship Between the Use of Antihypertensive Drugs and the Incidence of Dementia in General Practices in Germany. *J. Alzheimer's Dis.* **2019**, *70*, 91–97. [CrossRef] [PubMed]

93. Stutzmann, G.E.; Smith, I.; Caccamo, A.; Oddo, S.; Parker, I.; LaFerla, F. Enhanced Ryanodine-Mediated Calcium Release in Mutant PS1-Expressing Alzheimer's Mouse Models. *Ann. New York Acad. Sci.* **2007**, *1097*, 265–277. [CrossRef] [PubMed]

94. Paula-Lima, A.C.; Adasme, T.; Sanmartín, C.; Sebollela, A.; Hetz, C.; Carrasco, M.A.; Ferreira, J.; Hidalgo, C. Amyloid β-Peptide Oligomers Stimulate RyR-Mediated Ca2+ Release Inducing Mitochondrial Fragmentation in Hippocampal Neurons and Prevent RyR-Mediated Dendritic Spine Remodeling Produced by BDNF. *Antioxid. Redox Signal.* **2011**, *14*, 1209–1223. [CrossRef]

95. DeMuro, A.; Parker, I. Cytotoxicity of intracellular aβ42 amyloid oligomers involves Ca2+ release from the endoplasmic reticulum by stimulated production of inositol trisphosphate. *J. Neurosci.* **2013**, *33*, 3824–3833. [CrossRef]

96. Cheung, K.-H.; Shineman, D.; Müller, M.; Cárdenas, C.; Mei, L.; Yang, J.; Tomita, T.; Iwatsubo, T.; Lee, V.M.-Y.; Foskett, J.K. Mechanism of Ca2+ Disruption in Alzheimer's Disease by Presenilin Regulation of InsP3 Receptor Channel Gating. *Neuron* **2008**, *58*, 871–883. [CrossRef]

97. Bojarski, L.; Pomorski, P.; Szybinska, A.; Drab, M.; Skibinska-Kijek, A.; Gruszczynska-Biegala, J.; Kuźnicki, J. Presenilin-dependent expression of STIM proteins and dysregulation of capacitative Ca2+ entry in familial Alzheimer's disease. *Biochim. Biophys. Acta* **2009**, *1793*, 1050–1057. [CrossRef]

98. Tong, B.C.-K.; Lee, C.S.-K.; Cheng, W.-H.; Lai, K.-O.; Foskett, J.K.; Cheung, K.-H. Familial Alzheimer's disease–associated presenilin 1 mutants promote γ-secretase cleavage of STIM1 to impair store-operated Ca2+entry. *Sci. Signal.* **2016**, *9*, ra89. [CrossRef]

99. Ye, J.; Yin, Y.; Yin, Y.; Zhang, H.; Wan, H.; Wang, L.; Zuo, Y.; Gao, D.; Li, M.; Li, J.; et al. Tau-induced upregulation of C/EBPβ-TRPC1-SOCE signaling aggravates tauopathies: A vicious cycle in Alzheimer neurodegeneration. *Aging Cell* **2020**, e13209. [CrossRef]

100. Mata, A.M. Functional interplay between plasma membrane Ca 2+ -ATPase, amyloid β-peptide and tau. *Neurosci. Lett.* **2018**, *663*, 55–59. [CrossRef]

101. Berrocal, M.; Sepulveda, M.R.; Vazquez-Hernandez, M.; Mata, A.M. Calmodulin antagonizes amyloid-β peptides-mediated inhibition of brain plasma membrane Ca2+-ATPase. *Biochim. Et Biophys. Acta (BBA)—Mol. Basis Dis.* **2012**, *1822*, 961–969. [CrossRef] [PubMed]

102. Wu, A.; A Derrico, C.; Hatem, L.; A Colvin, R. Alzheimer's amyloid-beta peptide inhibits sodium/calcium exchange measured in rat and human brain plasma membrane vesicles. *Neuroscience* **1997**, *80*, 675–684. [CrossRef]

103. Pannaccione, A.; Piccialli, I.; Secondo, A.; Ciccone, R.; Molinaro, P.; Boscia, F.; Annunziato, L. The Na+/Ca2+exchanger in Alzheimer's disease. *Cell Calcium* **2020**, *87*, 102190. [CrossRef] [PubMed]

104. Atherton, J.; Kurbatskaya, K.; Bondulich, M.; Croft, C.L.; Garwood, C.J.; Chhabra, R.; Wray, S.; Jeromin, A.; Hanger, D.P.; Noble, W. Calpain cleavage and inactivation of the sodium calcium exchanger-3 occur downstream of Aβ in Alzheimer's disease. *Aging Cell* **2013**, *13*, 49–59. [CrossRef]

105. Kip, S.N.; Strehler, E.E. Rapid downregulation of NCX and PMCA in hippocampal neurons following H2O2 oxidative stress. *Ann. New York Acad. Sci.* **2007**, *1099*, 436–439. [CrossRef]

106. Ferreira, A.; Bigio, E.H. Calpain-Mediated Tau Cleavage: A Mechanism Leading to Neurodegeneration Shared by Multiple Tauopathies. *Mol. Med.* **2011**, *17*, 676–685. [CrossRef]

107. Liang, B.; Duan, B.-Y.; Zhou, X.-P.; Gong, J.-X.; Luo, Z.-G. Calpain Activation Promotes BACE1 Expression, Amyloid Precursor Protein Processing, and Amyloid Plaque Formation in a Transgenic Mouse Model of Alzheimer Disease. *J. Boil. Chem.* **2010**, *285*, 27737–27744. [CrossRef]

108. Rao, M.; McBrayer, M.K.; Campbell, J.; Kumar, A.; Hashim, A.; Sershen, H.; Stavrides, P.H.; Ohno, M.; Hutton, M.; Nixon, R.A. Specific Calpain Inhibition by Calpastatin Prevents Tauopathy and Neurodegeneration and Restores Normal Lifespan in Tau P301L Mice. *J. Neurosci.* **2014**, *34*, 9222–9234. [CrossRef]

109. Reinecke, J.B.; Devos, S.L.; McGrath, J.P.; Shepard, A.M.; Goncharoff, D.K.; Tait, D.N.; Fleming, S.R.; Vincent, M.; Steinhilb, M.L. Implicating Calpain in Tau-Mediated Toxicity In Vivo. *PLoS ONE* **2011**, *6*, e23865. [CrossRef]

110. O'Day, D.H.; Eshak, K.; Myre, M.A. Calmodulin Binding Proteins and Alzheimer's Disease. *J. Alzheimer's Dis.* **2015**, *46*, 553–569. [CrossRef]

111. Esteras, N.; Alquézar, C.; De La Encarnación, A.; Villarejo, A.; Bermejo-Pareja, F.; Requero, A.M. Calmodulin levels in blood cells as a potential biomarker of Alzheimer's disease. *Alzheimer's Res. Ther.* **2013**, *5*, 55. [CrossRef] [PubMed]

112. Esteras, N.; Dinkova-Kostova, A.T.; Abramov, A.Y. Nrf2 activation in the treatment of neurodegenerative diseases: A focus on its role in mitochondrial bioenergetics and function. *Boil. Chem.* **2016**, *397*, 383–400. [CrossRef] [PubMed]

113. Petersen, C.A.H.; Alikhani, N.; Behbahani, H.; Wiehager, B.; Pavlov, P.F.; Alafuzoff, I.; Leinonen, V.; Ito, A.; Winblad, B.; Glaser, E.; et al. The amyloid -peptide is imported into mitochondria via the TOM import machinery and localized to mitochondrial cristae. *Proc. Natl. Acad. Sci. USA* **2008**, *105*, 13145–13150. [CrossRef] [PubMed]

114. Pavlov, P.F.; Wiehager, B.; Sakai, J.; Frykman, S.; Behbahani, H.; Winblad, B.; Ankarcrona, M. Mitochondrial γ-secretase participates in the metabolism of mitochondria-associated amyloid precursor protein. *FASEB J.* **2010**, *25*, 78–88. [CrossRef] [PubMed]

115. Cieri, D.; Vicario, M.; Vallese, F.; D'Orsi, B.; Berto, P.; Grinzato, A.; Catoni, C.; De Stefani, D.; Rizzuto, R.; Brini, M.; et al. Tau localises within mitochondrial sub-compartments and its caspase cleavage affects ER-mitochondria interactions and cellular Ca2+ handling. *Biochim. Et Biophys. Acta (Bba)* **2018**, *1864*, 3247–3256. [CrossRef]

116. Amadoro, G.; Corsetti, V.; Stringaro, A.; Colone, M.; D'Aguanno, S.; Meli, G.; Ciotti, M.; Sancesario, G.; Cattaneo, A.; Bussani, R.; et al. A NH2 Tau Fragment Targets Neuronal Mitochondria at AD Synapses: Possible Implications for Neurodegeneration. *J. Alzheimer's Dis.* **2010**, *21*, 445–470. [CrossRef]

117. Abramov, A.Y.; Duchen, M.R. Mechanisms underlying the loss of mitochondrial membrane potential in glutamate excitotoxicity. *Biochim. Biophys. Acta (BBA)* **2008**, *1777*, 953–964. [CrossRef]

118. Duchen, M. Mitochondria and Ca2+in cell physiology and pathophysiology. *Cell Calcium* **2000**, *28*, 339–348. [CrossRef]

119. Abramov, A.Y.; Canevari, L.; Duchen, M.R. β-Amyloid Peptides Induce Mitochondrial Dysfunction and Oxidative Stress in Astrocytes and Death of Neurons through Activation of NADPH Oxidase. *J. Neurosci.* **2004**, *24*, 565–575. [CrossRef]

120. Stout, A.K.; Raphael, H.M.; Kanterewicz, B.I.; Klann, E.; Reynolds, I.J. Glutamate-induced neuron death requires mitochondrial calcium uptake. *Nat. Neurosci.* **1998**, *1*, 366–373. [CrossRef]

121. Pivovarova, N.B.; Nguyen, H.V.; Winters, C.A.; Brantner, C.A.; Smith, C.L.; Andrews, S.B. Excitotoxic Calcium Overload in a Subpopulation of Mitochondria Triggers Delayed Death in Hippocampal Neurons. *J. Neurosci.* **2004**, *24*, 5611–5622. [CrossRef] [PubMed]

122. Qiu, J.; Tan, Y.-W.; Hagenston, A.M.; Martel, M.-A.; Kneisel, N.; Skehel, P.A.; Wyllie, D.J.; Bading, H.; Hardingham, G.E. Mitochondrial calcium uniporter Mcu controls excitotoxicity and is transcriptionally repressed by neuroprotective nuclear calcium signals. *Nat. Commun.* **2013**, *4*, 4. [CrossRef] [PubMed]

123. Angelova, P.R.; Vinogradova, D.; Neganova, M.E.; Serkova, T.P.; Sokolov, V.V.; Bachurin, S.O.; Shevtsova, E.F.; Abramov, A.Y. Pharmacological Sequestration of Mitochondrial Calcium Uptake Protects Neurons Against Glutamate Excitotoxicity. *Mol. Neurobiol.* **2018**, *56*, 2244–2255. [CrossRef] [PubMed]

124. Sanz-Blasco, S.; Valero, R.A.; Rodríguez-Crespo, I.; Villalobos, C.; Nuñez, L. Mitochondrial Ca2+ Overload Underlies Aβ Oligomers Neurotoxicity Providing an Unexpected Mechanism of Neuroprotection by NSAIDs. *PLoS ONE* **2008**, *3*, e2718. [CrossRef] [PubMed]

125. Sanz-Blasco, S.; Calvo-Rodriguez, M.; Caballero, E.; García-Durillo, M.; Nuñez, L.; Villalobos, C. Is it All Said for NSAIDs in Alzheimer's Disease? Role of Mitochondrial Calcium Uptake. *Curr. Alzheimer Res.* **2018**, *15*, 504–510. [CrossRef]

126. Calvo-Rodriguez, M.; Hou, S.S.; Snyder, A.C.; Kharitonova, E.K.; Russ, A.N.; Das, S.; Fan, Z.; Muzikansky, A.; Garcia-Alloza, M.; Serrano-Pozo, A.; et al. Increased mitochondrial calcium levels associated with neuronal death in a mouse model of Alzheimer's disease. *Nat. Commun.* **2020**, *11*, 1–17. [CrossRef]

127. Jadiya, P.; Kolmetzky, D.W.; Tomar, D.; Di Meco, A.; Lombardi, A.A.; Lambert, J.; Luongo, T.S.; Ludtmann, M.H.; Praticò, M.; Elrod, J.W. Impaired mitochondrial calcium efflux contributes to disease progression in models of Alzheimer's disease. *Nat. Commun.* **2019**, *10*, 3885. [CrossRef]

128. Csordás, G.; Várnai, P.; Golenár, T.; Roy, S.; Purkins, G.; Schneider, T.G.; Balla, T.; Hajnóczky, G. Imaging Interorganelle Contacts and Local Calcium Dynamics at the ER-Mitochondrial Interface. *Mol. Cell* **2010**, *39*, 121–132. [CrossRef]

129. Del Prete, D.; Suski, J.M.; Oulès, B.; Debayle, D.; Gay, A.S.; Lacas-Gervais, S.; Bussiere, R.; Bauer, C.; Pinton, P.; Paterlini-Bréchot, P.; et al. Localization and Processing of the Amyloid-β Protein Precursor in Mitochondria-Associated Membranes. *J. Alzheimer's Dis.* **2016**, *55*, 1549–1570. [CrossRef]

130. Area-Gomez, E.; Schon, E.A. On the Pathogenesis of Alzheimer's Disease: The MAM Hypothesis. *FASEB J.* **2017**, *31*, 864–867. [CrossRef]

131. Hedskog, L.; Pinho, C.M.; Filadi, R.; Rönnbäck, A.; Hertwig, L.; Wiehager, B.; Larssen, P.; Gellhaar, S.; Sandebring, A.; Westerlund, M.; et al. Modulation of the endoplasmic reticulum-mitochondria interface in Alzheimer's disease and related models. *Proc. Natl. Acad. Sci. USA* **2013**, *110*, 7916–7921. [CrossRef] [PubMed]

132. Zampese, E.; Fasolato, C.; Kipanyula, M.J.; Bortolozzi, M.; Pozzan, T.; Pizzo, P. Presenilin 2 modulates endoplasmic reticulum (ER)-mitochondria interactions and Ca2+ cross-talk. *Proc. Natl. Acad. Sci. USA* **2011**, *108*, 2777–2782. [CrossRef] [PubMed]

133. Kipanyula, M.J.; Contreras, L.; Zampese, E.; Lazzari, C.; Wong, A.K.C.; Pizzo, P.; Fasolato, C.; Pozzan, T. Ca2+dysregulation in neurons from transgenic mice expressing mutant presenilin. *Aging Cell* **2012**, *11*, 885–893. [CrossRef] [PubMed]

134. Tu, H.; Nelson, O.; Bezprozvanny, A.; Wang, Z.; Lee, S.-F.; Hao, Y.-H.; Serneels, L.; De Strooper, B.; Yu, G.; Bezprozvanny, I. Presenilins Form ER Ca2+ Leak Channels, a Function Disrupted by Familial Alzheimer's Disease-Linked Mutations. *Cell* **2006**, *126*, 981–993. [CrossRef] [PubMed]

135. Shilling, D.; Mak, D.-O.D.; Kang, D.E.; Foskett, J.K. Lack of Evidence for Presenilins as Endoplasmic Reticulum Ca2+Leak Channels. *J. Boil. Chem.* **2012**, *287*, 10933–10944. [CrossRef]

136. Parks, J.K.; Smith, T.S.; A Trimmer, P.; Bennett, J.P.; Parker, W.D. Neurotoxic Aβ peptides increase oxidative stress in vivo through NMDA-receptor and nitric-oxide-synthase mechanisms, and inhibit complex IV activity and induce a mitochondrial permeability transition in vitro. *J. Neurochem.* **2001**, *76*, 1050–1056. [CrossRef]

137. Abramov, A.Y.; Fraley, C.; Diao, C.T.; Winkfein, R.; Colicos, M.A.; Duchen, M.R.; French, R.J.; Pavlov, E. Targeted polyphosphatase expression alters mitochondrial metabolism and inhibits calcium-dependent cell death. *Proc. Natl. Acad. Sci. USA* **2007**, *104*, 18091–18096. [CrossRef]

138. Britti, E.; Ros, J.; Esteras, N.; Abramov, A.Y. Tau inhibits mitochondrial calcium efflux and makes neurons vulnerable to calcium-induced cell death. *Cell Calcium* **2020**, *86*, 102150. [CrossRef]

139. Du, H.; Guo, L.; Fang, F.; Chen, D.; A Sosunov, A.; McKhann, G.M.; Yan, Y.; Wang, C.; Zhang, H.; Molkentin, J.D.; et al. Cyclophilin D deficiency attenuates mitochondrial and neuronal perturbation and ameliorates learning and memory in Alzheimer's disease. *Nat. Med.* **2008**, *14*, 1097–1105. [CrossRef]

140. Kopach, O.; Esteras, N.; Wray, S.; Rusakov, D.A.; Abramov, A.Y. Maturation and phenotype of pathophysiological neuronal excitability of human cells in tau-related dementia. *J. Cell Sci.* **2020**, *133*, jcs241687. [CrossRef]

141. Esteras, N.; Rohrer, J.D.; Hardy, J.; Wray, S.; Abramov, A.Y. Mitochondrial hyperpolarization in iPSC-derived neurons from patients of FTDP-17 with 10+16 MAPT mutation leads to oxidative stress and neurodegeneration. *Redox Boil.* **2017**, *12*, 410–422. [CrossRef] [PubMed]

142. Bartolome, F.; Esteras, N.; Martin-Requero, A.; Boutoleau-Bretonnière, C.; Vercelletto, M.; Gabelle, A.; Le Ber, I.; Honda, T.; Dinkova-Kostova, A.T.; Hardy, J.; et al. Pathogenic p62/SQSTM1 mutations impair energy metabolism through limitation of mitochondrial substrates. *Sci. Rep.* **2017**, *7*, 1666. [CrossRef] [PubMed]

143. Delgado-Camprubi, M.; Esteras, N.; Soutar, M.P.M.; Plun-Favreau, H.; Abramov, A.Y. Deficiency of Parkinson's disease-related gene Fbxo7 is associated with impaired mitochondrial metabolism by PARP activation. *Cell Death Differ.* **2016**, *24*, 120–131. [CrossRef] [PubMed]

144. Abramov, A.Y. Actions of ionomycin, 4-BrA23187 and a novel electrogenic Ca2+ ionophore on mitochondria in intact cells. *Cell Calcium* **2003**, *33*, 101–112. [CrossRef]

145. Zamaraeva, M.V.; Hagelgans, A.I.; Abramov, A.Y.; Ternovsky, V.I.; Merzlyak, P.G.; Tashmukhamedov, B.A.; Saldkhodzjaev, A.I. Ionophoretic properties of ferutinin. *Cell Calcium* **1997**, *22*, 235–241. [CrossRef]

Review

Presenilin-2 and Calcium Handling: Molecules, Organelles, Cells and Brain Networks

Paola Pizzo [1,2,*], Emy Basso [1,2], Riccardo Filadi [1,2], Elisa Greotti [1,2], Alessandro Leparulo [1], Diana Pendin [1,2], Nelly Redolfi [1], Michela Rossini [1], Nicola Vajente [1,2], Tullio Pozzan [1,2,3] and Cristina Fasolato [1,*]

[1]   Department of Biomedical Sciences, University of Padua, Via U. Bassi 58/B, 35131 Padua, Italy; emy.basso@cnr.it (E.B.); riccardo.filadi@unipd.it (R.F.); elisa.greotti@cnr.it (E.G.); alessandro.leparulo@unipd.it (A.L.); diana.pendin@unipd.it (D.P.); nelly.redolfi@unipd.it (N.R.); michela.rossini@studenti.unipd.it (M.R.); nicola.vajente@studenti.unipd.it (N.V.); tullio.pozzan@unipd.it (T.P.)

[2]   Neuroscience Institute, Italian National Research Council (CNR), Via U. Bassi 58/B, 35131 Padua, Italy

[3]   Venetian Institute of Molecular Medicine (VIMM), Via G. Orus 2B, 35131 Padua, Italy

*   Correspondence: paola.pizzo@unipd.it (P.P.); cristina.fasolato@unipd.it (C.F.); Tel.: +39-049-827-6067 (P.P.); +39-049-827-6065 (C.F.)

Received: 31 August 2020; Accepted: 18 September 2020; Published: 25 September 2020

Abstract: Presenilin-2 (PS2) is one of the three proteins that are dominantly mutated in familial Alzheimer's disease (FAD). It forms the catalytic core of the γ-secretase complex—a function shared with its homolog presenilin-1 (PS1)—the enzyme ultimately responsible of amyloid-β (Aβ) formation. Besides its enzymatic activity, PS2 is a multifunctional protein, being specifically involved, independently of γ-secretase activity, in the modulation of several cellular processes, such as Ca^{2+} signalling, mitochondrial function, inter-organelle communication, and autophagy. As for the former, evidence has accumulated that supports the involvement of PS2 at different levels, ranging from organelle Ca^{2+} handling to Ca^{2+} entry through plasma membrane channels. Thus FAD-linked PS2 mutations impact on multiple aspects of cell and tissue physiology, including bioenergetics and brain network excitability. In this contribution, we summarize the main findings on PS2, primarily as a modulator of Ca^{2+} homeostasis, with particular emphasis on the role of its mutations in the pathogenesis of FAD. Identification of cell pathways and molecules that are specifically targeted by PS2 mutants, as well as of common targets shared with PS1 mutants, will be fundamental to disentangle the complexity of memory loss and brain degeneration that occurs in Alzheimer's disease (AD).

Keywords: presenilin-2; calcium signalling; Alzheimer's disease mouse models; SOCE; mitochondria; autophagy; brain networks; oscillations; slow-waves; functional connectivity

---

1. Presenilin-2 in Physiology and Pathology

Presenilin-2 (PS2)—and its homolog presenilin-1 (PS1)—is a 50-kDa multi-pass membrane protein with nine helical transmembrane (TM) domains, and in humans it is encoded by a gene present on chromosome 1 (*PSEN2*) [1]. Both presenilins (PSs) mainly localize to the endoplasmic reticulum (ER) and Golgi apparatus (GA) membranes but also, although less abundantly, in plasma membrane (PM) and endosomes [2]. Their mRNAs are expressed in different human and mouse tissues, with the highest levels in the hippocampus and cerebellum [3].

Both PSs represent the catalytic core of the γ-secretase complex, the enzyme ultimately responsible for generation of Aβ peptides; they were both discovered in genetic analyses of families in which Alzheimer's disease (AD) is transmitted as an autosomal dominant trait. In fact, as of now, about 300 mutations in *PSEN1* and 58 mutations in *PSEN2* have been described (https://www.alzforum.org/mutations), the majority of which are dominant, mostly missense, and have been associated with the inherited forms of the disease (familial Alzheimer's disease (FAD)) [4,5]. Mutations in the gene for one of the substrate of the γ-secretase complex, the amyloid precursor protein (APP), are also responsible for FAD cases [6]. It has been proposed that FAD-PS mutations lead to a less precise γ-secretase cleavage of APP, in some cases decreasing the total production of Aβ but increasing the relative amount of the more amyloidogenic Aβ42 peptide, the seeding core of extracellular amyloid plaques, over the more soluble Aβ40 peptide [7,8].

The γ-secretase complex is part of the family of intramembrane-cleaving proteases (I-CliPs), which perform hydrolysis of protein domains embedded in the hydrophobic environment of the membrane. The family includes SP2 metalloproteases, serine proteases of the rhomboid family, and the aspartyl proteases to which γ-secretase belongs.

The γ-secretase has a central role in cellular biology, with about 150 different integral membrane proteins recognized as substrates [9]; the most studied are the Notch family of receptors, with a crucial role in signalling and cell differentiation, and APP [4,9]. The γ-secretase complex is composed of four subunits: PS1 or PS2; nicastrin, an integral membrane protein concerned with substrate recognition and selection [10]; PS enhancer-2 (PEN-2) that stabilizes the PS complex and has a role in its endoproteolytic cleavage [11–13]; and anterior pharynx defective 1 (APH1), which interacts with nicastrin, providing the initial scaffold to which PS1/2 and PEN-2 are added [14,15]. In humans, APH1 is encoded by two paralogous genes (APH1A and APH1B), and each protein can interact with either PS, resulting in the existence of four different γ-secretase complexes that might have slightly different specificities [16]. After its enclosure within the complex, PS undergoes an endoproteolytic cleavage that produces N- and C-terminal fragments; the two fragments remain associated and represent the biologically active form of the complex, each carrying one of the two key aspartic acid residues on TM6 and TM7, respectively [17,18].

PS1 and PS2 share about 66% of amino acidic sequence; one key difference is a motif in PS2 that interacts with activating protein-1 (AP-1) complexes in a phosphorylation-dependent manner and targets PS2 to the late endosome/lysosome compartment, leading to a different subcellular distribution of PS2 and perhaps to subtly different functions [19,20]. For example, it has been demonstrated that PS2-containing γ-secretase complexes are involved in the processing of premelanosome (PMEL) protein, which is involved in melanosome maturation and melanin deposition [19]. Indeed, PS2-null zebrafish showed defects in skin pigmentation [21]. Importantly, melanosome biogenesis seems to be Ca^{2+}-dependent [22] (see also below).

Several γ-secretase-independent functions of PSs have emerged in the recent years, enriching the overall importance of these proteins in cell biology. For example, PSs bind to glycogen synthase kinase 3β (GSK3β), a key protein of the Wnt signalling pathway, and to its substrate β-catenin, a transcription regulator [23,24]. The interaction of PSs with GSK3β and β-catenin is independent of γ-secretase activity [25] and influences β-catenin phosphorylation and turnover [26], as well as the activity of kinesin-1 and dynein and thus axonal transport of type 1 transmembrane receptors [27]. PSs have been implicated also in autophagy (see below) and protein trafficking [28].

Last, but not least, the regulation of cellular Ca^{2+} homeostasis has emerged as a key PS function, independent of γ-secretase activity, with relevant implications in multiple Ca^{2+}-regulated cell processes. In the present review, we summarize the central role played by PS2 in cellular Ca^{2+} homeostasis, highlighting divergent and convergent aspects of PS2 vs. PS1 pathophysiology.

2. PS2 and Ca^{2+} Homeostasis

2.1. Alterations of Ca^{2+} Homeostasis in FAD-PS2 Cell Models

According to the so-called "Ca^{2+} overload" hypothesis for AD, FAD-PS mutations increase the ER Ca^{2+} content and cause excessive cytosolic Ca^{2+} release upon cell stimulations that, in turn, alters APP processing and sensitizes neurons to Ca^{2+}-dependent cell death mechanisms [29]. Indeed, an excessive release of Ca^{2+} from the ER has been reported in different cell models expressing various FAD-PS mutations, as well as in neurons from transgenic (Tg) mice carrying FAD-PS1 mutations [30–35]. FAD-linked mutations in PS2 have also been reported to potentiate ER Ca^{2+} release from both ryanodine receptors (RyRs) [36] and inositol trisphosphate (IP3) receptors (IP3Rs) in *Xenopus* oocytes [30] and neurons from Tg mice expressing the PS2-N141I mutation [37]. Moreover, it has been proposed that wild type (WT) PSs form constitutively active ER Ca^{2+} leak channels whereas FAD-PS mutations disrupt the channel functionality; as a result of the reduced leak, the ER Ca^{2+} level increases and more Ca^{2+} is released upon stimulation [38].

In contrast, we showed that FAD patient-derived fibroblasts carrying the PS2-M239I mutation, as well as HeLa and HEK293 cells stably or transiently expressing the same PS2 mutant, show a decreased ER Ca^{2+} release when stimulated by IP3-generating agonists [39] (Figure 1); this result was confirmed in FAD patient-derived fibroblasts carrying another PS2 mutation (T122R) [40]. Of note, in this study, we analysed two monozygotic twins, one with overt signs of disease at the time of biopsy, whereas the other one was still asymptomatic; nevertheless, both cell samples shared a similar Ca^{2+} handling defect, strongly suggesting that Ca^{2+} dysregulation represents an early event in the pathogenesis of AD [40].

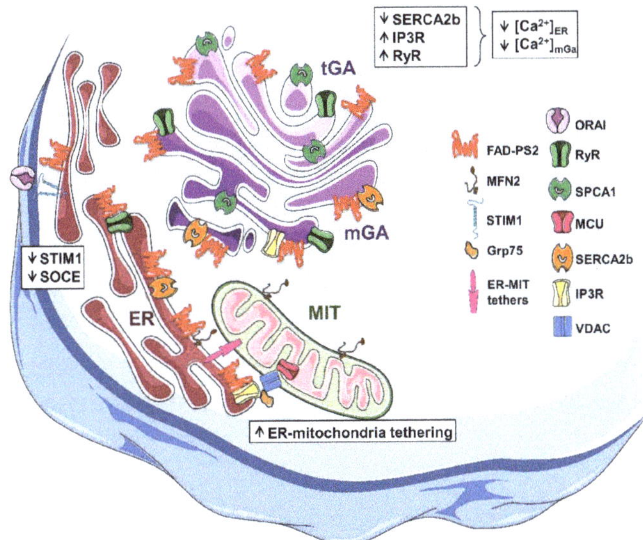

Figure 1. Familial Alzheimer's disease (FAD)-presenilin-2 (PS2) alters multiple Ca^{2+} signalling pathways. The cartoon represents different intracellular membrane localizations of FAD-PS2, its interactions with several components of the molecular Ca^{2+} toolkit, and multiple Ca^{2+} signalling pathways that are altered by its action. See text for details. ER, endoplasmic reticulum; mGA, medial-Golgi apparatus; tGA, trans-Golgi apparatus; MIT, mitochondrion.

To clarify the divergent results, we directly monitored Ca^{2+} dynamics within intracellular stores. We employed aequorin-based Ca^{2+} probes targeted to ER and GA in cells expressing different PS1 and PS2 mutants. In several cell lines [SH-SY5Y, HeLa, HEK293 and Mouse Embryonic Fibroblast (MEF) cells], expressing the ER (or GA)-targeted aequorin together with a number of PS1 (P117L, M146L, L286V, and A246E) or PS2 (M239I, T122R, and N141I) mutants, we analysed Ca^{2+} concentrations and dynamics in the two organelles. By this more specific approach, we confirmed lower ER and GA Ca^{2+} levels in the presence of all the analysed FAD-PS2 mutants, and unchanged or slightly decreased ER/GA Ca^{2+} concentrations when PS1 mutants were expressed [41]. Similar results were obtained in FAD patient-derived fibroblasts and rat primary neurons, expressing either PS1 or PS2 mutants and loaded with the Ca^{2+} sensor fura-2, confirming the capability of PS to modify Ca^{2+} homeostasis, but questioning the "Ca^{2+} overload" hypothesis for AD [41,42]. Indeed, it was also shown that FAD-PS are associated with IP3R hyperactivity [43,44], providing an alternative explanation to the "Ca^{2+} overload" hypothesis based on increased ER Ca^{2+} release findings previously reported in FAD-PS-expressing cells. In particular, Foskett and co-workers showed that FAD-PS1/2 mutants, by physically interacting with the IP3R, modulate the channel gating, causing an exaggerated ER Ca^{2+} release regardless of its Ca^{2+} content [43,44]. Furthermore, by employing ER- and GA-targeted Ca^{2+} indicators, the same group subsequently confirmed that cells expressing FAD-PS1 mutants do not present ER Ca^{2+} overload, arguing against the previously proposed role of PS as ER Ca^{2+} leak channels [45].

Similarly, by using newly developed genetically encoded Ca^{2+} indicators, the Förster resonance energy transfer (FRET)-based probe targeted to ER (D4ER [46]), medial-GA [47], and trans-GA [48], we showed that (i) in SH-SY5Y and Baby Hamster Kidney (BHK) cells, expressing the FAD-PS2-T122R mutant, and in PS2-N141I patient-derived fibroblasts, there is a clear reduction in ER Ca^{2+} concentration; (ii) in cells expressing the FAD-PS1-A246E mutant, instead, no change was observed [46,49]; (iii) the expression of FAD-PS2 mutants induced a selective decrease in the medial-GA Ca^{2+} content, but not in that of the trans-GA; (iv) in contrast, the expression of the FAD-PS1 mutant was ineffective on both GA sub-compartments [49] (Figure 1).

Concerning the molecular mechanism through which FAD-PS2 alters intracellular Ca^{2+} store dynamics, researchers have shown that the Ca^{2+} phenotype is caused by the holoprotein and that it is independent of γ-secretase activity [38,39,42,44,49–51]. Moreover, it has been shown that the protein directly interacts with the IP3R, sensitizing it to lower IP3 concentrations [43], the RyR [36], the RyR-regulating protein sorcin [52], and the SERCA2b [34], inhibiting its activity [51]. This latter result is consistent with the differential effect of FAD-PS2 mutants on GA sub-compartments (see above), given that the trans-GA, where FAD-PS2 mutants are ineffective, relies only on the secretory pathway Ca^{2+} ATPase 1 (SPCA1) for Ca^{2+} uptake [48]. Finally, in the presence of FAD-PS1/2 mutations, increased expression levels and activity of RyRs have been reported [53–55], suggesting a RyR-dependent Ca^{2+} hyperexcitability in AD that is antagonized by the channel inhibitor dantrolene (see [56] for an extensive discussion of this issue; see also [57] for the involvement of IP3Rs).

Intracellular Ca^{2+} stores, mainly the ER, are functionally and physically coupled to mitochondria with which they jointly operate modulating several cell functionalities, such as lipid synthesis and Ca^{2+} homeostasis. Specific ER membrane domains tightly juxtaposed to mitochondria, called mitochondria-associated membranes (MAM; [58]), represent signalling platforms and play a key role in these processes [59]. Interestingly, MAM appear to be altered in AD samples [42,54,59–64]; in addition PS1/2, as well as the other components of the γ-secretase complex and APP, are enriched in these domains [63,65,66]. Only FAD-PS2 mutants, however, are able to increase the interaction between the two organelles, facilitating ER–mitochondria Ca^{2+} transfer [42,54,63] by binding to mitofusin-2 [63] and thus removing its negative modulation on organelle tethering [67] (Figure 1).

The other Ca^{2+} signalling pathway affected by FAD-PS2 is the store-operated Ca^{2+} entry (SOCE) [68,69]. In particular, it has been shown that several FAD-PS2 mutants reduce SOCE activity in different cell types [40,41,49,70]. Interestingly, this effect is shared with FAD-PS1 mutants, which similarly reduce this Ca^{2+} influx [41,49,70,71] (Figure 1). Accordingly, SOCE is potentiated in cells where PS levels are reduced [70,72]. In PS double knock out (KO) MEFs and in B-lymphocytes derived from patients expressing FAD-PS mutants [73], researchers have found that the levels of the key SOCE components Stromal interaction molecule (STIM) STIM1 and STIM2 [68,69] are reduced. Of note, alterations in SOCE and STIM1 protein level have also been reported in sporadic AD (SAD) patients [74]. It has been proposed that SOCE is regulated by a γ-secretase-dependent mechanism, with STIM1 being a substrate of PS1-containing γ-secretase complexes [71]. Nevertheless, we found lower SOCE and STIM1 protein levels in both FAD-PS1- and FAD-PS2-expressing cells treated with the γ-secretase inhibitor DAPT [49] (Figure 1).

Of note, the overexpression of WT-PS2 often mimics the effect of its FAD mutants on Ca^{2+} homeostasis, although higher levels of WT-PS2 are required to obtain the alterations in Ca^{2+} homeostasis elicited by FAD-PS2 mutants [40]. This latter finding could be relevant for SAD forms of the disease, where an upregulation of the endogenous PS2 has been reported in brain AD samples due to the loss of repressor element 1-silencing transcription factor (REST) [75]. It can be speculated that an abnormal accumulation of PS2 holoprotein could cause Ca^{2+} signalling dysregulation, also typically observed in SAD cases.

2.2. Calcium Handling in AD Mouse Models Expressing PS2-N141I

The findings reported above led us to investigate Ca^{2+} handling and brain network functionality in Tg mouse lines based on FAD-PS2 mutants by means of in vitro and in vivo approaches. We took advantage of two homozygous mouse lines expressing the PS2-N141I mutant, as described in detail in Box 1: the double Tg line B6.152H, also known as B6.PS2APP, and the single Tg line PS2.30H [76]. Here, we simply refer to these two lines as 2TG and TG, respectively.

We were firstly interested to verify whether the same Ca^{2+} changes found in FAD-PS2-expressing cell lines were also detectable in primary neuronal cultures and in acute hippocampal slices from 2-week-old animals. At this age, total brain Aβ levels in 2TG mice are still very low, but are already detectable and higher when compared to WT and TG mice [54]. Both TG and 2TG neurons, in culture or in situ, show a reduction in the ER Ca^{2+} content, when estimated indirectly, through Ca^{2+} release induced by IP3-generating agonists, or directly, with the Cameleon probe D4ER [46,54].

In acute hippocampal slices, upon stimulation with IP3-generating agonists, Ca^{2+} release was dramatically reduced not only in neurons but also in astrocytes of TG and 2TG mice, suggesting defective store Ca^{2+} content, as well as Ca^{2+} entry, in these latter cell types [54]. Importantly, these changes occur precociously and independently of APP overexpression and brain Aβ load, being found equally in TG and 2TG mice; thus, they reflect the intrinsic capability of modulating Ca^{2+} handling of FAD-PS2 mutants.

Studying neurons in vitro and in situ allowed us to also highlight relevant network properties brought about by the PS2 mutant. In both conditions, neuronal cells, when exposed to picrotoxin, a γ-aminobutyric acid (GABA)-A receptor antagonist, showed synchronous Ca^{2+} spiking activity that was higher in TG and 2TG mice with respect to WT [54]. This type of Ca^{2+} spiking is independent of Ca^{2+} stores and likely due to an imbalance between excitatory and inhibitory inputs that represent an early sign of network dysfunction [77,78].

3. Functional Effects of Ca^{2+} Dysregulation by FAD-PS2

3.1. Autophagy

Macroautophagy (hereafter autophagy) is a process in which double-membrane vesicles (called autophagosomes) engulf different cellular components (including misfolded proteins, portions of cytosol, and damaged organelles) and target them to lysosomes, where they are degraded into simpler molecular constituents.

In 2004, two seminal papers firstly suggested that PSs might be involved in autophagy modulation [79,80]. Specifically, Esselens and co-workers reported telencephalin accumulation within autophagosomes in PS1-KO hippocampal neurons as a result of a defective fusion of these vesicles with lysosomes. Similarly, Wilson and colleagues observed that PS1 deficiency, in fibroblasts and primary cortical neurons, resulted in the formation of enlarged lysosomes, with accumulation of α- and β-synuclein. Importantly, this phenomenon was likely associated with SOCE augmentation, suggesting that altered Ca^{2+} signalling may underpin the effect of PSs on the autophagy pathway [79]. Additional investigations, mostly focused on PS1, consistently reported that PSs modulate autophagosome-lysosome fusion. Nevertheless, consensus has not been reached on the underlying mechanism. Indeed, either a defective lysosomal acidification [81], a reduced lysosomal Ca^{2+} content [82], or altered expression of key genes belonging to the coordinated lysosomal expression and regulation (CLEAR) network [83,84] have been suggested as possible mechanisms (reviewed in [85]).

As far as FAD-PS mutants are concerned, some of the discrepancies might be linked to mutation-specific effects. Nevertheless, most studies converge on the lack of involvement of the γ-secretase activity, whereas Ca^{2+} signalling dysregulation has been frequently reported as a common feature among different FAD-PS mutants [85]. Recently, we observed that the reduced ER Ca^{2+} content, consistently observed in different FAD-PS2 cell models, affects the fusion of autophagosomes with lysosomes, thus inducing autophagosome accumulation [86] (Figure 2). Specifically, the phenomenon appears linked to the generation of lower cytosolic Ca^{2+} rises upon IP3-induced release of ER Ca^{2+}, given that it can be mimicked by increasing the cytosolic Ca^{2+}-buffering capacity (loading cells with the permeable forms of Ca^{2+} chelating agents). Mechanistically, we found that alterations of cytosolic Ca^{2+} dynamics affect the recruitment to autophagosomes of Ras-associated binding protein RAB7, a small GTPase whose association with both autophagosomes and lysosomes tunes their fusion in the final steps of the autophagy pathway [87]. Importantly, at variance with previous studies focused on FAD-PS1 [81,82,88], neither the pH of lysosomes nor their Ca^{2+} content were found to be affected by FAD-PS2 mutants [86]. Taken together these observations suggest that slightly different mechanisms might underlie the effects of FAD-PS1 and FAD-PS2 on the autophagy flux, with an altered Ca^{2+} signalling (though by distinct pathways) being a common feature.

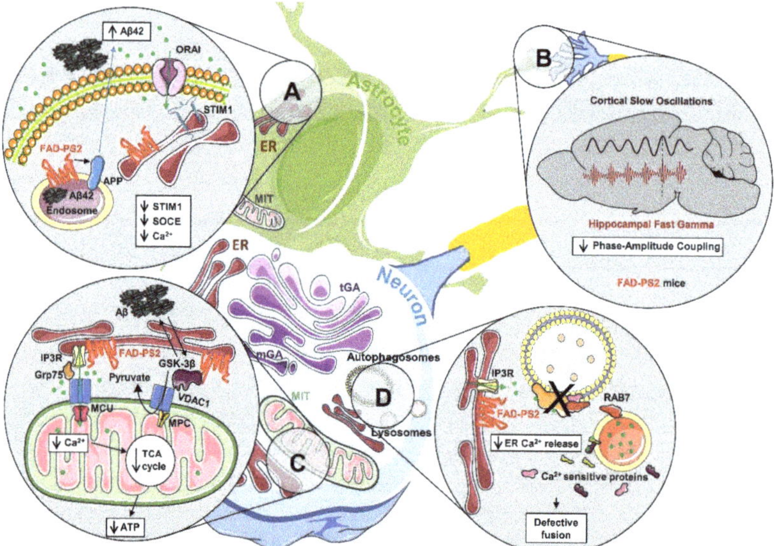

Figure 2. Functional consequences of dysregulated Ca^{2+} signalling induced by FAD-PS2. The cartoon represents the major dysfunctions linked to the expression of FAD-PS2 mutants at both the cellular and brain network levels. (**A**) Decreased store-operated Ca^{2+} entry (SOCE) potentiates amyloid precursor protein (APP) processing and Aβ42 production. (**B**) FAD-PS2-N141I-based mice show altered neuronal circuits (decreased phase-amplitude coupling between cortical slow oscillations and hippocampal fast gamma frequencies). (**C**) Decreased mitochondrial Ca^{2+} signalling and pyruvate uptake impair mitochondrial metabolism and cell bioenergetics. (**D**) Reduced endoplasmic reticulum (ER) Ca^{2+} release blocks the recruitment to autophagosomes of the Ras-associated binding protein RAB7 and their subsequent fusion with lysosomes. See text for details.

3.2. Cell Metabolism and Bioenergetics

The first piece of evidence that WT-PS2 modulates mitochondrial metabolism was found in 2006, when lower mitochondrial respiration and decreased mitochondrial membrane potential ($\Delta\psi$m) were observed in $PSEN2^{-/-}$, but not in $PSEN1^{-/-}$ MEFs [89]. Later, similar results were obtained by Contino and co-investigators [90], who found reduced basal and maximal mitochondrial oxygen consumption in $PSEN2^{-/-}$ and PS double KO MEFs (but not in $PSEN1^{-/-}$ MEFs), associated with an altered morphology of the mitochondrial cristae and a dampened expression of different subunits of the mitochondrial respiratory chain. Interestingly, in both studies, the ATP/ADP ratio was not significantly altered by *PSEN2* ablation, likely because of a compensatory upregulation of the glycolytic flux [90]. Recently, we obtained data suggesting that, in primary cortical neurons from PS2$^{-/-}$ mice (see Box 1), reduced mitochondrial respiration is associated with a defective mitochondrial Ca^{2+} signal (Rossi et al., in preparation). This finding suggests that the Ca^{2+}-mediated modulation of mitochondrial metabolism has a key role in the effects reported above [91].

It is well established that mitochondrial activity is critical for brain health—not only is the majority of neuronal ATP synthesized by mitochondria, but also the rate of ATP synthesis matches synaptic activity [92]. Therefore, the effects of FAD-PS on mitochondria metabolism might be relevant to FAD pathogenesis.

Alterations of mitochondrial activity have been reported in different AD models, mostly in Tg mouse models harboring FAD-PS1 and FAD-APP mutations. However, consensus has not been reached on the underlying molecular mechanisms. Indeed, either defective assembly/expression/activity of different subunits of the mitochondrial respiratory chain [93], altered mitochondrial Ca^{2+} signals [94,95], or organelle positioning/transport [96] have been suggested to contribute to the observed alterations. In contrast, only a few studies have focused on FAD-PS2 mutants. In primary cortical neurons from 2TG mice (see Box 1), we observed a reduced mitochondrial respiratory capacity [97]. This defect is not due to any intrinsic alteration of the respiratory chain, but rather depends on an impaired glycolytic flux, in turn affecting nutrient supply to mitochondria and thus organelle metabolism. However, considering that 2TG mice also express the FAD-APP mutant, it is not clear to what extent FAD-PS2 contributes to this phenotype. Recently, however, in different FAD-PS2-expressing cells, we observed a lower mitochondrial activity associated with a reduced ATP synthesis [98]. Mechanistically, these alterations depend in part on reduced mitochondrial Ca^{2+} signalling (due to partial depletion of ER Ca^{2+} content; see above), and in part on defective mitochondrial pyruvate uptake, caused by alterations in a signalling pathway driven by hyperactive GSK3β [98] (Figure 2), a feature commonly reported in AD [99]. Importantly, when compared to WT, in primary cortical neurons from TG mice (see Box 1), basal ATP levels are not significantly affected, whereas a faster ATP decrease is observed in cells exposed to ATP-consuming stimuli. In addition, we found that these metabolic alterations are associated with an increased susceptibility of FAD-PS2-N141I neurons to excitotoxicity induced by glutamate at physiological concentrations [98]. Overall, these results suggest that subtle mitochondrial alterations may be tolerated for a long time until specific stress conditions, imposing a high energy-demand, unveil their pathological potential. This might be relevant in neurological disorders characterized by a late onset, such as AD.

3.3. Brain Network Activity

Ca^{2+} dysregulation and altered APP processing, the two major hits linked to PS2-N141I expression, could affect neural circuit dynamics during the progression of amyloidosis. By studying brain oscillatory activity of adult 2TG mice under anesthesia, we observed that these mice develop a condition of hippocampal hyperactivity, with increased power in the gamma frequency range (45–90 Hz), as measured by spontaneous local field potential (LFP) signals. Curiously, age-matched TG mice also show a similar increase in the gamma power [100]. This hyperactivity is thus independent of Aβ production given that TG mice, unlike the 2TG animals, show neither plaque deposition nor gliosis, and Aβ42 levels are not significantly different from those found in WT mice [100]. This also suggests that, in 2TG mice, network hyperactivity is not due to compensatory, protective mechanisms and likely exerts a pathogenic role in the disease [78]. Of note, in humans, mild cognitive impairment (MCI) is marked by hyperactivity in the hippocampus, as well as in other cortical regions, that disappears with overt AD [101]. 2TG mice also present hyper-synchronicity, which is detectable as early as 3 months of age [100]. This aspect is likely attributable to the early phase of Aβ accumulation and represents a common feature in AD, often in the form of silent seizures, especially in FAD cases that show a higher incidence of epilepsy [102–104].

Studies on the brain electrical activity of both AD patients and mouse models have recently been focused on slow oscillations, which are directly involved in memory consolidation during sleep and unconsciousness [105]. By detecting mesoscale Ca^{2+} signals at the mouse brain level, Busche and coworkers elegantly demonstrated that functional connectivity in the slow-wave range (0.1–3 Hz) is severely reduced in the neocortex, thalamus, and hippocampus of different AD mouse models, also on the basis of PS1 [106]. Interestingly, slow-wave manipulation restores the functionality of brain circuits, rescues neuronal Ca^{2+} [107], and enhances memory consolidation in both types of mice [106,107].

Given that PS2-N141I alters neuronal and astrocytic Ca^{2+} homeostasis, it might also disturb hippocampal and cortical oscillatory activity in the slow-wave range. In mice under anesthesia, the oscillatory activity of different brain depths can be measured by simultaneously recording LFP signals with a multi-site linear probe. We used this approach to study brain rhythmicity at the cortical and hippocampal levels. In both TG and 2TG mice, the total power, which mostly reflects spontaneous activity in the low frequency range (0.1–5 Hz), is reduced, particularly at the hippocampal level, suggesting that the PS2 mutant by itself alters the brain electrical activity [108].

Another interesting feature shared by both TG and 2TG mice is the disruption of cortico-hippocampal oscillation coupling (Figure 2, [108]). The phenomenon, also known as phase-amplitude coupling (PAC), occurs when the phase of slower rhythms influences the amplitude of faster ones, and it has been found to be involved in memory consolidation and information transfer [105,109].

Unique features of 2TG mice help to mark the progression of Aβ accumulation and deposition—loss of functional connectivity in the slow-wave range marks the onset of Aβ accumulation, similarly to what reported in PS1-based AD mice [106], whereas low/high power imbalances characterize Aβ deposition in plaque-seeding mice [108]. Since Aβ42 oligomers are associated with Ca^{2+} homeostasis dysregulation [110–113], it is tempting to speculate that, in 2TG mice, Ca^{2+} handling alterations, due to PS2-N141I, sum up or synergize with defects linked to Aβ accumulation.

4. Concluding Remarks and Possible Therapeutic Targets

At the brain circuit level, the FAD-linked PS2-N141I mutant increases excitability [54,100] and disrupts the coupling of cortical slow-waves to hippocampal fast gamma frequencies [108]. Altogether, these findings are consistent with the high frequency of seizures and behavioral changes found in both FAD-PS2-N141I patients [1] and other mouse models expressing PS2-N141I [114,115]. From an pathogenic point of view, major alterations are expected in subpopulations of fast spiking interneurons that control the excitability of neuronal microcircuits, as reported in AD mouse models [103,116–118]. These highly active cells are likely more susceptible to the metabolic failure brought about by the aforementioned defective mitochondrial function [97,98]. One should also consider that, in these mouse models, only the PS2 mutant is expressed in both neurons and glial cells. In particular, astrocytes are good candidates to explain circuit dysfunctions given that, through spontaneous Ca^{2+} oscillations and intercellular Ca^{2+} waves, they can control the excitability of large neuronal networks [119,120], as well as modulate neighboring neurons by glio-transmission [121,122]. Furthermore, Ca^{2+} dysregulation and metabolic impairment in a cell type can also affect the closest cells, thus necessitating their investigation at the in situ and in vivo level.

It can be speculated that defects in metabolic and autophagic pathways, directly dependent on Ca^{2+} dysregulation (see above), are responsible for the described network hyperexcitability and excitation/inhibition imbalances, which has also been reported in other AD models [77,103,123].

As for Ca^{2+} dysregulation, a common denominator between FAD-PS1 and -PS2 mutations is SOCE. Nevertheless, up until now, only a few studies have addressed the role of this Ca^{2+} pathway in neurons, mainly because of technical problems, i.e., the difficulty of distinguishing between activation of SOCE and voltage-operated Ca^{2+} channels (VOCCs) [124,125]. Recently, STIM2 and ORAI2, two key players in SOCE machinery, have emerged as key components of neuronal SOCE, being implicated in SOCE impairment in mushroom spines of hippocampal neurons from FAD-PS1-M146V knock-in mice [126–128]. Although the role of SOCE in neurons is still unclear, it is important to stress that, in excitable cells, STIM and ORAI components might also play non-canonical roles—STIM1 binds to L-type VOCCs, inhibits their gating, and induces channel internalization [129,130] while ORAI1 increases neuronal excitability [131,132]. At variance with neuronal cells, it is now largely accepted that SOCE is crucial for the Ca^{2+}-based excitability that characterizes glial cells both in vitro [133,134] and in vivo [135]. Nonetheless, studies that specifically address the role of FAD-PS2 in glial SOCE modulation are still lacking. Considering also the complexity of microglia involvement in the onset and progression of AD [113,136], it is conceivable that these cells might also be primarily affected in the SOCE pathway, given that PS2 is the major core component of γ-secretase complexes expressed in this cell type [137].

We have recently shown that there is an inverse relationship between SOCE level and Aβ42 accumulation [138], consistent with data obtained in neurons [70] and other model cells [139]. These observations suggest the possibility of rescuing the SOCE defect in neural cells while antagonizing Aβ42 production. It has been demonstrated that, in mouse lymphocytes, SOCE is increased by knockout of ORAI2, a channel subunit and a negative modulator of SOCE that is responsible for the Ca^{2+} release-activated Ca^{2+} current [140]. In Aβ42-secreting neuroglioma cells, ORAI2 downregulation also increases SOCE and reduces the Aβ42/Aβ40 ratio [138] (Figure 2). Of note, astrocytes actively participate in Aβ production and clearance [141]. We do not know yet whether ORAI2 can play a similar role in neurons; up until now, it looks unlikely, given that recent data by Betzprozvanny's group favor the hypothesis that ORAI2 is a component of a specific type of neuronal SOCE that is based on transient receptor potential canonical 6 (TRPC6) channel and regulated by diacylglycerol [127]. What is clear is that investigating Ca^{2+} dysregulation in AD allows the design of alternative therapeutic approaches to this devastating disease.

Additional therapeutic approaches could be suggested on the basis of altered bioenergetic and autophagy pathways. Impaired mitochondria, unable to supply cellular ATP demand, cause alterations in neuronal excitability, eventually leading to Ca^{2+} overload and cell death [142]. Moreover, the accumulation of damaged mitochondria (and misfolded proteins), due to defective autophagy, further contributes to dysfunctional neurons, causing, over the long term, neurodegeneration. Indeed, mitochondrial alterations, and in particular defects in bioenergetic pathways, have been widely reported to be key factors not only in AD but also in other neurodegenerative diseases [142,143]. Importantly, bioenergetic alterations are reported in different SAD and FAD samples, appearing at the early stage of the disease, before Aβ plaque formation [144].

The bioenergetic state of neurons is a crucial determinant of their response to glutamate, with cells containing defective mitochondria undergoing bioenergetic crises, Ca^{2+} mishandling, and excitotoxicity. The Food and Drug Administration (FDA)-approved molecule memantine targets glutamate receptors and is among the few pharmacological treatments that provide modest benefits in AD patients, in addition to cholinesterase inhibitors [145]. Targeting Ca^{2+} defects, at multiple levels, was suggested as a possible therapeutic strategy, especially in the form of drug repurposing. Among the best candidates, there are dantrolene, a RyR modulator [146], and isradipine, a VOCC inhibitor, as reviewed by Chakraborty and Stutzmann [147]. Attention has to be payed to the fact that dihydropyridines, especially nimodipine, also increase Aβ42 secretion [148]. None of these drugs are in the pipeline yet, and thus additional interventions aiming at supporting other pathways, such as mitochondrial performance, are desirable. In line with this, we showed that GSK3β inhibition rescues the FAD-PS2-linked bioenergetic defect [98]. Interestingly, both PS and Aβ oligomers have been reported to interact with the kinase, favoring its activity [99]. Considering the fact that GSK3β activity has been observed at MAM [149], where PSs are also enriched and Aβ peptides are generated [63,65,66], a MAM-targeted intervention might represent a useful therapeutic strategy [98].

Finally, the impairment in mitochondrial bioenergetics described in AD models is likely linked to a metabolic rewiring, possibly resulting in systemic alterations in the concentration of specific metabolites. Thus, the detailed metabolic profiling of AD patient-derived peripheral samples (blood and cerebrospinal fluid) might offer the possibility to discover new biomarkers that are helpful for early AD diagnosis, as has been previously suggested [150,151].

5. Box 1: AD Mouse Models Based on PS2

Several mouse models have been developed to understand the pathogenesis of AD, however, none of them are capable of reproducing the full spectrum of the human disease. The large majority of the most used AD models are double-Tg mice based on human FAD-APP and -PS1 mutations, both required to obtain fast amyloid accumulation, plaque deposition, and gliosis between 2 and 8 months of age. These Tg mice are widely considered to be adequate models of Aβ amyloidosis and its inflammatory process; they allow us to study the initial stages of the disease, according to the vision that places Aβ toxicity among the first hits in the AD cascade [6,152,153]. Nonetheless, the latter appears necessary but not sufficient in terms of causing neurodegeneration, with other concomitant and downstream factors playing a key role [7]. Neurodegeneration, linked to tau aggregation, is in fact mainly present in 3xTg-AD mice, which host three human mutant genes encoding PS1, APP, and tau [154].

Curiously, only the PS2-N141I mutation has been used to generate AD mouse models based on *PSEN2*. In terms of the latter, we used two homozygous lines: the double Tg (2TG) B6.152H, also known as B6.PS2APP, and the single Tg (TG) PS2.30H [76]. The latter line expresses the human PS2-N141I under the *prion protein* promoter, with background C57Bl/6 > 90% [155]. The B6.152H line was instead obtained by co-injection of human *PSEN2*, carrying the N141I mutation—under the mouse *prion protein* promoter—and the human *APP* isoform 751, carrying the APP-KM670/671NL Swedish mutation—under the *Thy1.2* promoter—into zygotes of the C57Bl/6 strain (background C57Bl/6 100%) [76].

The PS2.30H line was originally used to obtain hemizygous PS2APP mice by crossing PS2.30H females with APP-Swedish males of the BD.AD147.71H line, with background C57Bl/6 > 90% [155]. Up to 12 months of age, TG mice show neither plaques nor Aβ accumulation in the brain [100]. The histopathological traits of PS2APP and B6.PS2APP are very similar, showing an exponential growth of Aβ accumulation and plaques at 3 and 6 months of age, respectively [76,155]. Plaque deposition starts in the frontal cortex, subiculum, and hippocampus; increases for up to 12–16 months of age; and correlates with the level of human APP transcript [76,155]. Behavioral deficits have only been characterized thoroughly in PS2APP mice, with spatial learning (Morris water maze) and memory defects appearing at 8 months [155]. Biochemical and functional differences between the two closely related models are also present [156,157]. In our studies, TG and 2TG mice are maintained and used in homozygosity, a condition that allows for the reduction of the variability of APP expression [76]. The two lines express PS2 at a similar level, about twice that found in C57Bl/6 WT mice, used as controls [54].

B6.152H mice have also been used in hemizygosity (B6.152) to study different aspects of the AD phenotype [158], or to generate TauPS2APP triple Tg mice, upon crossing the B6.152H line with the Tau-overexpressing pR5 line [159,160] that expresses the human tau-40 isoform under the *Thy1.2* promoter [161]. Of note, a PS2-/- mouse line has been obtained by neomycin insertion in the C57Bl/6 × 129Sv genetic background [162,163]. This line does not show alterations of the endogenous APP processing and it is useful in terms of studying the physiological role of PS2 and possible loss-of-function defects associated with PS2-N141I expression [100,108]. It was also used to produce

PSEN conditional double KO mice and study neurodegeneration and memory impairments due to PS deficiency [164].

Other AD mouse models, based on PS2-N141I, have been generated under different promoters and genetic backgrounds. Comparison of their histopathological, functional, and behavioral properties is beyond the scope of this review. The latest generation of AD mouse models avoids the overexpression of FAD mutations and is focused on risk genes, such as Triggering receptor expressed on myeloid cells 2 (*TREM2*) [165] and Apolipoprotein E4 (*APOE4*) [166]. More than 200 AD mouse models are now available; detailed information about these AD animal models is available at the Alzforum website (https://www.alzforum.org/). It is also necessary to mention that doubts have recently been raised on the use of Tg mice to study human AD, given that, at variance with the trascriptomic profiles of physiological human and rodent brain aging, which appear very similar, those of AD brains are largely different between humans and rodents, and even between different Tg AD mouse lines [167].

Author Contributions: Writing—original draft preparation, P.P., E.B., R.F., E.G., A.L., D.P., N.R., M.R., N.V., T.P., C.F. Review and editing P.P., T.P., C.F. All authors have read and agreed to the published version of the manuscript.

Funding: Authors' work was funded by the following grants: Consiglio Nazionale delle Ricerche (CNR) "Premio Scientifico 2018 DSB-CNR" to R.F.; the University of Padova, Italy (SID 2019), the Italian Ministry of University and Scientific Research (PRIN2017XA5J5N), and EU Joint Programme in Neurodegenerative Disease, 2015–2018 (grant CeBioND) to P.P.; Fondazione Cassa di Risparmio di Padua e Rovigo (CARIPARO Foundation) Excellence project 2017 (2018/113), Veneto Region ([RISIB Project]), Consiglio Nazionale delle Ricerche (CNR) Special Project Aging, Euro Bioimaging Project Roadmap/ESFRI from European Commission, and Telethon Italy Grant GGP16029A to TP; PRIN-20175C22WM to C.F.; UNIPD Funds for Research Equipment-2015. As post-docs, N.R. and A.L. were supported by fellowships from PRIN20175C22WM to C.F. and BIRD 2017 to C.F., respectively.

Acknowledgments: PS2.30H and B6.152H mouse lines used for the authors' research were kindly donated by L. Ozmen and F. Hoffmann-La Roche Ltd. (Basel, Switzerland).

Conflicts of Interest: The authors declare no conflict of interest. The funders had no role in the design of the study; in the collection, analyses, or interpretation of data; in the writing of the manuscript; or in the decision to publish the results.

References

1. Jayadev, S.; Leverenz, J.B.; Steinbart, E.; Stahl, J.; Klunk, W.; Yu, C.E.; Bird, T.D. Alzheimer's disease phenotypes and genotypes associated with mutations in presenilin 2. *Brain* **2010**, *133*, 1143–1154. [CrossRef] [PubMed]

2. Brunkan, A.L.; Goate, A.M. Presenilin function and gamma-secretase activity. *J. Neurochem.* **2005**, *93*, 769–792. [CrossRef] [PubMed]

3. Lee, M.K.; Slunt, H.H.; Martin, L.J.; Thinakaran, G.; Kim, G.; Gandy, S.E.; Seeger, M.; Koo, E.; Price, D.L.; Sisodia, S.S. Expression of presenilin 1 and 2 (PS1 and PS2) in human and murine tissues. *J. Neurosci.* **1996**, *16*, 7513–7525. [CrossRef] [PubMed]

4. Chévez-Gutiérrez, L.; Bammens, L.; Benilova, I.; Vandersteen, A.; Benurwar, M.; Borgers, M.; Lismont, S.; Zhou, L.; Van Cleynenbreugel, S.; Esselmann, H.; et al. The Mechanism of γ-Secretase Dysfunction in Familial Alzheimer Disease. *EMBO J.* **2012**, *31*, 2261–2274. [CrossRef]

5. Chávez-Gutiérrez, L.; Szaruga, M. Mechanisms of neurodegeneration-insights from familial alzheimer's disease. In *Seminars in Cell and Developmental Biology*; Elsevier Ltd.: Amsterdam, The Netherlands, 2020; pp. 75–85.

6. Selkoe, D.J.; Hardy, J. The amyloid hypothesis of alzheimer's disease at 25years. *EMBO Mol. Med.* **2016**, *8*, 595–608. [CrossRef]

7. Long, J.M.; Holtzman, D.M. Alzheimer disease: An update on pathobiology and treatment strategies. *Cell* **2019**, *179*, 312–339. [CrossRef]

8. Walker, E.S.; Martinez, M.; Brunkan, A.L.; Goate, A. Presenilin 2 familial alzheimer's disease mutations result in partial loss of function and dramatic changes in abeta 42/40 ratios. *J. Neurochem.* **2005**, *92*, 294–301. [CrossRef]

9. Güner, G.; Lichtenthaler, S.F. The substrate repertoire of γ-secretase/presenilin. In *Seminars in Cell and Developmental Biology*; Elsevier Ltd.: Amsterdam, The Netherlands, 2020; pp. 27–42.

10. Bolduc, D.M.; Montagna, D.R.; Gu, Y.; Selkoe, D.J.; Wolfe, M.S. Nicastrin functions to sterically hinder γ-secretase-substrate interactions driven by substrate transmembrane domain. *Proc. Natl. Acad. Sci. USA* **2016**, *113*, E509–E518. [CrossRef]

11. Holmes, O.; Paturi, S.; Selkoe, D.J.; Wolfe, M.S. Pen-2 is essential for γ-secretase complex stability and trafficking but partially dispensable for endoproteolysis. *Biochemistry* **2014**, *53*, 4393–4406. [CrossRef]

12. Kim, S.H.; Sisodia, S.S. Evidence that the "NF" motif in transmembrane domain 4 of presenilin 1 is critical for binding with PEN-2. *J. Biol. Chem.* **2005**, *280*, 41953–41996. [CrossRef]

13. Prokop, S.; Shirotani, K.; Edbauer, D.; Haass, C.; Steiner, H. Requirement of PEN-2 for stabilization of the presenilin n-/c-terminal fragment heterodimer within the γ-secretase complex. *J. Biol. Chem.* **2004**, *279*, 23255–23261. [CrossRef] [PubMed]

14. Pardossi-Piquard, R.; Yang, S.P.; Kanemoto, S.; Gu, Y.; Chen, F.; Böhm, C.; Sevalle, J.; Li, T.; Wong, P.C.; Checler, F.; et al. APH1 polar transmembrane residues regulate the assembly and activity of presenilin complexes. *J. Biol. Chem.* **2009**, *284*, 16298–16307. [CrossRef] [PubMed]

15. Gu, Y.; Chen, F.; Sanjo, N.; Kawarai, T.; Hasegawa, H.; Duthie, M.; Li, W.; Ruan, X.; Luthra, A.; Mount, H.T.; et al. APH-1 interacts with mature and immature forms of presenilins and nicastrin and may play a role in maturation of presenilin.nicastrin complexes. *J. Biol. Chem.* **2003**, *278*, 7374–7380. [CrossRef] [PubMed]

16. Serneels, L.; Van Biervliet, J.; Craessaerts, K.; Dejaegere, T.; Horré, K.; Van Houtvin, T.; Esselmann, H.; Paul, S.; Schäfer, M.K.; Berezovska, O.; et al. γ-Secretase heterogeneity in the aph1 subunit: Relevance for alzheimer's disease. *Science* **2009**, *324*, 629–632. [CrossRef] [PubMed]

17. Wolfe, M.S.; Xia, W.; Ostaszewski, B.L.; Diehl, T.S.; Kimberly, W.T.; Selkoe, D.J. Two transmembrane aspartates in presenilin-1 required for presenilin endoproteolysis and g-secretase activity. *Nature* **1999**, *398*, 513–517. [CrossRef]

18. Wolfe, M.S. Substrate recognition and processing by γ-secretase. *Biochim. Biophys. Acta Biomembr.* **2020**, *1862*, 183016. [CrossRef]

19. Sannerud, R.; Esselens, C.; Ejsmont, P.; Mattera, R.; Rochin, L.; Tharkeshwar, A.K.; De Baets, G.; De Wever, V.; Habets, R.; Baert, V.; et al. Restricted location of PSEN2/gamma-secretase determines substrate specificity and generates an intracellular abeta pool. *Cell* **2016**, *166*, 193–208. [CrossRef]

20. Meckler, X.; Checler, F. Presenilin 1 and presenilin 2 target gamma-secretase complexes to distinct cellular compartments. *J. Biol. Chem.* **2016**, *291*, 12821–12837. [CrossRef]

21. Jiang, H.; Newman, M.; Lardelli, M. The zebrafish orthologue of familial alzheimer's disease gene PRESENILIN 2 is required for normal adult melanotic skin pigmentation. *PLoS ONE* **2018**, *13*, e0206155. [CrossRef]

22. Zhang, Z.; Gong, J.; Sviderskaya, E.V.; Wei, A.; Li, W. Mitochondrial NCKX5 regulates melanosomal biogenesis and pigment production. *J. Cell Sci.* **2019**, *132*, jcs232009. [CrossRef]

23. Murayama, M.; Tanaka, S.; Palacino, J.; Murayama, O.; Honda, T.; Sun, X.; Yasutake, K.; Nihonmatsu, N.; Wolozin, B.; Takashima, A. Direct association of presenilin-1 with β-catenin. *FEBS Lett.* **1998**, *433*, 73–77. [CrossRef]

24. Zhang, Z.; Hartmann, H.; Do, V.M.; Abramowski, D.; Sturchler-Pierrat, C.; Staufenbiel, M.; Sommer, B.; Van De Wetering, M.; Clevers, H.; Saftig, P.; et al. Destabilization of β-Catenin by mutations in Presenilin-1 potentiates neuronal apoptosis. *Nature* **1998**, *395*, 698–702. [CrossRef]

25. Soriano, S.; Kang, D.E.; Fu, M.; Pestell, R.; Chevallier, N.; Zheng, H.; Koo, E.H. Presenilin 1 negatively regulates β-Catenin/T cell factor/Lymphoid enhancer factor-1 signaling independently of β-Amyloid precursor protein and notch processing. *J. Cell Biol.* **2001**, *52*, 785–794. [CrossRef] [PubMed]

26. Kang, D.E.; Soriano, S.; Frosch, M.P.; Collins, T.; Naruse, S.; Sisodia, S.S.; Leibowitz, G.; Levine, F.; Koo, E.H. Presenilin 1 facilitates the constitutive turnover of β-Catenin: Differential activity of alzheimer's disease-linked PS1 mutants in the β-catenin-signaling pathway. *J. Neurosci.* **1999**, *19*, 4229–4237. [CrossRef] [PubMed]

27. Dolma, K.; Iacobucci, G.J.; Hong Zheng, K.; Shandilya, J.; Toska, E.; White, J.A.; Spina, E.; Gunawardena, S. Presenilin influences glycogen synthase kinase-3 beta (GSK-3beta) for Kinesin-1 and dynein function during axonal transport. *Hum. Mol. Genet.* **2014**, *23*, 1121–1133. [CrossRef] [PubMed]

28. Barthet, G.; Dunys, J.; Shao, Z.; Xuan, Z.; Ren, Y.; Xu, J.; Arbez, N.; Mauger, G.; Bruban, J.; Georgakopoulos, A.; et al. Presenilin mediates neuroprotective functions of ephrinb and brain-derived neurotrophic factor and regulates ligand-induced internalization and metabolism of EphB2 and TrkB receptors. *Neurobiol. Aging* **2013**, *34*, 499–510. [CrossRef] [PubMed]

29. LaFerla, F.M. Calcium dyshomeostasis and intracellular signalling in alzheimer's disease. *Nat. Rev. Neurosci.* **2002**, *3*, 862–872. [CrossRef]

30. Leissring, M.A.; Paul, B.A.; Parker, I.; Cotman, C.W.; LaFerla, F.M. Alzheimer's presenilin-1 mutation potentiates inositol 1,4,5-trisphosphate-mediated calcium signaling in xenopus oocytes. *J. Neurochem.* **1999**, *72*, 1061–1068. [CrossRef]

31. Chan, S.L.; Mayne, M.; Holden, C.P.; Geiger, J.D.; Mattson, M.P. Presenilin-1 Mutations increase levels of ryanodine receptors and calcium release in PC12 cells and cortical neurons. *J. Biol. Chem.* **2000**, *275*, 18195–18200. [CrossRef]

32. Leissring, M.A.; Akbari, Y.; Fanger, C.M.; Cahalan, M.D.; Mattson, M.P.; LaFerla, F.M. Capacitative calcium entry deficits and elevated luminal calcium content in mutant presenilin-1 knockin mice. *J. Cell Biol.* **2000**, *149*, 793–798. [CrossRef]

33. Smith, I.F.; Boyle, J.P.; Vaughan, P.F.; Pearson, H.A.; Cowburn, R.F.; Peers, C.S. Ca^{2+} stores and capacitative Ca^{2+} entry in human neuroblastoma (SH- SY5Y) cells expressing a familial alzheimer's disease presenilin-1 mutation. *Brain Res.* **2002**, *949*, 105–111. [CrossRef]

34. Green, K.N.; Demuro, A.; Akbari, Y.; Hitt, B.D.; Smith, I.F.; Parker, I.; LaFerla, F.M. SERCA pump activity is physiologically regulated by presenilin and regulates amyloid β production. *J. Cell Biol.* **2008**, *181*, 1107–1116. [CrossRef] [PubMed]

35. Cai, D.; Netzer, W.J.; Zhong, M.; Lin, Y.; Du, G.; Frohman, M.; Foster, D.A.; Sisodia, S.S.; Xu, H.; Gorelick, F.S.; et al. Presenilin-1 uses phospholipase d1 as a negative regulator of β-amyloid formation. *Proc. Natl. Acad. Sci. USA* **2006**, *103*, 1941–1946. [CrossRef] [PubMed]

36. Hayrapetyan, V.; Rybalchenko, V.; Rybalchenko, N.; Koulen, P. The N-terminus of presenilin-2 increases single channel activity of brain ryanodine receptors through direct protein-protein interaction. *Cell Calcium* **2008**, *44*, 507–518. [CrossRef] [PubMed]

37. Schneider, I.; Reverse, D.; Dewachter, I.; Ris, L.; Caluwaerts, N.; Kuiperi, C.; Gilis, M.; Geerts, H.; Kretzschmar, H.; Godaux, E.; et al. Mutant presenilins disturb neuronal calcium homeostasis in the brain of transgenic mice, decreasing the threshold for excitotoxicity and facilitating long-term potentiation. *J. Biol. Chem.* **2001**, *276*, 11539–11544. [CrossRef] [PubMed]

38. Tu, H.; Nelson, O.; Bezprozvanny, A.; Wang, Z.; Lee, S.F.; Hao, Y.H.; Serneels, L.; De Strooper, B.; Yu, G.; Bezprozvanny, I. Presenilins form ER Ca^{2+} leak channels, a function disrupted by familial alzheimer's disease-linked mutations. *Cell* **2006**, *126*, 981–993. [CrossRef]

39. Zatti, G.; Ghidoni, R.; Barbiero, L.; Binetti, G.; Pozzan, T.; Fasolato, C.; Pizzo, P. The presenilin 2 M239I mutation associated with familial alzheimer's disease reduces Ca^{2+} release from intracellular stores. *Neurobiol. Dis.* **2004**, *15*, 269–278. [CrossRef]

40. Giacomello, M.; Barbiero, L.; Zatti, G.; Squitti, R.; Binetti, G.; Pozzan, T.; Fasolato, C.; Ghidoni, R.; Pizzo, P. Reduction of Ca^{2+} stores and capacitative Ca^{2+} entry is associated with the familial alzheimer's disease presenilin-2 T122R mutation and anticipates the onset of dementia. *Neurobiol. Dis.* **2005**, *18*, 638–648. [CrossRef]

41. Zatti, G.; Burgo, A.; Giacomello, M.; Barbiero, L.; Ghidoni, R.; Sinigaglia, G.; Florean, C.; Bagnoli, S.; Binetti, G.; Sorbi, S.; et al. Presenilin mutations linked to familial alzheimer's disease reduce endoplasmic reticulum and golgi apparatus calcium levels. *Cell Calcium* **2006**, *39*, 539–550. [CrossRef]

42. Zampese, E.; Fasolato, C.; Kipanyula, M.J.; Bortolozzi, M.; Pozzan, T.; Pizzo, P. Presenilin 2 modulates endoplasmic reticulum (ER)-mitochondria interactions and Ca^{2+} cross-talk. *Proc. Natl. Acad. Sci. USA* **2011**, *108*, 2777–2782. [CrossRef]

43. Cheung, K.H.; Shineman, D.; Muller, M.; Cardenas, C.; Mei, L.; Yang, J.; Tomita, T.; Iwatsubo, T.; Lee, V.M.; Foskett, J.K. Mechanism of Ca^{2+} disruption in alzheimer's disease by presenilin regulation of InsP3 receptor channel gating. *Neuron* **2008**, *58*, 871–883. [CrossRef] [PubMed]

44. Cheung, K.H.; Mei, L.; Mak, D.O.; Hayashi, I.; Iwatsubo, T.; Kang, D.E.; Foskett, J.K. Gain-of-function enhancement of IP3 receptor modal gating by familial alzheimer's disease-linked presenilin mutants in human cells and mouse neurons. *Sci. Signal.* **2010**, *3*, ra22. [CrossRef] [PubMed]

45. Shilling, D.; Mak, D.O.; Kang, D.E.; Foskett, J.K. Lack of evidence for presenilins as endoplasmic reticulum Ca^{2+} leak channels. *J. Biol. Chem.* **2012**, *14*, 10933–10944. [CrossRef] [PubMed]

46. Greotti, E.; Wong, A.; Pozzan, T.; Pendin, D.; Pizzo, P. Characterization of the ER-targeted low affinity Ca^{2+} probe D4ER. *Sensors* **2016**, *16*, 1419. [CrossRef] [PubMed]

47. Lissandron, V.; Podini, P.; Pizzo, P.; Pozzan, T. Unique characteristics of Ca^{2+} homeostasis of the trans-golgi compartment. *Proc. Natl. Acad. Sci. USA* **2010**, *107*, 9198–9203. [CrossRef]

48. Wong, A.K.; Capitanio, P.; Lissandron, V.; Bortolozzi, M.; Pozzan, T.; Pizzo, P. Heterogeneity of Ca^{2+} handling among and within golgi compartments. *J. Mol. Cell Biol.* **2013**, *5*, 266–276. [CrossRef]

49. Greotti, E.; Capitanio, P.; Wong, A.; Pozzan, T.; Pizzo, P.; Pendin, D. Familial alzheimer's disease-linked presenilin mutants and intracellular Ca^{2+} handling: A single-organelle, FRET-based analysis. *Cell Calcium* **2019**, *79*, 44–56. [CrossRef]

50. Nelson, O.; Tu, H.; Lei, T.; Bentahir, M.; de Strooper, B.; Bezprozvanny, I. Familial alzheimer disease-linked mutations specifically disrupt Ca^{2+} leak function of presenilin 1. *J. Clin. Investig.* **2007**, *117*, 1230–1239. [CrossRef]

51. Brunello, L.; Zampese, E.; Florean, C.; Pozzan, T.; Pizzo, P.; Fasolato, C. Presenilin-2 dampens intracellular Ca^{2+} stores by increasing Ca^{2+} leakage and reducing Ca^{2+} uptake. *J. Cell. Mol. Med.* **2009**, *13*, 3358–3369. [CrossRef]

52. Pack-Chung, E.; Meyers, M.B.; Pettingell, W.P.; Moir, R.D.; Brownawell, A.M.; Cheng, I.; Tanzi, R.E.; Kim, T.W. Presenilin 2 interacts with sorcin, a modulator of the ryanodine receptor. *J. Biol. Chem.* **2000**, *275*, 14440–14445. [CrossRef]

53. Chakroborty, S.; Goussakov, I.; Miller, M.B.; Stutzmann, G.E. Deviant ryanodine receptor-mediated calcium release resets synaptic homeostasis in presymptomatic 3×Tg-AD mice. *J. Neurosci.* **2009**, *29*, 9458–9470. [CrossRef] [PubMed]

54. Kipanyula, M.J.; Contreras, L.; Zampese, E.; Lazzari, C.; Wong, A.K.C.; Pizzo, P.; Fasolato, C.; Pozzan, T. Ca^{2+} dysregulation in neurons from transgenic mice expressing mutant presenilin 2. *Aging Cell* **2012**, *11*, 885–893. [CrossRef] [PubMed]

55. Smith, I.F.; Hitt, B.; Green, K.N.; Oddo, S.; LaFerla, F.M. Enhanced caffeine-induced Ca^{2+} release in the 3×Tg-AD mouse model of alzheimer's disease. *J. Neurochem.* **2005**, *94*, 1711–1718. [CrossRef]

56. Del Prete, D.; Checler, F.; Chami, M. Ryanodine receptors: Physiological function and deregulation in alzheimer disease. *Mol. Neurodegener.* **2014**, *9*, 21. [CrossRef] [PubMed]

57. Shilling, D.; Muller, M.; Takano, H.; Mak, D.O.; Abel, T.; Coulter, D.A.; Foskett, J.K. Suppression of InsP3 receptor-mediated Ca^{2+} signaling alleviates mutant presenilin-linked familial alzheimer's disease pathogenesis. *J. Neurosci.* **2014**, *34*, 6910–6923. [CrossRef]

58. Vance, J.E. Phospholipid synthesis in a membrane fraction associated with mitochondria. *J. Biol. Chem.* **1990**, *265*, 7248–7256.

59. Filadi, R.; Theurey, P.; Pizzo, P. The endoplasmic reticulum-mitochondria coupling in health and disease: Molecules, functions and significance. *Cell Calcium* **2017**, *62*, 1–15. [CrossRef]

60. Hedskog, L.; Pinho, C.M.; Filadi, R.; Ronnback, A.; Hertwig, L.; Wiehager, B.; Larssen, P.; Gellhaar, S.; Sandebring, A.; Westerlund, M.; et al. Modulation of the endoplasmic reticulum-mitochondria interface in alzheimer's disease and related models. *Proc. Natl. Acad. Sci. USA* **2013**, *110*, 7916–7921. [CrossRef]

61. Area-Gomez, E.; Del Carmen Lara Castillo, M.; Tambini, M.D.; Guardia-Laguarta, C.; de Groof, A.J.; Madra, M.; Ikenouchi, J.; Umeda, M.; Bird, T.D.; Sturley, S.L.; et al. Upregulated function of mitochondria-associated ER membranes in alzheimer disease. *EMBO J.* **2012**, *31*, 4106–4123. [CrossRef]

62. Sepulveda-Falla, D.; Barrera-Ocampo, A.; Hagel, C.; Korwitz, A.; Vinueza-Veloz, M.F.; Zhou, K.; Schonewille, M.; Zhou, H.; Velazquez-Perez, L.; Rodriguez-Labrada, R.; et al. Familial alzheimer's disease-associated presenilin-1 alters cerebellar activity and calcium homeostasis. *J. Clin. Investig.* **2014**, *124*, 1552–1567. [CrossRef]

63. Filadi, R.; Greotti, E.; Turacchio, G.; Luini, A.; Pozzan, T.; Pizzo, P. Presenilin 2 modulates endoplasmic reticulum-mitochondria coupling by tuning the antagonistic effect of mitofusin 2. *Cell Rep.* **2016**, *15*, 2226–2238. [CrossRef] [PubMed]

64. Liu, Y.; Zhu, X. Endoplasmic reticulum-mitochondria tethering in neurodegenerative diseases. *Transl. Neurodegener.* **2017**, *6*, 1–8. [CrossRef] [PubMed]

65. Area-Gomez, E.; de Groof, A.J.; Boldogh, I.; Bird, T.D.; Gibson, G.E.; Koehler, C.M.; Yu, W.H.; Duff, K.E.; Yaffe, M.P.; Pon, L.A.; et al. Presenilins are enriched in endoplasmic reticulum membranes associated with mitochondria. *Am. J. Pathol.* **2009**, *175*, 1810–1816. [CrossRef] [PubMed]

66. Schreiner, B.; Hedskog, L.; Wiehager, B.; Ankarcrona, M. Amyloid-beta peptides are generated in mitochondria-associated endoplasmic reticulum membranes. *J. Alzheimer's Dis.* **2015**, *43*, 369–374. [CrossRef]

67. Filadi, R.; Greotti, E.; Turacchio, G.; Luini, A.; Pozzan, T.; Pizzo, P. Mitofusin 2 ablation increases endoplasmic reticulum-mitochondria coupling. *Proc. Natl. Acad. Sci. USA* **2015**, *112*, E2174–E2181. [CrossRef]

68. Hogan, P.G.; Rao, A. Store-operated calcium entry: Mechanisms and modulation. *Biochem. Biophys. Res. Commun.* **2015**, *460*, 40–49. [CrossRef]

69. Putney, J.W. Forms and functions of store-operated calcium entry mediators, stim and orai. *Adv. Biol. Regul.* **2018**, *68*, 88–96. [CrossRef]

70. Yoo, A.S.; Cheng, I.; Chung, S.; Grenfell, T.Z.; Lee, H.; Pack-Chung, E.; Handler, M.; Shen, J.; Xia, W.; Tesco, G.; et al. Presenilin-mediated modulation of capacitative calcium entry. *Neuron* **2000**, *27*, 561–572. [CrossRef]

71. Tong, B.C.; Lee, C.S.; Cheng, W.H.; Lai, K.O.; Foskett, J.K.; Cheung, K.H. Familial alzheimer's disease-associated presenilin 1 mutants promote gamma-secretase cleavage of STIM1 to impair store-operated Ca^{2+} entry. *Sci. Signal.* **2016**, *9*, ra89. [CrossRef]

72. Herms, J.; Schneider, I.; Dewachter, I.; Caluwaerts, N.; Kretzschmar, H.; Van Leuven, F. Capacitive calcium entry is directly attenuated by mutant presenilin-1, independent of the expression of the amyloid precursor protein. *J. Biol. Chem.* **2003**, *278*, 2484–2489. [CrossRef]

73. Bojarski, L.; Pomorski, P.; Szybinska, A.; Drab, M.; Skibinska-Kijek, A.; Gruszczynska-Biegala, J.; Kuznicki, J. Presenilin-dependent expression of STIM proteins and dysregulation of capacitative Ca^{2+} entry in familial Alzheimer's disease. *Biochim. Biophys. Acta* **2009**, *1793*, 1050–1057. [CrossRef] [PubMed]

74. Pascual-Caro, C.; Berrocal, M.; Lopez-Guerrero, A.M.; Alvarez-Barrientos, A.; Pozo-Guisado, E.; Gutierrez-Merino, C.; Mata, A.M.; Martin-Romero, F.J. STIM1 deficiency is linked to Alzheimer's disease and triggers cell death in SH-SY5Y cells by upregulation of L-Type voltage-operated Ca^{2+} entry. *J. Mol. Med.* **2018**, *96*, 1061–1079. [CrossRef] [PubMed]

75. Lu, T.; Aron, L.; Zullo, J.; Pan, Y.; Kim, H.; Chen, Y.; Yang, T.H.; Kim, H.M.; Drake, D.; Liu, X.S.; et al. Rest and stress resistance in ageing and Alzheimer's disease. *Nature* **2014**, *507*, 448–454. [CrossRef] [PubMed]

76. Ozmen, L.; Albientz, A.; Czech, C.; Jacobsen, H. Expression of transgenic APP MRNA is the key determinant for beta-amyloid deposition in PS2APP transgenic mice. *Neurodegener. Dis.* **2009**, *6*, 29–36. [CrossRef]

77. Zott, B.; Busche, M.A.; Sperling, R.A.; Konnerth, A. What happens with the circuit in Alzheimer's disease in mice and humans? *Annu. Rev. Neurosci.* **2018**, *41*, 277–297. [CrossRef]

78. Stargardt, A.; Swaab, D.F.; Bossers, K. Storm before the quiet: Neuronal hyperactivity and ab in the presymptomatic stages of alzheimer's Disease. *Neurobiol. Aging* **2015**, *36*, 1–11. [CrossRef]

79. Wilson, C.A.; Murphy, D.D.; Giasson, B.I.; Zhang, B.; Trojanowski, J.Q.; Lee, V.M. Degradative organelles containing mislocalized alpha-and beta-synuclein proliferate in presenilin-1 null neurons. *J. Cell Biol.* **2004**, *165*, 335–346. [CrossRef]

80. Esselens, C.; Oorschot, V.; Baert, V.; Raemaekers, T.; Spittaels, K.; Serneels, L.; Zheng, H.; Saftig, P.; De Strooper, B.; Klumperman, J.; et al. Presenilin 1 mediates the turnover of telencephalin in hippocampal neurons via an autophagic degradative pathway. *J. Cell Biol.* **2004**, *166*, 1041–1054. [CrossRef]

81. Lee, J.H.; Yu, W.H.; Kumar, A.; Lee, S.; Mohan, P.S.; Peterhoff, C.M.; Wolfe, D.M.; Martinez-Vicente, M.; Massey, A.C.; Sovak, G.; et al. Lysosomal proteolysis and autophagy require presenilin 1 and are disrupted by alzheimer-related PS1 mutations. *Cell* **2010**, *141*, 1146–1158. [CrossRef]

82. Coen, K.; Flannagan, R.S.; Baron, S.; Carraro-Lacroix, L.R.; Wang, D.; Vermeire, W.; Michiels, C.; Munck, S.; Baert, V.; Sugita, S.; et al. Lysosomal calcium homeostasis defects, not proton pump defects, cause endo-lysosomal dysfunction in PSEN-deficient cells. *J. Cell Biol.* **2012**, *198*, 23–35. [CrossRef]

83. Zhang, X.; Garbett, K.; Veeraraghavalu, K.; Wilburn, B.; Gilmore, R.; Mirnics, K.; Sisodia, S.S. A role for presenilins in autophagy revisited: Normal acidification of lysosomes in cells lacking PSEN1 and PSEN2. *J. Neurosci.* **2012**, *32*, 8633–8648. [CrossRef] [PubMed]

84. Reddy, K.; Cusack, C.L.; Nnah, I.C.; Khayati, K.; Saqcena, C.; Huynh, T.B.; Noggle, S.A.; Ballabio, A.; Dobrowolski, R. Dysregulation of nutrient sensing and CLEARance in presenilin deficiency. *Cell Rep.* **2016**, *14*, 2166–2179. [CrossRef] [PubMed]

85. Filadi, R.; Pizzo, P. Defective autophagy and alzheimer's disease: Is calcium the key? In *Neural Regeneration Research*; Wolters Kluwer Medknow Publications: Mumbai, India, 2019; pp. 2081–2082. [CrossRef]

86. Fedeli, C.; Filadi, R.; Rossi, A.; Mammucari, C.; Pizzo, P. PSEN2 (Presenilin 2) mutants linked to familial alzheimer disease impair autophagy by altering Ca^{2+} homeostasis. *Autophagy* **2019**, 1–19. [CrossRef] [PubMed]

87. Gutierrez, M.G.; Munafo, D.B.; Beron, W.; Colombo, M.I. Rab7 is required for the normal progression of the autophagic pathway in mammalian cells. *J. Cell Sci.* **2004**, *117*, 2687–2697. [CrossRef]

88. Lee, J.H.; McBrayer, M.K.; Wolfe, D.M.; Haslett, L.J.; Kumar, A.; Sato, Y.; Lie, P.P.; Mohan, P.; Coffey, E.E.; Kompella, U.; et al. Presenilin 1 maintains lysosomal Ca^{2+} homeostasis via TRPML1 by regulating VATPase-mediated lysosome acidification. *Cell Rep.* **2015**, *12*, 1430–1444. [CrossRef]

89. Behbahani, H.; Shabalina, I.G.; Wiehager, B.; Concha, H.; Hultenby, K.; Petrovic, N.; Nedergaard, J.; Winblad, B.; Cowburn, R.F.; Ankarcrona, M. Differential role of presenilin-1 and -2 on mitochondrial membrane potential and oxygen consumption in mouse embryonic fibroblasts. *J. Neurosci. Res.* **2006**, *84*, 891–902. [CrossRef]

90. Contino, S.; Porporato, P.E.; Bird, M.; Marinangeli, C.; Opsomer, R.; Sonveaux, P.; Bontemps, F.; Dewachter, I.; Octave, J.N.; Bertrand, L.; et al. Presenilin 2-dependent maintenance of mitochondrial oxidative capacity and morphology. *Front. Physiol.* **2017**, *8*, 796. [CrossRef]

91. Rossi, A.; Pizzo, P.; Filadi, R. Calcium, mitochondria and cell metabolism: A functional triangle in bioenergetics. *Biochim. Biophys. Acta Mol. Cell Res.* **2019**, *1866*, 1068–1078. [CrossRef]

92. Rangaraju, V.; Calloway, N.; Ryan, T.A. Activity-driven local ATP synthesis is required for synaptic function. *Cell* **2014**, *156*, 825–835. [CrossRef]

93. Weidling, I.W.; Swerdlow, R.H. Mitochondria in alzheimer's disease and their potential role in alzheimer's proteostasis. In *Experimental Neurology*; Academic Press Inc.: Cambridge, MA, USA, 2020.

94. Jadiya, P.; Kolmetzky, D.W.; Tomar, D.; Di Meco, A.; Lombardi, A.A.; Lambert, J.P.; Luongo, T.S.; Ludtmann, M.H.; Praticò, D.; Elrod, J.W. Impaired mitochondrial calcium efflux contributes to disease progression in models of alzheimer's disease. *Nat. Commun.* **2019**, *10*, 3885. [CrossRef]

95. Toglia, P.; Cheung, K.H.; Mak, D.O.D.; Ullah, G. Impaired mitochondrial function due to familial alzheimer's disease-causing presenilins mutants via Ca^{2+} disruptions. *Cell Calcium* **2016**, *59*, 240–250. [CrossRef] [PubMed]

96. Flannery, P.J.; Trushina, E. Mitochondrial dynamics and transport in alzheimer's disease. In *Molecular and Cellular Neuroscience*; Academic Press Inc.: Cambridge, MA, USA, 2019; pp. 109–120.

97. Theurey, P.; Connolly, N.M.C.; Fortunati, I.; Basso, E.; Lauwen, S.; Ferrante, C.; Moreira Pinho, C.; Joselin, A.; Gioran, A.; Bano, D.; et al. Systems biology identifies preserved integrity but impaired metabolism of mitochondria due to a glycolytic defect in alzheimer's disease neurons. *Aging Cell* **2019**, *18*, e12924. [CrossRef] [PubMed]

98. Rossi, A.; Rigotto, G.; Valente, G.; Giorgio, V.; Basso, E.; Filadi, R.; Pizzo, P. Defective mitochondrial pyruvate flux affects cell bioenergetics in alzheimer's disease-related models. *Cell Rep.* **2020**, *30*, 2332–2348. [CrossRef] [PubMed]

99. Llorens-Martin, M.; Jurado, J.; Hernandez, F.; Avila, J. GSK-3beta, a pivotal kinase in alzheimer disease. *Front. Mol. Neurosci.* **2014**, *7*, 46. [CrossRef] [PubMed]

100. Fontana, R.; Agostini, M.; Murana, E.; Mahmud, M.; Scremin, E.; Rubega, M.; Sparacino, G.; Vassanelli, S.; Fasolato, C. Early hippocampal hyperexcitability in PS2APP mice: Role of mutant PS2 and APP. *Neurobiol. Aging* **2017**, *50*, 64–76. [CrossRef]

101. Dickerson, B.C.; Salat, D.H.; Greve, D.N.; Chua, E.F.; Rand-Giovannetti, E.; Rentz, D.M.; Bertram, L.; Mullin, K.; Tanzi, R.E.; Blacker, D.; et al. Increased hippocampal activation in mild cognitive impairment compared to normal aging and AD. *Neurology* **2005**, *65*, 404–411. [CrossRef]

102. Bakker, A.; Albert, M.S.; Krauss, G.; Speck, C.L.; Gallagher, M. Response of the medial temporal lobe network in amnestic mild cognitive impairment to therapeutic intervention assessed by FMRI and memory task performance. *Neuroimage Clin.* **2015**, *7*, 688–698. [CrossRef]

103. Palop, J.J.; Mucke, L. Network abnormalities and interneuron dysfunction in alzheimer disease. *Nat. Rev. Neurosci.* **2016**, *17*, 777–792. [CrossRef]

104. Lam, A.D.; Deck, G.; Goldman, A.; Eskandar, E.N.; Noebels, J.; Cole, A.J. Silent hippocampal seizures and spikes identified by foramen ovale electrodes in alzheimer's disease. *Nat. Med.* **2017**, *23*, 678–680. [CrossRef]

105. Mitra, A.; Snyder, A.Z.; Hacker, C.D.; Pahwa, M.; Tagliazucchi, E.; Laufs, H.; Leuthardt, E.C.; Raichle, M.E. Human cortical–hippocampal dialogue in wake and slow-wave sleep. *Proc. Natl. Acad. Sci. USA* **2016**, *113*, E6868–E6876. [CrossRef]

106. Busche, M.A.; Grienberger, C.; Keskin, A.D.; Song, B.; Neumann, U.; Staufenbiel, M.; Forstl, H.; Konnerth, A. Decreased amyloid-beta and increased neuronal hyperactivity by immunotherapy in alzheimer's models. *Nat. Neurosci.* **2015**, *18*, 1725–1727. [CrossRef] [PubMed]

107. Kastanenka, K.V.; Hou, S.S.; Shakerdge, N.; Logan, R.; Feng, D.; Wegmann, S.; Chopra, V.; Hawkes, J.M.; Chen, X.; Bacskai, B.J. Optogenetic restoration of disrupted slow oscillations halts amyloid deposition and restores calcium homeostasis in an animal model of alzheimer's disease. *PLoS ONE* **2017**, *12*, e0170275. [CrossRef] [PubMed]

108. Leparulo, A.; Mahmud, M.; Scremin, E.; Pozzan, T.; Vassanelli, S.; Fasolato, C. Dampened slow oscillation connectivity anticipates amyloid deposition in the PS2APP mouse model of alzheimer's disease. *Cells* **2019**, *9*, 54. [CrossRef] [PubMed]

109. Born, J. Slow-wave sleep and the consolidation of long-term memory. *World J. Biol. Psychiatry* **2010**, *11*, 16–21. [CrossRef] [PubMed]

110. Agostini, M.; Fasolato, C. When, Where and How? Focus on neuronal calcium dysfunctions in alzheimer's disease. *Cell Calcium* **2016**, *60*, 289–298. [CrossRef]

111. Lazzari, C.; Kipanyula, M.J.; Agostini, M.; Pozzan, T.; Fasolato, C. Ab42 oligomers selectively disrupt neuronal calcium release. *Neurobiol. Aging* **2015**, *36*, 877–885. [CrossRef]

112. Tong, B.C.; Wu, A.J.; Li, M.; Cheung, K.H. Calcium signaling in alzheimer's disease & therapies. *Biochim. Biophys. Acta Mol. Cell Res.* **2018**, *1865*, 1745–1760. [CrossRef] [PubMed]

113. Brawek, B.; Garaschuk, O. Network-wide dysregulation of calcium homeostasis in alzheimer's disease. In *Cell and Tissue Research*; Springer: Berlin/Heidelberg, Germany, 2014; pp. 427–438.

114. Lee, Y.J.; Choi, I.S.; Park, M.H.; Lee, Y.M.; Song, J.K.; Kim, Y.H.; Kim, K.H.; Hwang, D.Y.; Jeong, J.H.; Yun, Y.P.; et al. 4-O-methylhonokiol attenuates memory impairment in presenilin 2 mutant mice through reduction of oxidative damage and inactivation of astrocytes and the ERK pathway. *Free Radic. Biol. Med.* **2011**, *50*, 66–77. [CrossRef]

115. Yuk, D.Y.; Lee, Y.K.; Nam, S.Y.; Yun, Y.W.; Hwang, D.Y.; Choi, D.Y.; Oh, K.W.; Hong, J.T. Reduced anxiety in the mice expressing mutant (N141I) presenilin 2. *J. Neurosci. Res.* **2009**, *87*, 522–531. [CrossRef] [PubMed]

116. Verret, L.; Mann, E.O.; Hang, G.B.; Barth, A.M.; Cobos, I.; Ho, K.; Devidze, N.; Masliah, E.; Kreitzer, A.C.; Mody, I.; et al. Inhibitory interneuron deficit links altered network activity and cognitive dysfunction in alzheimer model. *Cell* **2012**, *149*, 708–721. [CrossRef]

117. Hahn, T.T.G.; Sakmann, B.; Mehta, M.R. Phase-locking of hippocampal interneurons' membrane potential to neocortical up-down states. *Nat. Neurosci.* **2006**, *9*, 1359–1361. [CrossRef] [PubMed]

118. Hijazi, S.; Heistek, T.S.; van der Loo, R.; Mansvelder, H.D.; Smit, A.B.; van Kesteren, R.E. Hyperexcitable parvalbumin interneurons render hippocampal circuitry vulnerable to amyloid beta. *Iscience* **2020**, *23*, 101271. [CrossRef] [PubMed]

119. Fellin, T.; Halassa, M.M.; Terunuma, M.; Succol, F.; Takano, H.; Frank, M.; Moss, S.J.; Haydon, P.G. Endogenous nonneuronal modulators of synaptic transmission control cortical slow oscillations in vivo. *Proc. Natl. Acad. Sci. USA* **2009**, *106*, 15037–15042. [CrossRef] [PubMed]

120. Poskanzer, K.E.; Yuste, R. Astrocytic regulation of cortical Up states. *Proc. Natl. Acad. Sci. USA* **2011**, *108*, 18453–18458. [CrossRef] [PubMed]

121. Araque, A.; Carmignoto, G.; Haydon, P.G.; Oliet, S.H.; Robitaille, R.; Volterra, A. Gliotransmitters travel in time and space. *Neuron* **2014**, *81*, 728–739. [CrossRef]

122. Durkee, C.A.; Araque, A. Diversity and specificity of astrocyte-neuron communication. *Neuroscience* **2019**, *396*, 73–78. [CrossRef]

123. Busche, M.A.; Chen, X.; Henning, H.A.; Reichwald, J.; Staufenbiel, M.; Sakmann, B.; Konnerth, A. Critical role of soluble amyloid-beta for early hippocampal hyperactivity in a mouse model of alzheimer's disease. *Proc. Natl. Acad. Sci. USA* **2012**, *109*, 8740–8745. [CrossRef]

124. Lu, B.; Fivaz, M. Neuronal SOCE: Myth or reality? In *Trends in Cell Biology*; Elsevier Ltd.: Amsterdam, The Netherlands, 2016; pp. 890–893.

125. Majewski, L.; Kuznicki, J. SOCE in Neurons: Signaling or just refilling? In *Biochimica et Biophysica Acta-Molecular Cell Research*; Elsevier: Amsterdam, The Netherlands, 2014; pp. 1940–1952.

126. Sun, S.; Zhang, H.; Liu, J.; Popugaeva, E.; Xu, N.J.; Feske, S.; White, C.L.; Bezprozvanny, I. Reduced synaptic STIM2 expression and impaired store-operated calcium entry cause destabilization of mature spines in mutant presenilin mice. *Neuron* **2014**, *82*, 79–93. [CrossRef]

127. Zhang, H.; Sun, S.; Wu, L.; Pchitskaya, E.; Zakharova, O.; Tacer, K.F.; Bezprozvanny, I. Store-operated calcium channel complex in postsynaptic spines: A new therapeutic target for alzheimer's disease treatment. *J. Neurosci.* **2016**, *36*, 11837–11850. [CrossRef]

128. Zhang, H.; Wu, L.; Pchitskaya, E.; Zakharova, O.; Saito, T.; Saido, T.; Bezprozvanny, I. Neuronal store-operated calcium entry and mushroom spine loss in amyloid precursor protein knock-in mouse model of alzheimer's disease. *J. Neurosci.* **2015**, *35*, 13275–13286. [CrossRef]

129. Park, C.Y.; Shcheglovitov, A.; Dolmetsch, R. The CRAC channel activator STIM1 binds and inhibits L-Type voltage-gated calcium channels. *Science* **2010**, *330*, 101–105. [CrossRef] [PubMed]

130. Wang, Y.J.; Deng, X.X.; Mancarella, S.; Hendron, E.; Eguchi, S.; Soboloff, J.; Tang, X.A.D.; Gill, D.L. The calcium store sensor, STIM1, reciprocally controls orai and Ca(V)1.2 channels. *Science* **2010**, *330*, 105–109. [CrossRef] [PubMed]

131. Moccia, F.; Zuccolo, E.; Soda, T.; Tanzi, F.; Guerra, G.; Mapelli, L.; Lodola, F.; D'Angelo, E. Stim and orai proteins in neuronal Ca^{2+} signaling and excitability. *Front. Cell. Neurosci.* **2015**, *9*, 153. [CrossRef] [PubMed]

132. Dou, Y.; Ia, J.; Gao, R.; Gao, I.; Munoz, F.M.; Wei, D.; Tian, Y.; Barrett, J.E.; Ajit, S.; Meucci, O.; et al. Orai1 plays a crucial role in central sensitization by modulating neuronal excitability. *J. Neurosci.* **2018**, *38*, 887–900. [CrossRef]

133. Pizzo, P.; Burgo, A.; Pozzan, T.; Fasolato, C. Role of capacitative calcium entry on glutamate-induced calcium influx in Type-I rat cortical astrocytes. *J. Neurochem.* **2001**, *79*, 98–109. [CrossRef]

134. Okubo, Y.; Iino, M.; Hirose, K. Store-operated Ca^{2+} entry-dependent Ca^{2+} refilling in the endoplasmic reticulum in astrocytes. *Biochem. Biophys. Res. Commun.* **2020**, *522*, 1003–1008. [CrossRef]

135. Toth, A.B.; Hori, K.; Novakovic, M.M.; Bernstein, N.G.; Lambot, L.; Prakriya, M. CRAC channels regulate astrocyte Ca^{2+} signaling and gliotransmitter release to modulate hippocampal GABAergic transmission. *Sci. Signal.* **2019**, *12*, 5450. [CrossRef]

136. Verkhratsky, A.; Rodríguez-Arellano, J.J.; Parpura, V.; Zorec, R. Astroglial calcium signalling in alzheimer's disease. In *Biochemical and Biophysical Research Communications*; Elsevier, B.V.: Amsterdam, The Netherlands, 2017; pp. 1005–1012.

137. Jayadev, S.; Case, A.; Eastman, A.J.; Nguyen, H.; Pollak, J.; Wiley, J.C.; Möller, T.; Morrison, R.S.; Garden, G.A. Presenilin 2 is the predominant γ-secretase in microglia and modulates cytokine release. *PLoS ONE* **2010**, *5*, e15743. [CrossRef]

138. Scremin, E.; Agostini, M.; Leparulo, A.; Pozzan, T.; Greotti, E.; Fasolato, C. ORAI2 down-regulation potentiates SOCE and decreases Aβ42 accumulation in human neuroglioma cells. *Int. J. Mol. Sci.* **2020**, *21*, 5288. [CrossRef]

139. Zeiger, W.; Vetrivel, K.S.; Buggia-Prevot, V.; Nguyen, P.D.; Wagner, S.L.; Villereal, M.L.; Thinakaran, G. Ca^{2+} influx through store-operated Ca^{2+} channels reduces alzheimer disease b-amyloid peptide secretion. *J. Biol. Chem.* **2013**, *288*, 26955–26966. [CrossRef]

140. Vaeth, M.; Yang, J.; Yamashita, M.; Zee, I.; Eckstein, M.; Knosp, C.; Kaufmann, U.; Karoly Jani, P.; Lacruz, R.S.; Flockerzi, V.; et al. ORAI2 modulates store-operated calcium entry and T cell-mediated immunity. *Nat. Commun.* **2017**, *8*, 14714. [CrossRef] [PubMed]

141. Frost, G.R.; Li, Y.M. The role of astrocytes in amyloid production and alzheimer's disease. In *Open Biology*; Royal Society Publishing: London, UK, 2017.

142. Filadi, R.; Pizzo, P. Mitochondrial calcium handling and neurodegeneration: When a good signal goes wrong. *Curr. Opin. Physiol.* **2020**. [CrossRef]

143. Golpich, M.; Amini, E.; Mohamed, Z.; Azman Ali, R.; Mohamed Ibrahim, N.; Ahmadiani, A. Mitochondrial dysfunction and biogenesis in neurodegenerative diseases: Pathogenesis and treatment. *CNS Neurosci. Ther.* **2017**, *23*, 5–22. [CrossRef] [PubMed]

144. Swerdlow, R.H.; Burns, J.M.; Khan, S.M. The alzheimer's disease mitochondrial cascade hypothesis: Progress and perspectives. *Biochim. Biophys. Acta* **2014**, *1842*, 1219–1231. [CrossRef] [PubMed]

145. Bazzari, F.H.; Abdallah, D.M.; El-Abhar, H.S. Pharmacological interventions to attenuate alzheimer's disease progression: The story so far. *Curr. Alzheimer Res.* **2019**, *16*, 261–277. [CrossRef]

146. Oules, B.; Del Prete, D.; Greco, B.; Zhang, X.; Lauritzen, I.; Sevalle, J.; Moreno, S.; Paterlini-Brechot, P.; Trebak, M.; Checler, F.; et al. Ryanodine receptor blockade reduces amyloid-b load and memory impairments in Tg2576 mouse model of alzheimer disease. *J. Neurosci.* **2012**, *32*, 11820–11834. [CrossRef]

147. Chakroborty, S.; Stutzmann, G.E. Calcium channelopathies and alzheimer's disease: Insight into therapeutic success and failures. *Eur. J. Pharmacol.* **2014**, *739*, 83–95. [CrossRef]

148. Facchinetti, F.; Fasolato, C.; Del Giudice, E.; Burgo, A.; Furegato, S.; Fusco, M.; Basso, E.; Seraglia, R.; D'Arrigo, A.; Leon, A. Nimodipine selectively stimulates b-amyloid 1-42 secretion by a mechanism independent of calcium influx blockage. *Neurobiol. Aging* **2006**, *27*, 218–227. [CrossRef]

149. Bantug, G.R.; Fischer, M.; Grahlert, J.; Balmer, M.L.; Unterstab, G.; Develioglu, L.; Steiner, R.; Zhang, L.; Costa, A.S.H.; Gubser, P.M.; et al. Mitochondria-endoplasmic reticulum contact sites function as immunometabolic hubs that orchestrate the rapid recall response of memory CD8(+) T cells. *Immunity* **2018**, *48*, 542–555. [CrossRef]

150. Trushina, E.; Dutta, T.; Persson, X.M.; Mielke, M.M.; Petersen, R.C. Identification of altered metabolic pathways in plasma and CSF in mild cognitive impairment and alzheimer's disease using metabolomics. *PLoS ONE* **2013**, *8*, e63644. [CrossRef]

151. Parnetti, L.; Gaiti, A.; Polidori, M.C.; Brunetti, M.; Palumbo, B.; Chionne, F.; Cadini, D.; Cecchetti, R.; Senin, U. Increased cerebrospinal fluid pyruvate levels in alzheimer's disease. *Neurosci. Lett.* **1995**, *199*, 231–233. [CrossRef]

152. Sasaguri, H.; Nilsson, P.; Hashimoto, S.; Nagata, K.; Saito, T.; De Strooper, B.; Hardy, J.; Vassar, R.; Winblad, B.; Saido, T.C. APP mouse models for alzheimer's disease preclinical studies. *EMBO J.* **2017**, *36*, 2473–2487. [CrossRef] [PubMed]

153. Jankowsky, J.L.; Zheng, H. Practical considerations for choosing a mouse model of alzheimer's disease. *Mol. Neurodegener.* **2017**, *12*, 1–22. [CrossRef] [PubMed]

154. Oddo, S.; Caccamo, A.; Shepherd, J.D.; Murphy, M.P.; Golde, T.E.; Kayed, R.; Metherate, R.; Mattson, M.P.; Akbari, Y.; LaFerla, F.M. Triple-transgenic model of alzheimer's disease with plaques and tangles: Intracellular abeta and synaptic dysfunction. *Neuron* **2003**, *39*, 409–421. [CrossRef]

155. Richards, J.G.; Higgins, G.A.; Ouagazzal, A.M.; Ozmen, L.; Kew, J.N.; Bohrmann, B.; Malherbe, P.; Brockhaus, M.; Loetscher, H.; Czech, C.; et al. PS2APP transgenic mice, coexpressing HPS2mut and HAPPswe, show age-related cognitive deficits associated with discrete brain amyloid deposition and inflammation. *J. Neurosci.* **2003**, *23*, 8989–9003. [CrossRef]

156. Poirier, R.; Veltman, I.; Pflimlin, M.C.; Knoflach, F.; Metzger, F. Enhanced dentate gyrus synaptic plasticity but reduced neurogenesis in a mouse model of amyloidosis. *Neurobiol. Dis.* **2010**, *40*, 386–393. [CrossRef]

157. Weidensteiner, C.; Metzger, F.; Bruns, A.; Bohrmann, B.; Kuennecke, B.; Von Kienlin, M. Cortical hypoperfusion in the B6.PS2APP mouse model for alzheimer's disease: Comprehensive phenotyping of vascular and tissular parameters by MRI. *Magn. Reson. Med.* **2009**, *62*, 35–45. [CrossRef]

158. Brendel, M.; Jaworska, A.; Grießinger, E.; Rötzer, C.; Burgold, S.; Gildehaus, F.J.; Carlsen, J.; Cumming, P.; Baumann, K.; Haass, C.; et al. Cross-sectional comparison of small animal [18f]-florbetaben amyloid-PET between transgenic AD mouse models. *PLoS ONE* **2015**, *10*, e0116678. [CrossRef]

159. Grueninger, F.; Bohrmann, B.; Czech, C.; Ballard, T.M.; Frey, J.R.; Weidensteiner, C.; von Kienlin, M.; Ozmen, L. Phosphorylation of Tau at S422 is enhanced by Aβ in TauPS2APP triple transgenic mice. *Neurobiol. Dis.* **2010**, *37*, 294–306. [CrossRef]

160. Loreth, D.; Ozmen, L.; Revel, F.G.; Knoflach, F.; Wetzel, P.; Frotscher, M.; Metzger, F.; Kretz, O. Selective degeneration of septal and hippocampal GABAergic neurons in a mouse model of amyloidosis and tauopathy. *Neurobiol. Dis.* **2012**, *47*, 1–12. [CrossRef]

161. Kins, S.; Crameri, A.; Evans, D.R.H.; Hemmings, B.A.; Nitsch, R.M.; Götz, J. Reduced protein phosphatase 2A activity induces hyperphosphorylation and altered compartmentalization of Tau in transgenic mice. *J. Biol. Chem.* **2001**, *276*, 38193–38200. [CrossRef] [PubMed]

162. Nyabi, O.; Pype, S.; Mercken, M.; Herreman, A.; Saftig, P.; Craessaerts, K.; Serneels, L.; Annaert, W.; De Strooper, B. No endogenous ab production in presenilin-deficient fibroblasts. *Nat. Cell Biol.* **2002**, *4*, E164. [CrossRef] [PubMed]

163. Herreman, A.; Hartmann, D.; Annaert, W.; Saftig, P.; Craessaerts, K.; Serneels, L.; Umans, L.; Schrijvers, V.; Checler, F.; Vanderstichele, H.; et al. Presenilin 2 deficiency causes a mild pulmonary phenotype and no changes in amyloid precursor protein processing but enhances the embryonic lethal phenotype of presenilin 1 deficiency. *Proc. Natl. Acad. Sci. USA* **1999**, *96*, 11872–11877. [CrossRef] [PubMed]

164. Lee, S.H.; Lutz, D.; Mossalam, M.; Bolshakov, V.Y.; Frotscher, M.; Shen, J. Presenilins regulate synaptic plasticity and mitochondrial calcium homeostasis in the hippocampal mossy fiber pathway. *Mol. Neurodegener.* **2017**, *12*, 48. [CrossRef] [PubMed]

165. Gratuze, M.; Leyns, C.E.G.; Sauerbeck, A.D.; St-Pierre, M.-K.; Xiong, M.; Kim, N.; Remolina Serrano, J.; Tremblay, M.-È.; Kummer, T.T.; Colonna, M.; et al. Impact of TREM2R47H variant on tau pathology-induced gliosis and neurodegeneration. *J. Clin. Investig.* **2020**. [CrossRef]

166. Balu, D.; Karstens, A.J.; Loukenas, E.; Maldonado Weng, J.; York, J.M.; Valencia-Olvera, A.C.; LaDu, M.J. The role of APOE in transgenic mouse models of AD. *Neurosci. Lett.* **2019**, *707*, 134285. [CrossRef]

167. Hargis, K.E.; Blalock, E.M. Transcriptional signatures of brain aging and alzheimer's disease: What are our rodent models telling us? *Behav. Brain Res.* **2017**, *322*, 311–328. [CrossRef]

 *cells*

Review

Potential Drug Candidates to Treat TRPC6 Channel Deficiencies in the Pathophysiology of Alzheimer's Disease and Brain Ischemia

Veronika Prikhodko [1,2,3], Daria Chernyuk [1], Yurii Sysoev [1,2,3,4], Nikita Zernov [1], Sergey Okovityi [2,3] and Elena Popugaeva [1,*]

[1] Laboratory of Molecular Neurodegeneration, Peter the Great St. Petersburg Polytechnic University, 195251 St. Petersburg, Russia; veronika.prihodko@pharminnotech.com (V.P.); dasha0703@mail.ru (D.C.); susoyev92@mail.ru (Y.S.); quakenbush97@gmail.com (N.Z.)

[2] Department of Pharmacology and Clinical Pharmacology, Saint Petersburg State Chemical Pharmaceutical University, 197022 St. Petersburg, Russia; okovityy@mail.ru

[3] N.P. Bechtereva Institute of the Human Brain of the Russian Academy of Sciences, 197376 St. Petersburg, Russia

[4] Institute of Translational Biomedicine, Saint Petersburg State University, 199034 St. Petersburg, Russia

* Correspondence: lena.popugaeva@gmail.com

Received: 31 August 2020; Accepted: 20 October 2020; Published: 24 October 2020

Abstract: Alzheimer's disease and cerebral ischemia are among the many causative neurodegenerative diseases that lead to disabilities in the middle-aged and elderly population. There are no effective disease-preventing therapies for these pathologies. Recent in vitro and in vivo studies have revealed the TRPC6 channel to be a promising molecular target for the development of neuroprotective agents. TRPC6 channel is a non-selective cation plasma membrane channel that is permeable to Ca^{2+}. Its Ca^{2+}-dependent pharmacological effect is associated with the stabilization and protection of excitatory synapses. Downregulation as well as upregulation of TRPC6 channel functions have been observed in Alzheimer's disease and brain ischemia models. Thus, in order to protect neurons from Alzheimer's disease and cerebral ischemia, proper TRPC6 channels modulators have to be used. TRPC6 channels modulators are an emerging research field. New chemical structures modulating the activity of TRPC6 channels are being currently discovered. The recent publication of the cryo-EM structure of TRPC6 channels should speed up the discovery process even more. This review summarizes the currently available information about potential drug candidates that may be used as basic structures to develop selective, highly potent TRPC6 channel modulators to treat neurodegenerative disorders, such as Alzheimer's disease and cerebral ischemia.

Keywords: TRPC6; Alzheimer's disease; cerebral ischemia; pharmaceutical agents

---

1. Introduction

Due to increased life expectancies, neurodegenerative diseases (NDD), such as Alzheimer's disease (AD), dementia, and cerebrovascular diseases, are considered by WHO the main cause of disability in the coming decades. Currently, there are no effective disease-modifying or preventing therapies for those NDDs.

AD is caused by the progressive loss of neurons in brain structures that are responsible for memory acquisition and preservation, such as the hippocampus and cortical areas. The exact mechanism that causes neurons to die is not known. Among the most studied toxic changes of proteins that cause neuronal degeneration in AD are extracellular amyloid-beta (Aβ) aggregates and intracellular hyperphosphorylated tau (p-tau) that forms neurofibrillary tangles. Recently, disruption of immune system signaling in glial cells has started to receive growing attention due to the appearance of genome-wide association study (GWAS) data for late-onset AD patients [1].

Brain ischemia is a cerebrovascular disease that is caused by a restriction in blood supply, leading to oxygen deprivation and the rapid death of neurons. Several mechanisms, including excitotoxicity, ionic imbalance, oxidative and nitrosative stress, and apoptosis, have been implicated in ischemic neuronal death [2,3]. Acute brain ischemia can be treated successfully in modern healthcare settings, although treatment success depends on how quickly the patient receives medical care as well as the brain volume affected. However, there are currently no effective treatment options for chronic brain ischemia, which is usually caused by cerebral atherosclerosis.

Although the pathophysiological mechanisms causing AD and cerebral ischemia may differ, cerebral ischemia serves as a risk factor for AD development [4], and vice versa [5,6], indicating that a common intracellular mechanism may be disrupted in these two distinct pathologies. Such a common mechanism may be associated with Ca^{2+} dyshomeostasis. The N-methyl-D-aspartate (NMDA) receptor, an important excitatory neurotransmitter receptor, has been reported as the key player in the Ca^{2+} signaling in AD [7] and cerebral ischemia [8]. However, the NMDA receptor blocker memantine only relieves the symptoms temporarily in the early to moderate stage of AD patients [9,10]. Moreover, blocking NMDA receptors to prevent ischemic neuronal damage in clinical trials has caused severe adverse effects [11,12]. Although multiple factors might have contributed to the unsuccessful clinical trials, it is possible that the disruption of neuroprotective pathways, which precedes NMDA receptors hyperactivation, could be responsible for either AD and/or ischemic brain damage.

One of such neuroprotective pathways is the transient receptor potential cation channel, subfamily C, member 6 (TRPC6)-dependent regulation of excitatory synapse formation. TRPC6 overexpression has been shown to increase dendritic spine density [13] and rescue mushroom spine loss in mouse AD models [14], as well as protect neurons from ischemic brain damage [15,16]. TRPC6 upregulates the cAMP-response element-binding protein (CREB) pathway that is important for dendritic growth [17] and promotes synapse and dendritic spine formation, spatial memory, and learning [13]. In addition, TRPC6 also acts as a negative regulator that suppresses NMDA-induced Ca^{2+} influx in hippocampal neurons [15], which may protect neurons from excitotoxicity in the first stages of the disease. TRPC6 overexpression has been observed in breast cancer cells [18]. Overactivation of TRPC6 is toxic to immune cells (reviewed here [19]). Gain-of-function mutations of TRPC6 have been associated with familial forms of focal segmental glomerular sclerosis [20]. The mentioned studies indicate that despite the positive effect of TRPC6 activation in the brain, excessive TRPC6 activation has toxic effects on other cellular systems in the body.

Current prevalent evidence suggests that the TRPC6 channel function is downregulated in AD and cerebral ischemia [14,21–23]. However, there are reports in the literature that observe TRPC6 overactivation in AD [24] and cerebral ischemia models [25,26]. Existing contradictions on TRPC6 channel function in AD and ischemia indicate that these NDDs might be heterogenic, meaning that in one group of patients, the disease leads to hypofunction of the TRPC6 channel; however, there is another group (most likely smaller than the first one) where TRPC6 is hyperactivated. Thus, in order to preserve brain function, TRPC6 activity has to be in its physiological state, and deviations towards hypoactivity as well as hyperactivity are toxic to the cells. In terms of patient treatment, a careful investigation of the dysfunction of the TRPC6-dependent molecular pathway has to be performed in order to develop appropriate pharmacological treatments for said pathologies.

This review is devoted to the description of the role of TRPC6 channels in AD and brain ischemia with a particular focus on the dysfunction of them as Ca^{2+}-dependent channels. Potential drug candidates that have shown their therapeutic effects in different cellular and animal models are discussed. When available, the pros and cons of each particular TRPC6 channel modulator are mentioned. TRPC6 is also involved in certain Ca^{2+}-independent processes, such as amyloid precursor protein (APP) interaction [27]. However, in order to keep the review focused, this Ca^{2+}-independent process observed in AD and brain ischemia pathogenesis is omitted.

2. TRPC Channels and Their Regulation in Cells

Transient receptor potential (TRP) cation channels form a large family of multifunctional cell sensors. There are 29 TRP channels that can be described and divided into six subfamilies based on sequence homology: seven canonical channels (TRPC), six vanilloid channels (TRPV), eight melastatin-related channels (TRPM), three polycystic channels (TRPP), three mucolipins (TPRML), and one ankyrin channel (TRPA) [28].

TRPCs participate in different physiological processes in the development of the nervous system [29]. TRPC subfamily consists of seven members: TRPC1, TRPC2, TRPC3, TRPC4, TRPC5, TRPC6, and TRPC7. TRPC2 is a pseudogene in a number of vertebrates, including humans [30].

Analysis of the expression of different types of TRPC channels in situ using the Allen Brain Atlas and of gene expression in different regions of the mouse brain has demonstrated that TRPC6 is highly expressed in the hippocampus [14]. Western blot analysis of the rat hippocampus has shown that TRPC6 levels are enhanced in postsynaptic structures compared with synaptosomes. Moreover, electron microscopy has shown that TRPC6 is mostly located in the postsynaptic sites [13].

TRPs are cloned and identified, assuming that they are calcium-selective and activated by the emptying of internal Ca^{2+} stores (store-operated channels (SOC)) [31]. After their initial functional characterization, it has turned out that both assumptions do not hold true, especially for TRPCs [31]. These ion channels only show a moderate Ca^{2+} selectivity (PCa/PNa from ~0.5 to 9), and the TRPC3/6/7 subfamily of TRPC channels can be activated by the second messenger diacylglycerol (DAG) produced by receptor-activated phospholipase C (PLC) without any involvement of internal stores (receptor-operated channels (ROC)) [32].

Store-operated Ca^{2+} entry (SOCE) via TRPCs happens when 1,4,5-trisphosphate (IP3) or some other intracellular mechanism empties Ca^{2+} stored in the endoplasmic reticulum (ER). The fall in the ER Ca^{2+} concentration signals to the plasma membrane to open store-operated channels [33]. The major breakthrough in the understanding of SOCE physiology happened when Ca^{2+}-sensing stromal interaction molecule (STIM) ER proteins and plasma membrane (PM) Orai 1-3 channels have been described [34–37]. STIM1 and 2 proteins reside in the ER and monitor ER Ca^{2+} concentration with their EF-hand motif. When the ER Ca^{2+} concentration drops, Ca^{2+} dissociates from EF-hand, thus allowing STIM proteins to oligomerize and move to ER-PM tight junctions where they interact with PM Orai and TRP channels in order to facilitate Ca^{2+} flow.

3. Role of TRPC6 in the Formation of Excitatory Synapses

TRPC6 plays a certain physiological role in the formation of excitatory synapses. Particularly, TRPC6 overexpression increases dendritic spine density [13] and attenuates mushroom spine loss in presenilin 1 knock-in (PS1-KI) and amyloid precursor protein knock in (APP-KI) hippocampal neurons to the level of wild type neurons [14]. At the same time, overexpression of TRPC1, TRPC3-5, and TRPC7 does not affect spine morphology [14]. Meanwhile, the downregulation of TRPC6 expression leads to spine density reduction, lowers the frequency of spontaneous miniature excitatory postsynaptic currents (mEPSCs) [13], decreases the expression of postsynaptic density protein 95 (PSD95), and inhibits phosphorylation of calcium-calmodulin-dependent protein kinase II (pCaMKII) [14].

Activation of TRPC6 can promote spine formation via the STIM2-neuronal SOCE-CaMKII pathway. The expression level of STIM2 is downregulated in hippocampal neurons from PS1-M146V-KI [38] and APP-KI [14] mice as well as in a cellular model of amyloid synaptotoxicity [39]. The STIM2 reduction may be a compensatory response to ER Ca^{2+} overload in these models and may lead to the loss of mushroom spines in PS1-M146V-KI and APP-KI hippocampal neurons. TRPC6 and Orai2 channels are suggested to be the key components of neuronal SOCE in hippocampal cells [14]. It is hypothesized that the downstream molecule for TRPC6-mediated SOCE in hippocampal neurons is CaMKII since upregulation of neuronal SOCE activity recovers phosphorylation of CaMKII, restores the number of mushroom spines in hippocampal neurons in mouse models of AD [38,39], and induces long-term potentiation in PS1-KI, APP-KI, and 5xFAD hippocampal slices [14,21].

Alternative TRPC6-downstream signaling pathways are the Ca^{2+}/calmodulin-dependent kinase IV (CaMKIV) pathway and the cAMP-response element-binding protein (CREB) pathway, which is important for dendritic growth in hippocampal neurons [17]; another study has shown that the CaMKIV-CREB pathway is important in promoting synapse and dendritic spine formation, spatial memory, and learning [13]. TRPC6 also acts as a negative regulator that suppresses NMDA-induced Ca^{2+} influx in hippocampal neurons [15]. In addition, NMDAR has been shown to regulate transcription and degradation of TRPC6 in neurons in a bidirectional manner through NMDAR subunit 2A (NR2A) or NR2B activation [40].

4. Hypo- and Hyperactivation of TRPC6 Channels in Different Pathogenetic Forms of AD

There is evidence that different genetically inherited familial forms of AD (fAD) can cause TRPC6 dysfunction [14,21,24]. Both hypo- [14,21,22] and hyperactivation [24] of TRPC6 channels have been reported for different fAD-associated mutations in *APP* and *PS* genes.

fAD-associated PS2-N141I, M239V mutations cause downregulation of TRPC6-mediated Ca^{2+} entry in transiently transfected HEK cells. Lessard et al. suggested that TRPC6 downregulation by PS2-N141I, M239V does not depend on ER Ca^{2+} content but rather involves the interaction of PS2 with an intermediate protein of unknown origin [22]. Later on, contradictory results were obtained on ER Ca^{2+} content in PS2-N141I expressing cells [41,42]. In double knockout fibroblasts, PS2-N141I increases ER Ca^{2+} content [41] but lowers it in Hela cells [42]. According to the classical understanding of the SOCE physiology, overloaded ER Ca^{2+} stores downregulate SOCE [33]; thus, the data provided by Tu et al. [41] seem to be more relevant.

PS1-M146V is an fAD-associated mutation that has been reported to downregulate TRPC6-dependent Ca^{2+} entry in hippocampal neurons in store-operated mode [14]. PS1-M146V has been shown to increase ER Ca^{2+} content in mouse embryonic fibroblasts (MEFs) [41] and in neurons [43]. It has been demonstrated in previous studies [38,44] that ER Ca^{2+} stores are overloaded in neurons from AD mouse models. Furthermore, it has been discovered that overloaded ER Ca^{2+} stores cause compensatory downregulation of TRPC6 channels in PS1-M146V neurons [14]. A similar impact on TRPC6 function has been reported for fAD-associated mutations in APP (KM670/671NL and I716F) [14] as well as for Aβ toxicity in a cell culture model of AD [21]. There are fAD mutations in PSEN1 (PSEN1M146L, PSEN1S170F, PSEN1I213F, PSEN1E318G, PSEN1P117R, PSEN1L226F, PSENA246E) [45,46] and PSEN2 (PSEN2M239I, PSEN2T122R) [42], which have been reported to downregulate SOCE, although their role in the regulation of TRPC6 function has not been investigated yet. TRPC6 activators have been shown to recover the percentage of mushroom spines in cell culture models of fAD and induce long-term potentiation in hippocampal brain slices taken from AD mouse models [14,21]. Based on these results, it is suggested that activators of TRPC6 may have a therapeutic value for the treatment of fAD with TRPC6 hypofunction [14,47–49].

PS1-ΔE9 mutation has been reported to empty ER Ca^{2+} stores [41] and enhance SOCE [24,50]. There are other fAD mutations in PSEN1 (PSEN1D257A, PSEN1D385A), which have been reported to enhance SOCE [51]; however, their role in the regulation of TRPC6 function has not been investigated yet. Today, there is only one fAD-associated PS1-ΔE9 mutation that has been shown to upregulate the TRPC6 function in store-operated mode [24]. A TRPC6 inhibitor has been shown to recover mushroom spine percentage in a cell culture model of fAD [24]. Inhibitors of TRPC6 have been proposed to have therapeutic effects in fAD with TRPC6 hyperfunction [24,49].

To conclude the section, in order to normalize TRPC6 function in neurons and preserve the stability of excitatory synaptic contacts, suitable pharmacological agents have to be used for distinct genetic forms of AD.

5. Cerebral Ischemia as a Risk Factor for AD Development

Recent experimental and clinical findings have demonstrated a high degree of correlation between cerebral ischemia and AD [4,52,53]. While some studies have indicated that ischemic stroke significantly increases the risk of AD [4], others, in turn, have associated AD with a higher risk of stroke [5,6]. Previous studies have suggested that almost 30% of AD subjects bear evidence of cerebral infarction at autopsy [54,55]. A meta-analysis that comprises seven cohort studies and two nested case-control studies has found that a history of stroke is associated with the development of AD [4]. Notably, several lines of evidence suggest that AD patients have a high risk of cerebral ischemia [5,6]. For example, among AD patients with no history of previous stroke, vascular dementia, or other cerebral degenerative diseases, the incidence of ischemic stroke amounts to 37.8 per 1000 persons (versus 23.2 in non-AD controls) [5]. In another register-based matched cohort study [6], patients with Alzheimer's dementia have a higher risk of hemorrhagic stroke, while there is no difference in ischemic stroke incidence. When the results are analyzed within different age groups, the risk of ischemic stroke is found to be increased among AD patients younger than 80 years [6].

6. Role of TRPC6 in the Development of Ischemia

Ca^{2+} overload is one of the main molecular mechanisms involved in ischemic cell damage and death [56]. TRPC6, along with a few other prominent members of the family, has recently gained considerable attention as a promising target for the prevention of Ca^{2+} overload [23]. Dysregulation of TRPC6 activity has been implicated in ischemic stroke [23,57], as well as retinal ischemia [58], and renal hypoxia following cerebral ischemia [59].

On the one hand, upregulation and maintenance of TRPC6 activity prevent NMDAR hyperactivation and the subsequent Ca^{2+} influx, development of excitotoxicity, and neuronal death. In supporting this notion, both direct activation by 1-oleoyl-2-acetyl-sn-glycerol (a synthetic analog of diacylglycerol, the main endogenous TRPC6 agonist) and overexpression of TRPC6 inhibit NMDA-induced currents in cultured hippocampal neurons [15], and TRPC overexpression attenuates excitotoxic damage in hippocampal and cortical neurons [16]. *Trpc6*-transgenic mice with an elevated basal level of TRPC6 expression are less susceptible to cerebral ischemia than their wild-type littermates and have lower mortality rates, reduced infarct volumes, and better neurological outcomes after middle cerebral artery occlusion (MCAO) [16]. Recent studies have shown that TRPC6-mediated signaling promotes neuronal survival [60], the brain-derived neurotrophic factor-mediated axonal growth cone guidance [61], dendritic outgrowth and branching [17], and excitatory synapse formation [13]. In addition, positive modulation of TRPC6 activity allows for the sustained activation of the CREB/CaMK-IV and Ras/MEK/ERK pathways, which is vital for neuronal development, survival, and proper functioning [23]. Blocking CREB signaling hinders post-stroke recovery [62], and CaMK-IV inhibition impairs blood–brain barrier integrity and exacerbates ischemic injury [63]. In turn, elevated CREB and CaMK-IV activity is associated with improved post-stroke outcomes in a number of animal studies [62–66].

On the other hand, some experimental data suggest that the upregulation of TRPC6 activity increases intracellular Ca^{2+} concentrations concomitantly with NMDAR activation, further exacerbating excitotoxic damage to neurons [16]. Oxygen-glucose deprivation in cultured cortical neurons and MCAO in wild-type mice are associated with elevated TRPC6 expression and activity, while TRPC6 deletion attenuates glutamate- and NMDA-induced cytotoxicity and reduced infarct volumes [25]. Knockdown of TRPC3, 6, and 7 prevents apoptosis in cultured astrocytes and ameliorates ischemic brain injury in mice [26]. Moreover, prevention of TRPC6 hyperactivation results in increased neuronal viability, reduced infarct volumes and brain edema, and improved functional recovery following acute ischemic stroke in rats [67–69] and crab-eating macaques [67]. Other TRPC6 inhibitors have also been reported to exert beneficial effects in experimental models to some extent relevant to ischemic brain injury (e.g., acute renal ischemia/reperfusion injury) [70–74].

Existing controversies regarding the TRPC6 function in the development of brain ischemia might be due to different experimental settings (i.e., rodent model, sex, age, the method used to model ischemia). However, similarly to AD, brain ischemia seems to be heterogenic, meaning that one group of patients has TRPC6 hypofunction, and another one has TRPC6 hyperfunction. This indicates the need to develop pathology-dependent strategies to treat different NDD patients.

7. Available Drug Candidates to Modulate TRPC6 Activity

To date, two strategies have been proposed for the pharmacological modulation of TRPC6 activity for the treatment of Alzheimer's disease and cerebral ischemia: (1) TRPC6 activation to allow Ca^{2+} influx via neuronal SOCE and sustain the stability of postsynaptic contacts (for AD) and to attenuate NMDAR activity and prevent calcium-dependent excitotoxicity [15,16] (for ischemia); (2) inhibition of TRPC6 in order to prevent calcium overload and the subsequent cell damage [23,64] (for AD and ischemia). Although apparently mutually exclusive, both of these strategies are aimed to keep the intracellular Ca^{2+} concentration within the normal range, which requires limiting its entry via transmembrane channels and/or release from intracellular stores [75]. Given the pivotal role of NMDAR in excitotoxic neuronal damage, NMDAR blockers have been proposed as potential neuroprotective agents, although most of them have failed to show substantial effectiveness in human patients so far [76]. Due to that fact, TRPC6 has emerged as an alternative therapeutic target for AD and ischemic stroke [23].

7.1. TRPC6 Activators

TRPC6 activation can be induced by several endogenous diacylglycerols (DAGs) [32], lysophosphatidylcholines [75], and 20-hydroxyeicosatetraenoic acid, which is a metabolite of arachidonic acid [77]. A number of DAG analogs, including 1,2-dioctanoyl-*sn*-glycerol, DAG-containing arachidonic and docosahexaenoic acids [78,79], and the docosanoid neuroprotectin D1 [80], have also been reported as TRPC6 agonists. This channel can also be activated by agents of synthetic or natural origin that are structurally different from DAG. Direct TRPC6 agonists acting in receptor-operated mode include synthetic compounds, such as flufenamic acid [81] and several pyrazolopyrimidine [82] and piperazine [83] derivatives. The benzimidazole-based small molecule agonist GSK1702934A, its azobenzene derivative OptoBI-1 [84], and the chromone-containing compound C20 [85] are also thought to be direct (ROC) stimulators of TRPC6 activity. In contrast, certain naturally occurring chemicals are known to potentiate TRPC6 effects in an indirect manner. These include the stilbenoid resveratrol [66], the isoflavone calycosin [86], and (−)-epigallocatechin-3-gallate, a catechin-type polyphenol [87]. Recently, a novel potent ethanolamine derivative, bis-{2-[(2E)-4-hydroxy-4-oxobut-2-enoyloxy]-N,N-diethylethanaminium} butandioate (FDES), has been demonstrated to exert neuroprotective effects due to selective TRPC6 channel activation in store-operated mode [88,89]. The aminoquinazoline derivative, NSN21778, was demonstrated by Zhang et al. to have a store-dependent mechanism of action where DAG is required as a co-factor for TRPC6 activation [14]. The piperazine derivative, 51164, has been shown to activate TRPC6 in store-operated mode requiring DAG as a co-factor [21].

7.1.1. Endogenous Ligands and Analogs

DAG

A compound containing both DAG and arachidonic acid fragments, 1-stearoyl-2-arachidonyl-sn-glycerol (SAG), elicits a rapid Ca^{2+} flux into HEK293 cells, while 1-stearoyl-2-docosahexaenoyl-sn-glycerol (SDG), which does not contain an arachidonic acid moiety, has significantly lower potency [78]. 1-oleoyl-2-acetyl-*sn*-glycerol (OAG), a diacylglycerol analog and a TRPC3/6/7 channel modulator, is found to cross the plasma membrane and intracellularly activate the channels [32,90]. OAG has been shown to activate Ca^{2+}-permeable channels, displaying TRPC6-like properties in cultured cortical neurons [91]. In addition, OAG has been demonstrated to increase field excitatory postsynaptic potential (fEPSP) levels in a TRPC-dependent manner in hippocampal slices from wild-type mice [92], indicating that it might have neuroprotective effects. However, as far as we are aware, those compounds have not yet been evaluated in in vivo models of AD and cerebral ischemia.

Lysophosphatidylcholine

Lysophosphatidylcholine (LPC) is produced from phosphatidylcholines via partial hydrolysis generally catalyzed by phospholipase A2. Increased LPC production has been observed in various disorders of the central nervous system, including stroke and AD [93], and is associated with acute and chronic brain ischemia [94,95]. Results of a cohort study have suggested that LPC levels could be used as a tool for ischemic stroke risk stratification in patients who have suffered a transitory ischemic attack before [96]. LPC is shown to activate TRPC6 channels and promote Ca^{2+} flux into endothelial cells, hampering their migration and preventing endothelial healing, thus contributing to atherogenesis [97,98].

20-Hydroxy-5Z,8Z,11Z,14Z-Eicosatetraenoic Acid

20-hydroxy-5Z,8Z,11Z,14Z-eicosatetraenoic acid (20-HETE) is the main eicosanoid metabolite of arachidonic acid and a potent inflammatory vasoconstrictor. In HEK293 cells, 20-HETE (half maximal effective concentration, EC_{50} of 0.8 μM) elicits a three-fold increase in TRPC6 activity (as indicated by an increased inward, the non-selective current observed in whole-cell patch-clamp recordings) but does not affect intracellular Ca^{2+} concentrations [77]. In isolated guinea pig airway smooth muscle cells, it induces a dose-dependent inotropic effect via TRPC6 activation and the subsequent promotion of Ca^{2+} entry [99]. Nevertheless, 20-HETE has been shown to have detrimental effects in ischemic and traumatic brain injury, which might be explained by its vasoconstrictor properties, and has even been proposed as a predictor of poor prognosis in stroke patients [100]. To our knowledge, 20-HETE has not yet been tested in AD models, but it has been shown to activate TRPV1 channels in dorsal root ganglia cultures [101].

10*R*,17*R*-Dihydroxydocosa-4Z,7Z,11E,13E,15Z,19Z-Hexaenoic Acid

10*R*,17*R*-dihydroxydocosa-4Z,7Z,11E,13E,15Z,19Z-hexaenoic acid (Neuroprotectin D1, NPD1, Table 1) is a docosahexaenoic acid ((4Z,7Z,10Z,13Z,16Z,19Z)-docosa-4,7,10,13,16,19-hexaenoic acid, DHA)-derived endogenous anti-inflammatory mediator commonly found in fish oil [102]. NPD1, among other neuroprotectins, is synthesized in ischemic brain tissue as a result of DHA enzymatic lipoxygenation. In a rat model of ischemic stroke, NPD1 administration is associated with a significantly elevated TRPC6 and CREB activity, while the inhibition of the MEK/ERK pathway results in a decrease in NPD1 neuroprotective activity. Continuous intracerebroventricular administration of NPD1 over 10 min at 2 h after reperfusion sustains TRPC6/CREB activity, reduces infarct volumes, and promotes functional recovery [80]. The role of NPD1 in TRPC6-mediated neuroprotection in AD has not been described yet.

Table 1. Pharmacological modulators of TRPC6 in cerebral ischemia and AD models.

No.	Compound Name and Structural Formula	Experimental Model	Mode of Administration	Effect(s)	Reference(s)
		TRPC6 Activators			
		Endogenous ligands			
1	Neuroprotectin D1 10*R*,17*R*-Dihydroxydocosa-4*Z*,7*Z*,11*E*,13*E*,15*Z*,19*Z*-hexaenoic acid	Rat transient MCAO	Icv injection at 2 h after reperfusion	reduced infarct volume reduced sensory and motor deficits	[80]
		Phytochemicals			
2	Hyperforin (1*R*,5*S*,6*R*,7*S*)-4-Hydroxy-6-methyl-1,3,7-tris(3-methylbut-2-en-1-yl)-6-(4-methylpent-3-en-1-yl)-5-(2-methylpropanoyl) bicyclo[3.3.1]non-3-ene-2,9-dione	Lipopolysaccharide stimulation in 28 d post-MCAO isolated mouse astrocytes	Co-incubation for 16 h	increased viability	[103]
		Lipopolysaccharide stimulation in 28 d post-MCAO isolated mouse cortical neurons		increased viability	[103]
		NMDA toxicity in rat hippocampal slices	Co-incubation for 30 min	reduced edema	[104]
		Male Sprague-Dawley rats injected with fibrillary Aβ	Intrahippocampal co-injection of Aβ with the drug for 14 days	amyloid deposits disaggregation reduced spatial memory deficit	[105]
		Rat hippocampal slice cultures	Co-incubation for 24 h	increased proportion of mature stubby spines	[106]
		Hippocampal cultures from PS1-M146VKI and APPKI transgenic mice	Incubation for 16 h	increased percentage of mushroom spines in TRPC6-dependent manner increased neuronal SOCE in postsynaptic spines	[14]
		Hippocampal cultures treated with synthetic Aβ42 peptides	Co-incubation for 16 h	increased percentage of mushroom spines increased neuronal SOCE in postsynaptic spines	[21]
		Mouse transient MCAO	Intranasal administration q.d. for 7 d starting at day 7 post-MCAO	increased hippocampal neurogenesis improved post-stroke depression and anxiety reduced memory deficit	[107]
			Icv injections at 1, 24, and 48 h after MCAO	reduced microglial activation reduced infarct volume reduced neurological deficit	[108]
			Icv injections q.d. for 14 d starting at day 14 post-MCAO	increased angiogenesis reduced motor deficit	[109]
				increased angiogenesis increased subventricular neurogenesis reduced motor deficit	[103]
		Mouse permanent MCAO	Ip injection before ischemia onset	no effect on infarct volume or brain edema	[104]
		Mouse water intoxication			
		Rat transient MCAO	Icv injection at 6, 12, or 24 h after reperfusion	prevented neuronal apoptosis reduced infarct volume reduced neurological deficit	[110]

Table 1. *Cont.*

No.	Compound Name and Structural Formula	Experimental Model	Mode of Administration	Effect(s)	Reference(s)
3	Tetrahydrohyperforin (1S,5S,7S,8R)-4,9-dihydroxy-1-(1-hydroxy-2-methylpropyl)-8-methyl-3,5,7-tris(3-methylbut-2-enyl)-8-(4-methylpent-3-enyl)bicyclo[3.3.1]non-3-en-2-one	APPSEN1ΔE9 mice	Ip injections for 4 weeks	reduced memory deficit reduced amyloid deposition attenuated neuroinflammation and oxidative stress	[111]
4	Resveratrol 5-[(E)-2-(4-Hydroxyphenyl)ethenyl] benzene-1,3-diol	Oxygen/glucose deprivation in isolated rat brain endothelial cells	Preincubation for 3 d	increased viability	[112]
		Mouse transient MCAO	Ip injection at 48 h before MCAO	reduced infarct volume no effect on cerebral blood flow	[113]
		Mouse transient MCAO	Oral gavage q.d. for 7 d starting from 24 h after MCAO, or for 5 d starting from 72 h after MCAO	increased vascular density in the basal ganglia region and cortex reduced infarct volume reduced neurological deficit	[114]
		Rat transient MCAO	Ip injections q.d. for 7 d before MCAO	reduced infarct volume reduced neurological deficit	[66]
		Rat recurrent transient MCAO	Oral gavage q.d. for 3 d between strokes	reduced infarct volume following an initial and recurrent stroke reduced blood–brain barrier disruption following a recurrent stroke reduced brain edema following a recurrent stroke reduced astrogliosis following a recurrent stroke no effect on cerebral blood flow during or after recurrent stroke	[112]
		Rat asphyxial cardiac arrest	Ip injection at 48 h before cardiac arrest	enhanced ATP synthesis efficiency in hippocampal mitochondria prevented hippocampal neuronal apoptosis	[115]
		Gerbil transient bilateral common carotid artery occlusion	Ip injections during occlusion or at reperfusion + at 24 h after reperfusion	reduced hippocampal microglial activation prevented hippocampal-delayed neuronal death	[116]
		Clinical trials in patients	diverse	affected neuroinflammation, Aβ deposition, and adaptive immunity in patients with mild to moderate Alzheimer's disease	for a review, see [117]

Table 1. *Cont.*

No.	Compound Name and Structural Formula	Experimental Model	Mode of Administration	Effect(s)	Reference(s)
	Synthetic compounds				
5	Flufenamic acid 2-[[3-(Trifluoromethyl)phenyl]amino]benzoic acid 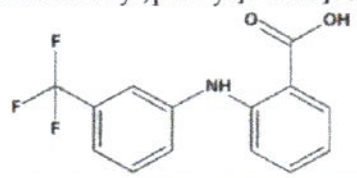	Oxygen/glucose deprivation in isolated mouse embryonic cortical neurons	Incubation from 15 min before oxygen/glucose deprivation until 24 h of reoxygenation	increased viability of male neurons no effect on female neurons	[118]
		Glutamate toxicity in isolated rat embryonic hippocampal neurons	Co-incubation for 10 min	increased viability	[119]
6	PPZ1 [4-(5-Chloro-2-methylphenyl)piperazin-1-yl]-3-fluorophenylmethanone 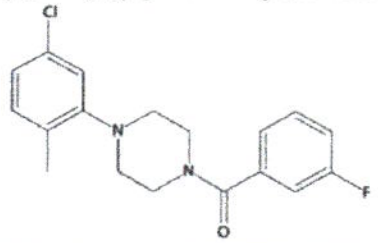				
7	PPZ2 2-[4-(2,3-Dimethylphenyl)piperazin-1-yl]-N-(2-ethoxyphenyl)acetamide 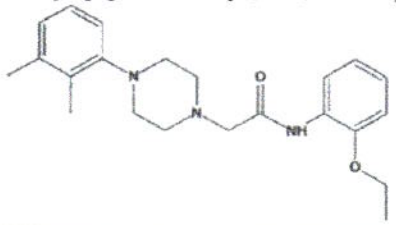	Serum deprivation in isolated rat cerebellar granule neurons	Incubation for 24 h before and 24 h after serum deprivation	increased neurite outgrowth increased cell viability	[83]
8	51164 N-(2-chlorophenyl)-2-(4-phenylpiperazin-1-yl)acetamide 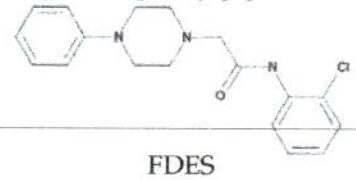	Aβ42-induced toxicity in primary hippocampal neurons	Co-incubation for 16 h	restored mushroom spines percentage induced neuronal SOCE in postsynaptic spines	[21]
		6 month-old 5xFAD mouse hippocampal slices	30 min incubation	restored LTP induction	
9	FDES Bis-{2-[(2E)-4-hydroxy-4-oxobut-2-enoyloxy]-N,N-diethylethanaminium} butandioate 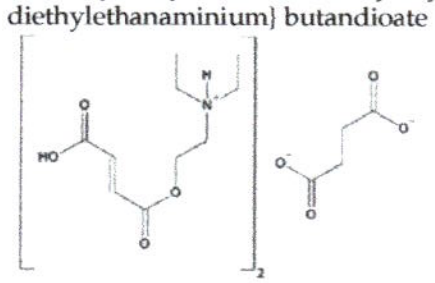	Aβ42-induced toxicity in primary hippocampal neurons	Incubation for 16 h	restored mushroom spines percentage induced neuronal SOCE in postsynaptic spines	[89]
		Rat transient MCAO	Ip injections at 1 h after reperfusion + q.d. for 7 d after reperfusion	improved spatial memory retention no effect on mortality	
		Rat permanent bilateral common carotid artery ligation	Oral gavage at 30 min before MCAO + q.d. for 21 d after reperfusion	reduced mortality reduced motor deficit reduced aggressiveness reduced emotional lability increased exploratory behavior	[120]

Table 1. *Cont.*

No.	Compound Name and Structural Formula	Experimental Model	Mode of Administration	Effect(s)	Reference(s)
10	NSN21778 N-[4-[2-[(6-aminoquinazolin-4-yl)amino]ethyl]phenyl]acetamide	Primary hippocampal cell culture models from PS1-M146V and APPKI mice	Incubation for 16 h	increased percentage of mushroom spines in TRPC6-dependent manner increased neuronal SOCE in postsynaptic spines	[14]
		Hippocampal brain slices from PS1-M146V and APPKI mice	Pretreatment for 30 min	recovered LTP induction	
TRPC6 inhibitors					
11	HET0016 N-Hydroxy-N′-(4-n-butyl-2-methylphenyl)formamidine	Rat pediatric asphyxial cardiac arrest	Iv injection at reperfusion	increased cortical CBF reduced brain edema	[69]
			Iv injection at reperfusion + ip injections every 6 h for 24 h after reperfusion	reduced neurological deficit reduced neurodegeneration	
		Rat transient MCAO	Iv injection immediately before reperfusion	reduced infarct volume	[68]
			Ip injections q.d. for 3 d before and 3 d after MCAO	increased CBF reduced infarct volume	[121]
		Piglet neonatal hypoxia/ischemia	5 min-infusion at 5 min after reperfusion + hypothermia at 3 h after reperfusion	increased neuronal viability in the putamen, cortex, and thalamus prevention of seizures	[122]
			Iv injection at 5 min after reperfusion	increased neuronal viability in the putamen reduced neurological deficit no effect on cerebral blood flow (CBF)	[123]
12	TS-011 N-(3-Chloro-4-morpholin-4-yl)phenyl-N′-hydroxyimidoformamide	Rat transient MCAO	Iv injection at 30 min before MCAO + 1 or 2 h-infusion during MCAO	reduced cortical, subcortical, and total infarct volumes reduced the delayed drop in CBF no effect on volume at risk	[68]
			Iv injection at 20 min after MCAO + 2 h-infusion at reperfusion	reduced cortical and total infarct volumes no effect on volume at risk no effect on CBF	
			1 h-infusion at reperfusion, 1, 2, or 4 h after reperfusion	reduced infarct volumes	
			1 h-infusion at reperfusion + iv injections q.d. for 7 d	reduced infarct volumes reduced sensory and motor deficits	[67]
		Crab-eating macaque thrombotic internal carotid artery occlusion	Iv injection + 24 h-infusion after embolization	reduced infarct volume (when co-administered with tissue plasminogen activator) reduced neurological deficit	

Table 1. *Cont.*

No.	Compound Name and Structural Formula	Experimental Model	Mode of Administration	Effect(s)	Reference(s)
13	Mefenamic acid 2-(2,3-Dimethylphenyl)aminobenzoic acid 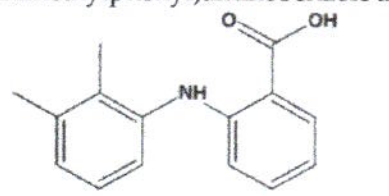	Glutamate toxicity in isolated rat embryonic hippocampal neurons	Co-incubation for 10 min	increased cell viability	[124]
		3xTg mice AD model	Administration by osmotic minipump over 28 days	reduced cognitive deficit	[125]
		Rat transient MCAO	Iv injection before MCAO	no effect on infarct and penumbra volumes and brain edema	[124]
			Iv injections at 1 h before + at 1, 2, and 3 h after MCAO	reduced infarct volume reduced brain edema	
			Icv 24 h-infusion starting at 1 h before MCAO		[119]
14	Meclofenamic acid 2-(2,6-Dichloro-3-methylanilino)benzoic acid 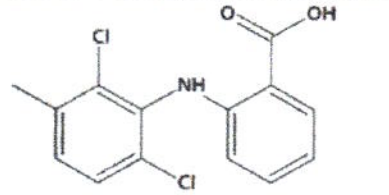	Glutamate toxicity in isolated rat embryonic hippocampal neurons	Co-incubation for 10 min	increased viability	[119]
15	Niflumic acid 2-{[3-(Trifluoromethyl)phenyl]amino}pyridine-3-carboxylic acid 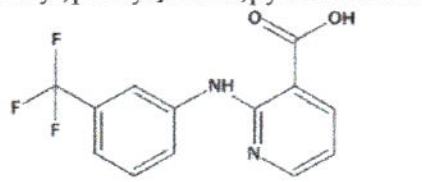				
16	SAR7334 4-[[(1R,2R)-2-[(3R)-3-Amino-1-piperidinyl]-2,3-dihydro-1H-inden-1-yl]oxy]-3-chlorobenzonitrile dihydrochloride 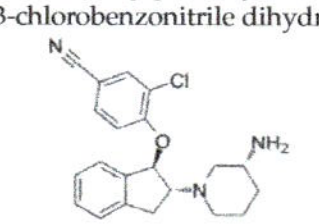	Primary cortical neurons	Treatment with 1 µM SAR7334 at the time of imaging	no effect on neuronal SOCE	[126]
17	EVP4593 4-N- [2- (4-phenoxyphenyl)ethyl]quinazoline-4,6-diamine 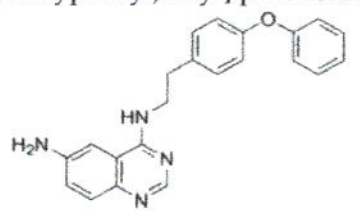	PSEN1ΔE9-hyperexpressing primary hippocampal neurons	Co-incubation for 16 h	reduced TRPC6-dependent neuronal SOCE in postsynaptic spines increased mushroom spines percentages	[24]

7.1.2. Hyperforin and Other Phytochemicals

Hyperforin

((1*R*,5*S*,6*R*,7*S*)-4-hydroxy-6-methyl-1,3,7-tris(3-methylbut-2-en-1-yl)-6-(4-methylpent-3-en-1-yl)-5-(2-methylpropanoyl)bicyclo[3.3.1]non-3-ene-2,9-dione) (Hyperforin, Table 1) is a phloroglucinol derivative and a major active constituent of St. John's wort (*Hypericum perforatum* L.). Hyperforin is a potent inhibitor of TRPC6 proteolysis and a positive modulator of TRPC6/CREB activity, acting in a manner similar to that of the brain-derived neurotrophic factor (BDNF) [110,127]. It is thought to bind to TRPC6 due to structural similarities to DAG and has higher selectivity because of the relative rigidity of the phloroglucinol pharmacophore moiety [128]. Neuroprotective and antidepressant-like properties of hyperforin and hyperforin-containing *H. perforatum* preparations involve the modulation of axonal growth, neurite growth and branching, dendritic spine formation, and the promotion of neuronal plasticity [106,127]. As proposed by Singer et al., the increase in Na^+ concentration resulting from TRPC6 activation by hyperforin might inhibit serotonin reuptake via the serotonin/Na^+ symporter, which, together with increased synaptic plasticity, could explain the antidepressant-like activity of *H. perforatum*-based drugs [129]. Confirming this hypothesis, larixyl acetate, a selective blocker of TRPC6, abolishes the antidepressant-like effects of hyperforin observed in mice in the tail suspension test [130].

In an ex vivo experiment, hyperforin (0.3 μM) promotes mature stubby spine formation and decreases the proportion of immature thin spine formation in rat hippocampal pyramidal neurons but does not affect mushroom spine density and morphology. Proper TRPC6 expression level and the presence of a fully functional TRPC6 channel are required for hyperforin to exert its effects, which suggests the key role of TRPC6 activation in its mechanism of action [106]. Using different rodent models of ischemic stroke, hyperforin has been shown to promote post-stroke neuro- and angiogenesis [103,109], inhibit microglial activation [108], attenuate brain edema [104], stimulate hippocampal neurogenesis, ameliorate post-stroke depression and anxiety, and restore memory function [107]. Chronic hyperforin treatment stimulates the expression of the tropomyosine receptor kinase B (TrkB) BDNF receptors as well as of TRPC6 in murine cortical neurons but has no effect on hippocampal neurogenesis [131]. When applied intracerebroventricularly to rats immediately after MCAO, hyperforin preserves TRPC6 activity, reduces infarct volumes, promotes functional recovery, and increases neurologic scores at 24 h after reperfusion [110].

There is a lot of evidence that hyperforin and its derivatives are highly selective towards the TRPC6 channel and do not exert similar effects on its closest relative, the TRPC3 channel [106,127,128,132]. Several studies have shown that hyperforin activates TRPC6 and increases its expression [106,133], leading to a decrease in the Aβ level and an improvement in cognitive performance in AD models [92,105,111].

The neuroprotective effect of hyperforin has been demonstrated in several rodent models of AD. In rats co-injected with amyloid fibrils and hyperforin in the hippocampus, hyperforin reduces amyloid deposit formation and, therefore, decreases the Aβ-induced neurotoxicity, reactive oxidative species formation, and attenuated behavioral impairments [105]. A more stable hyperforin derivative, tetrahydrohyperforin, also prevents the cognitive decline and synaptic impairment in double transgenic APPswe/PSEN1ΔE9 mice in a dose-dependent manner. It has been shown that the neuroprotective mechanism of tetrahydrohyperforin is associated with a reduced rate of proteolytic processing of APP, decreased the total amount of fibrillar and oligomeric forms of Aβ, reduced level of tau hyperphosphorylation, and attenuated astrogliosis [132]. Tetrahydrohyperforin has been shown to specifically target TRPC6 [92].

Hyperforin is also thought to be responsible for the induction of the cytochrome P450 enzyme CYP3A4 by binding to and activating the pregnane X receptor [134], indicating that it might have side effects and undesirable drug-to-drug interactions. Moreover, hyperforin is difficult to synthetize [135], unstable when exposed to light, and irritant to the gastrointestinal tract [136]. Such side effects might limit the use of hyperforin as a TRPC6 activator.

Resveratrol

Resveratrol (5-[(*E*)-2-(4-hydroxyphenyl)ethenyl]benzene-1,3-diol) (Table 1), which is found in grapes, berries, peanuts, and other plants, is among the best-known natural compounds with antioxidant and antihypoxic activity [66,137,138]. In the pioneering study of Wang et al. [116], resveratrol is shown to be able to cross the blood–brain barrier (BBB) and exert neuroprotective activity. Injected intraperitoneally either during or shortly after the induction of cerebral ischemia, it largely prevents delayed neuronal cell death and glial cell activation in a gerbil model of transient bilateral common carotid artery occlusion [116]. Based on the observed increase in TRPC6 and CREB activity, which has been prevented by PD98059 or KN62, inhibitors of MEK and CAMKIV/ CaMKII, respectively, it is suggested that resveratrol has exerted its effects via TRPC6 activation. Since its administration has been accompanied by a marked decrease in calpain activity, resveratrol is classified as an indirect positive modulator of TRPC6 activity [66]. When given to rats for 7 days before MCAO, resveratrol significantly reduces infarct volumes and enhances neurological scores at 24 h after reperfusion [66]. Low-dose oral resveratrol treatment for three consecutive days before and after an ischemic stroke induced in rats by middle cerebral artery clipping alleviates brain damage caused by the following recurrent stroke. Resveratrol normalizes BBB function and reduces cerebral edema without affecting regional cerebral blood flow and systemic blood pressure [112]. Resveratrol preconditioning (48 h before the induction of ischemia) effectively prevents neuronal cell loss in a mouse MCAO model of stroke [113]. In a rat model of global cerebral ischemia induced by asphyxial cardiac arrest, resveratrol (48 h before the induction of ischemia) is shown to protect the CA1 region of the hippocampus similarly to ischemic preconditioning [115].

Resveratrol is actively used in AD research [139–141]. Neuroprotective effects of resveratrol have been shown to be associated with the activation of silent mating type information regulation 2 homolog 1 (SIRT1) and vitagenes production [142]. In a phase II trial in AD patients, resveratrol is safe and well-tolerated, but its effectiveness is contradictory [143,144].

Although generally considered as a non-toxic therapeutic agent, high doses of resveratrol inhibit CYP3A4 activity in vitro [145] and in healthy volunteers [146], thus potentially inhibiting drug metabolic clearance, increasing bioavailability and toxicity of drugs taken concomitantly [147].

7.1.3. Synthetic Compounds

Flufenamic Acid

Flufenamic (2-[[3-(trifluoromethyl)phenyl]amino]benzoic) (Table 1) acid (FFA) is a member of the fenamate class of nonsteroidal anti-inflammatory drugs that have limited clinical applications due to their toxicity. FFA has been found to selectively activate TRPC6, at the same time, inhibiting TRPC3, 4, 5, and 7 [81]. Direct activation of TRPC6 by FFA has been confirmed in a number of in vitro studies [148], including those in glomerular podocytes [149] and ventricular cardiomyocytes [150]. Male (but not female) embryonic mice cortical neurons, which have been pre-treated with FFA 15 min before oxygen-glucose deprivation, have shown significantly higher viability, although this effect is not linked by the authors to TRPC6 activation [118]. Similar results are obtained in a recent in vitro glutamate toxicity assay using isolated rat embryonic hippocampal neurons [119].

Multiple experimental evidences suggest that FFA is a broad spectrum ion channel modulator, with a preference for non-selective cation channels and chloride channels (reviewed here [81]). Its activity seems to be dose-dependent since FFA inhibits TRPC6 with a half maximal inhibitory concentration (IC_{50}) of 17.1 µM [151] but activates the same channel at 100 µM [148]. TRPM8 is inhibited by 100 µM FFA but is slightly activated at higher concentrations [152]. A worse situation is reported for big calcium-activated potassium channels (BK_{Ca}) modulation since FFA activates the channel below 10 µM, inhibits the channel between 10 to 50 µM, and then activates the channel above 50 µM [153]. Such opposing effects on the same channels and a huge number of other ion channels that are modulated by FFA makes it an inappropriate drug for usage in humans.

Piperazines

Sawamura et al. discovered a group of piperazine-based potent TRPC3/6/7 agonists functioning in receptor-operated mode with varying selectivity for different channel subtypes [83]. Among that group, 2-[4-(2,3-dimethylphenyl)piperazin-1-yl]-N-(2-ethoxyphenyl)acetamide (PPZ2, Table 1) dose-dependently activates TRPC6 and TRPC6-like channels in HEK cells, vascular smooth muscle cells, and cultured rat cerebellar granule neurons. PPZ2 and PPZ1 ([4-(5-chloro-2-methylphenyl)piperazin-1-yl]-3-fluorophenylmethanone) (Table 1) promote neurite outgrowth in a manner similar to that of BDNF and provide protection against serum deprivation-induced neuronal death [83].

Later on, another piperazine derivative, N-(2-chlorophenyl)-2-(4-phenylpiperazin-1-yl)acetamide (51164, Table 1), has been shown to activate the TRPC6 channel in store-operated mode with DAG acting as a co-factor [21]. Nanomolar concentrations of 51164 protect mushroom spines from amyloid toxicity, induce TRPC6-dependent neuronal SOCE in postsynaptic spines, and restore the induction of long-term potentiation in hippocampal slices taken from 6 months old 5xFAD mice [21].

N-[3-[4-[3-[bis(2-methylpropyl)amino]propyl]piperazin-1-yl]propyl]-1*H*-benzimidazol-2-amine (AZP2006), another piperazine derivative, attenuates Aβ and tau toxicity and improves cognitive performance in various mouse models [154]. Currently, AZP2006 is in phase 2 clinical trial in patients with progressive supranuclear palsy [155].

Piperazine derivatives as the majority of TRPC6 agonists cross-react with TRPC3 and TRPC7 [83], limiting their use as specific TRPC6 modulators. Among other side effects of piperazines is their hepatotoxicity [137]; however, hepatotoxicity has not been predicted by bioinformatical analyses for a 51164 compound [21]. Gastrointestinal hemorrhage and multiple organ failure have been predicted by bioinformatical analyses; thus, there is a need to search for the lowest therapeutic dose for the 51164 compound, and most likely, there is a need to modify its structure in order to minimize the mentioned side effects [21].

Bis-{2-[(2*E*)-4-Hydroxy-4-Oxobut-2-Enoyloxy]-*N*,*N*-Diethylethanaminium} Butandioate

Bis-{2-[(2*E*)-4-hydroxy-4-oxobut-2-enoyloxy]-*N*,*N*-diethylethanaminium} butandioate (Table 1), abbreviated as FDES, is an ethanolamine derivative known to possess antihypoxic, anti-ischemic, and neuroprotective properties [88,89,120,156]. Chronic FDES administration decreases mortality and improves motor function and coordination following permanent bilateral common carotid artery ligation [120] and reduces spatial memory deficit following middle cerebral artery (MCA) occlusion/reperfusion [89] in rats. Later, FDES was demonstrated to ameliorate fore- and hindlimb motor disturbances and increase overall locomotor activity in rats with unilateral traumatic brain injury [88].

FDES is shown to potentiate neuronal SOCE into postsynaptic spines in mouse hippocampal neurons [89]. Since TRPC6 knockdown abolishes the effects of FDES on neuronal SOCE (similarly to hyperforin), TRPC6 activation is suggested to be the primary mechanism of FDES neuroprotective action. Nanomolar concentrations of FDES effectively protect mushroom dendritic spines from amyloid synaptotoxicity, stabilizing and enhancing synaptic transmission, and preserving synaptic contact density [89]. Similarly to hyperforin, FDES decreases the proportion of immature thin and stubby spines in hippocampal neurons. When administered intraperitoneally to rats subjected to MCAO for 7 consecutive days starting from 1 h after reperfusion, FDES improves short-term spatial memory retention, as observed in the Barnes maze [89].

FDES is a precursor of choline and has been shown to have nootropic and actoprotective properties [89]. We assume that FDES would cross-react with muscarinic acetylcholine receptors, causing phospholipase C activation and production of IP3 and DAG. This cross-reactivity of FDES could further enhance its TRPC6-agonistic properties, although this hypothesis remains to be experimentally proven.

N-[4-[2-[(6-Aminoquinazolin-4-yl)Amino]ethyl]phenyl]acetamide

N-[4-[2-[(6-aminoquinazolin-4-yl)amino]ethyl]phenyl]acetamide (NSN21778, Table 1) was proposed by Zhang et al. as a novel positive modulator of the TRPC6/neuronal SOCE pathway acting in a manner similar to that of 51164. Despite its effectiveness in terms of improving mushroom spine morphology and TRPC6-mediated SOCE in PS1-KI and APP-KI hippocampal neurons and rescuing long-term potentiation in the APP-KI mouse model of AD, NSN21778 is found to have a rather poor pharmacokinetic profile and a low penetration of the blood–brain barrier [14].

7.1.4. Other TRPC6 Agonists

Several compounds described below have been confirmed to activate the TRPC6 channel using in vitro assays. However, to the best of our knowledge, their specific neuroprotective properties remain unexplored. Given their ability to interact with TRPC6, these compounds can be considered as potential neuroprotective agents.

Pyrazolopyrimidines

A number of pyrazolopyrimidines obtained by Qu et al. are reported to be direct agonists of TRPC6, 3, and 7. Among the four pyrazolopyrimidines tested by Qu et al., ethyl 4-(7-hydroxy-2-methyl-3-(4-(trifluoromethyl)phenyl)-pyrazolo[1,5-a]pyrimidin-5-yl)piperidine-1-carboxylate (compound 4n) is most active towards TRPC6 (EC_{50} = 1.39 or 0.89 µM depending on the conditions). However, compound 4n has demonstrated a much higher affinity for TRPC3 and TRPC7 (EC_{50} = 0.019 and 0.090 µM, respectively) [82].

GSK1702934A and OptoBI-1

GSK1702934A (1,3-Dihydro-1-[1-[(5,6,7,8-tetrahydro-4H-cyclohepta[b]thien-2-yl)carbonyl]-4-piperidinyl]-2H-benzimidazol-2-one) has been reported by Xu et al. to activate TRPC3 and 6 (EC_{50} = 0.08 and 0.44 µM, respectively), acting directly and independent of protein lipase C signaling from the extracellular side [157]. OptoBI-1, an azobenzene moiety-containing photochromic derivative of GSK1702934A, is found to activate TRPC6 as well as TRPC3 and 7, although having a slightly higher affinity for TRPC3. Light treatment of cultured murine hippocampal neurons with OptoBI-1 suppresses action potential firing elicited by repetitive depolarizing current injections [84].

3-(6,7-Dimethoxy-3,3-Dimethyl-3,4-Dihydroisoquinolin-1-yl)-2H-Chromen-2-One

Recently, a novel small-molecule allosteric TRPC6 agent, 3-(6,7-dimethoxy-3,3-dimethyl-3,4-dihydroisoquinolin-1-yl)-2H-chromen-2-one (C20), has been reported by Häfner et al. [85]. C20 (EC_{50} = 2.37 µM) selectively activates TRPC6 channels in several HEK cell lines while only slightly reducing the basal activity of TRPC3 and increasing that of TRPC7 and not affecting TRPC4 and 5 activity at all. Higher concentrations of C20 (10 µM) potentiate the efficacy of OAG and GSK1702934A, low-selective TRPC6 agonists described above, in HEK cells and freshly prepared human platelets [85].

It can be assumed that the mechanism of action of C20 involves TRPC6 sensitization and not with its activation per se; that is, C20 allows TRPC6 to be activated at a low basal concentration of DAG [85].

7.2. TRPC6 Inhibitors

AD seems to be a multifactorial disease with different pathogenic cascades occurring in different patients. In terms of TRPC6 channel dysfunction, there are forms of fAD, which demonstrate

TRPC6 hyperfunction [24]. For those patients, TRPC6 inhibitors might be used in order to normalize intracellular Ca^{2+} homeostasis.

The pathogenesis of cerebral ischemia is not fully understood. Similarly to AD, cerebral ischemia involves several different pathological cascades. There are studies reporting that TRPC6-mediated Ca^{2+} and Na+ influx facilitates NMDAR activation and exacerbates excitotoxicity, while TRPC6 deletion attenuates neuronal damage and death following focal cerebral ischemia [25]. In such cases, where excessive TRPC6 activity seems to be present, the use of TRPC6 inhibitors might be beneficial.

Arachidonic acid, which is released following cerebral ischemia, can be metabolized to 20-hydroxyeicosatetraenoic acid (20-HETE). 20-HETE is a potent vasoconstrictor that may contribute to ischemic injury [67]. Synthetic 20-HETE has been shown to activate TRPC6 [77]. Thus, inhibition of 20-HETE production by HET0016 reduces TRPC6 activation [158]. Inhibition of 20-HETE synthesis by N-hydroxy-N′-(4-n-butyl-2-methylphenyl)formamidine (HET0016, Table 1) decreases infarct volumes and increases cortical cerebral blood flow in cerebral ischemia induced by asphyxia cardiac arrest in rat pups [69] and in transient MCAO-induced ischemia in adult rats [121]. In neonatal piglets subjected to 6 min of acute asphyxia, HET0016 potentiates the neuroprotective effects of delayed hypothermia, increasing neuronal viability, preventing seizure development, and reducing neurological deficit [122,123]. Another 20-HETE inhibitor known as N-(3-chloro-4-morpholin-4-yl)phenyl-N′-hydroxyimidoformamide (TS-011, Table 1) markedly decreases infarct volumes and improves functional recovery in rats and primates [67,68].

We are unaware of any direct investigations of neuroprotective effects of HET0016 and TS-011 (Table 1) in AD models. However, since patients with Alzheimer's disease show an accumulation of (2E)-4-hydroxy-2-nonenal (HNE) adducts [159], and HET0016 is a potent inhibitor of ω- and ω-1-hydroxylation of HNE/HNA (4-hydroxynonanoic acid) [160], HET0016 could be of interest regarding its potential neuroprotective properties in AD.

In mice with closed-head traumatic brain injury, larixyl acetate, a naturally occurring diterpene, ameliorates endothelial dysfunction [74], which is closely associated with cerebral ischemia as well [161]. In isolated mouse lungs, larixyl acetate prevents the development of the acute hypoxic ventilatory response [72]. TRPC6 inhibition is now considered to be the primary mechanism of larixyl's neuroprotective action [72,74].

Mefenamic (2-(2,3-dimethylphenyl)aminobenzoic acid) (MFA), meclofenamic (2-(2,6-dichloro-3-methylanilino)benzoic), and niflumic (2-{[3-(trifluoromethyl)phenyl]amino}pyridine-3-carboxylic) acids (Table 1), non-steroidal anti-inflammatory drugs structurally related to flufenamic acid, are potent inhibitors of TRPC6 and some other closely related ion channels [151]. These compounds have been shown to attenuate glutamate-evoked excitotoxicity in cultured rat embryonic hippocampal neurons similarly to flufenamic acid [119,124]. In a 3xTgAD mouse AD model, MFA ameliorates cognitive impairments [125].

Maier et al. discovered a novel TRPC3/6/7 inhibitor, the aminoindanol derivative 4-[[(1R,2R)-2-[(3R)-3-Amino-1-piperidinyl]-2,3-dihydro-1H-inden-1-yl]oxy]-3-chlorobenzonitrile dihydrochloride (SAR7334, Table 1), with a higher selectivity towards TRPC6 (IC_{50} of 7.9 nM, as indicated by patch-clamp data) [71]. Hou et al. found that TRPC6 knockout or inhibition by SAR7334 mitigates oxidative stress-induced apoptosis of renal proximal tubular cells, which is considered to play a major role in renal ischemia/reperfusion [73]. SAR7334 and the tryptoline derivative 8009-5364 are reported to diminish acute hypoxia-induced pulmonary vasoconstriction and pulmonary arterial pressure in isolated mouse lungs [70,71]. SAR7334 has no effect on SOCE in primary cortical neurons [126].

Some rare-earth metal ions, including La3+ and Gd3+, have been shown to inhibit TRPC6 activity [106,127,133]. But since those ions inhibit all channels of the TRP family (except TRPM2), they cannot be used as selective TRPC6 antagonists. Clotrimazole (1-[(2-chlorophenyl)diphenyl-methyl]-1H-imidazole) is an imidazole compound, which inhibits TRPM2, TRPM3, TRPV4, and TRPC6 channels [90]. In SOCE mechanism studies, 2-aminoethoxydiphenyl borate (2-APB) [90] and 1-[2-[3-(4-methoxyphenyl)propoxy]-2-(4-methoxyphenyl)ethyl]-1H-imidazole hydrochloride

(SKF-96365) [162] are identified as agents targeting TRPC6. 2-APB blocks TRPC6 [163] and was later found to impact Orai channel functioning [164]. 2-APB has also been shown to block Ca^{2+} influx induced by acetylcholine or thapsigargin application but not by DAG [163]. SKF-96365 is considered to be an inhibitor of receptor- and store-operated elevation of intracellular calcium levels via entry through voltage-independent channels [162,165]. Studies have demonstrated successful blocking of TRPC6 using SKF-96365 (IC_{50} = 2 μM), but a less pronounced effect on TRPC3 is also observed (IC_{50} = 12 μM) [90]. 8009-5364 is another highly specific TRPC6 antagonist, which has an IC_{50} of 3.2 μM, and is considered a promising agent for the treatment of pulmonary hypertension [70]. Investigation of neuronal SOCE mechanism in striatal neurons has revealed 4-N-[2-(4-phenoxyphenyl)ethyl]quinazoline-4,6-diamine (EVP4593, Table 1) (IC_{50} = 300 nM [126]) to be an inhibitor of TRPC1 channels [166]. EVP4593 has also been found to target heteromeric but not homomeric TRCP1 channels [166]. Later on, EVP4593 was demonstrated to block Orai channels at 300 nM concentration [126]. In cultured hippocampal neurons exhibiting PSEN1ΔE9 mutation, TRPC6 hyperactivation is blocked by 30 nM EVP4593 [24]. EVP4593 has been shown to inhibit the nuclear factor kappa-b (NF-Kb) [167]. It has been also observed that NF-Kb downregulates TRPC6 protein expression [168]; thus, EVP4593-mediated inhibition of the NF-Kb pathway might cause an increase in TRPC6 protein expression. Whether this increase in TRPC6 protein expression would compete with EVP4593-mediated blockade of TRPC6-dependent SOCE is an open question.

8. Conclusions

The present review summarizes current knowledge on the role of TRPC6 channels in the development of two neurological disorders: Alzheimer's disease and cerebral ischemia. Cerebral ischemia serves as a risk factor for AD, and vice versa. It is becoming evident that both diseases can be caused by either upregulation or downregulation of TRPC6 channels. Thus, understanding the nature of the disruption of this molecular pathway in each particular patient is extremely important for appropriate drug prescription.

TRPC6 is structurally similar to TRPC3 and 7, and therefore the majority of compounds do not act selectively on either of these three isoforms, making it difficult to develop selective TRPC6 pharmacological modulators. Moreover, a number of TRPC6 channel modulators are cross-reactive to other cellular targets, thus limiting their pharmacological potential. Toxicity profiles of some known TRPC6 modulators (such as FFA) require further structural optimization. The presence of a wide range of different chemical substances that are known to interact with TRPC6 channels as well as the availability of cryo-electron microscopy structures of TRPC6 and 3 [169] may allow determining the pharmacophore in order to design selective TRPC6 activators and inhibitors in the future. In turn, the development of selective TRPC6 channel modulators could help slow down the progression of AD, cerebral ischemia, and, most likely, other TRPC6-dependent diseases.

Author Contributions: D.C., V.P., Y.S., N.Z., and E.P. wrote the review; E.P. and S.O. edited the review. All authors have read and agreed to the published version of the manuscript.

Funding: This work was funded by the Russian Science Foundation grant No. 20-75-10026 (Chapters 1–5 and 7, 8) and in part by the Saint-Petersburg State University project No. 51134206 (Chapter 6).

Acknowledgments: Authors thank Anastasiya Bolshakova for administrative support.

Conflicts of Interest: The authors declare no conflict of interest.

References

1. Forabosco, P.; Ramasamy, A.; Trabzuni, D.; Walker, R.; Smith, C.; Bras, J.; Levine, A.P.; Hardy, J.; Pocock, J.M.; Guerreiro, R.; et al. Insights into TREM2 biology by network analysis of human brain gene expression data. *Neurobiol. Aging* **2013**, *34*, 2699–2714. [CrossRef] [PubMed]
2. Dirnagl, U.; Iadecola, C.; Moskowitz, M.A. Pathobiology of ischaemic stroke: An integrated view. *Trends Neurosci.* **1999**, *22*, 391–397. [CrossRef]

3. Lipton, P. Ischemic cell death in brain neurons. *Physiol. Rev.* **1999**, *79*, 1431–1568. [CrossRef] [PubMed]

4. Zhou, J.; Yu, J.; Wang, H.; Meng, X.; Tan, C.; Wang, J.; Wang, C.; Tan, L. Association between Stroke and Alzheimer's Disease: Systematic Review and Meta-Analysis. *J. Alzheimers Dis.* **2015**, *43*, 479–489. [CrossRef]

5. Chi, N.-F.; Chien, L.-N.; Ku, H.-L.; Hu, C.-J.; Chiou, H.-Y. Alzheimer disease and risk of stroke: A population-based cohort study. *Neurology* **2013**, *80*, 705–711. [CrossRef]

6. Tolppanen, A.; Lavikainen, P.; Solomon, A.; Kivipelto, M.; Esoininen, H.; Hartikainen, S. Incidence of stroke in people with Alzheimer disease: A national register-based approach. *Neurology* **2013**, *80*, 353–358. [CrossRef]

7. Wang, R.; Reddy, P.H. Role of Glutamate and NMDA Receptors in Alzheimer's Disease. *J. Alzheimers Dis.* **2017**, *57*, 1041–1048. [CrossRef]

8. Chen, M.; Lu, T.-J.; Chen, X.-J.; Zhou, Y.; Chen, Q.; Feng, X.-Y.; Xu, L.; Duan, W.-H.; Xiong, Z.-Q. Differential roles of NMDA receptor subtypes in ischemic neuronal cell death and ischemic tolerance. *Stroke* **2008**, *39*, 3042–3048. [CrossRef]

9. Grossberg, G.T.; Thomas, S.J. Memantine: A review of studies into its safety and efficacy in treating Alzheimer's disease and other dementias. *Clin. Interv. Aging* **2009**, *4*, 367–377. [CrossRef]

10. Schneider, L.S.; Dagerman, K.S.; Higgins, J.P.; McShane, R. Lack of Evidence for the Efficacy of Memantine in Mild Alzheimer Disease. *Arch. Neurol.* **2011**, *68*, 991. [CrossRef]

11. Hoyte, L.; Barber, P.A.; Buchan, A.M.; Hill, M.D. The rise and fall of NMDA antagonists for ischemic stroke. *Curr. Mol. Med.* **2004**, *4*, 131–136. [CrossRef] [PubMed]

12. Ginsberg, M.D. Neuroprotection for ischemic stroke: Past, present and future. *Neuropharmacology* **2008**, *55*, 363–389. [CrossRef] [PubMed]

13. Zhou, J.; Du, W.; Zhou, K.; Tai, Y.; Yao, H.; Jia, Y.; Ding, Y.; Wang, Y. Critical role of TRPC6 channels in the formation of excitatory synapses. *Nat. Neurosci.* **2008**, *11*, 741–743. [CrossRef]

14. Zhang, H.; Sun, S.; Wu, L.; Pchitskaya, E.; Zakharova, O.; Tacer, K.F.; Bezprozvanny, I. Store-Operated Calcium Channel Complex in Postsynaptic Spines: A New Therapeutic Target for Alzheimer's Disease Treatment. *J. Neurosci.* **2016**, *36*, 11837–11850. [CrossRef] [PubMed]

15. Shen, H.; Pan, J.; Pan, L.; Zhang, N. TRPC6 inhibited NMDA current in cultured hippocampal neurons. *Neuromol. Med.* **2013**, *15*, 389–395. [CrossRef]

16. Li, H.; Huang, J.; Du, W.; Jia, C.; Yao, H.; Wang, Y. TRPC6 inhibited NMDA receptor activities and protected neurons from ischemic excitotoxicity. *J. Neurochem.* **2012**, *123*, 1010–1018. [CrossRef]

17. Tai, Y.; Feng, S.; Ge, R.; Du, W.; Zhang, X.; He, Z.; Wang, Y. TRPC6 channels promote dendritic growth via the CaMKIV-CREB pathway. *J. Cell Sci.* **2008**, *121*, 2301–2307. [CrossRef]

18. Guilbert, A.; Dhennin-Duthille, I.; El Hiani, Y.; Haren, N.; Khorsi, H.; Sevestre, H.; Ahidouch, A.; Ouadid-Ahidouch, H. Expression of TRPC6 channels in human epithelial breast cancer cells. *BMC Cancer* **2008**, *8*, 125. [CrossRef]

19. Ramirez, G.A.; Coletto, L.A.; Sciorati, C.; Bozzolo, E.P.; Manunta, P.; Rovere-Querini, P.; Manfredi, A.A. Ion Channels and Transporters in Inflammation: Special Focus on TRP Channels and TRPC6. *Cells* **2018**, *7*, 70. [CrossRef]

20. Santín, S.; Ars, E.; Rossetti, S.; Salido, E.; Silva, I.; García-Maset, R.; Giménez, I.; Ruíz, P.; Mendizábal, S.; Nieto, J.L.; et al. TRPC6 mutational analysis in a large cohort of patients with focal segmental glomerulosclerosis. *Nephrol. Dial. Transplant.* **2009**, *24*, 3089–3096. [CrossRef]

21. Popugaeva, E.; Chernyuk, D.; Zhang, H.; Postnikova, T.Y.; Pats, K.; Fedorova, E.; Poroikov, V.; Zaitsev, A.V.; Bezprozvanny, I.; Cherniuk, D.; et al. Derivatives of Piperazines as Potential Therapeutic Agents for Alzheimer's Disease. *Mol. Pharmacol.* **2019**, *95*, 337–348. [CrossRef] [PubMed]

22. Lessard, C.B.; Lussier, M.P.; Cayouette, S.; Bourque, G.; Boulay, G. The overexpression of presenilin2 and Alzheimer's-disease-linked presenilin2 variants influences TRPC6-enhanced Ca^{2+} entry into HEK293 cells. *Cell. Signal.* **2005**, *17*, 437–445. [CrossRef]

23. Liu, L.; Gu, L.; Chen, M.; Zheng, Y.; Xiong, X.; Zhu, S. Novel Targets for Stroke Therapy: Special Focus on TRPC Channels and TRPC6. *Front. Aging Neurosci.* **2020**, *12*, 70. [CrossRef]

24. Chernyuk, D.; Zernov, N.; Kabirova, M.; Bezprozvanny, I.; Popugaeva, E. Antagonist of neuronal store-operated calcium entry exerts beneficial effects in neurons expressing PSEN1ΔE9 mutant linked to familial Alzheimer disease. *Neuroscience* **2019**, *410*, 118–127. [CrossRef] [PubMed]

25. Chen, J.; Li, Z.; Hatcher, J.T.; Chen, Q.-H.; Chen, L.; Wurster, R.D.; Chan, S.L.; Cheng, Z. Deletion of TRPC6 Attenuates NMDA Receptor-Mediated Ca(2+) Entry and Ca(2+)-Induced Neurotoxicity Following Cerebral Ischemia and Oxygen-Glucose Deprivation. *Front. Neurosci.* **2017**, *11*, 138. [CrossRef] [PubMed]

26. Chen, X.; Lu, M.; He, X.; Ma, L.; Birnbaumer, L.; Liao, Y. TRPC3/6/7 Knockdown Protects the Brain from Cerebral Ischemia Injury via Astrocyte Apoptosis Inhibition and Effects on NF-small ka, CyrillicB Translocation. *Mol. Neurobiol.* **2017**, *54*, 7555–7566. [CrossRef]

27. Wang, J.; Lu, R.; Yang, J.; Li, H.; He, Z.; Jing, N.; Wang, X.; Wang, Y. TRPC6 specifically interacts with APP to inhibit its cleavage by gamma-secretase and reduce Abeta production. *Nat. Commun.* **2015**, *6*, 8876. [CrossRef] [PubMed]

28. Venkatachalam, K.; Montell, C. TRP channels. *Annu. Rev. Biochem.* **2007**, *76*, 387–417. [CrossRef]

29. Sun, Y.; Sukumaran, P.; Bandyopadhyay, B.C.; Singh, B.B. Physiological Function and Characterization of TRPCs in Neurons. *Cells* **2014**, *3*, 455–475. [CrossRef]

30. Montell, C. The TRP superfamily of cation channels. *Sci. Signal.* **2005**, *2005*, re3. [CrossRef]

31. Dietrich, A.; Gudermann, T. TRPC6: Physiological function and pathophysiological relevance. *Handb. Exp. Pharmacol.* **2014**, *222*, 157–188. [PubMed]

32. Hofmann, T.; Obukhov, A.G.; Schaefer, M.; Harteneck, C.; Gudermann, T.; Schultz, G. Direct activation of human TRPC6 and TRPC3 channels by diacylglycerol. *Nature* **1999**, *397*, 259–263. [CrossRef] [PubMed]

33. Prakriya, M.; Lewis, R.S. Store-Operated Calcium Channels. *Physiol. Rev.* **2015**, *95*, 1383–1436. [CrossRef]

34. Roos, J.; Digregorio, P.J.; Yeromin, A.V.; Ohlsen, K.; Lioudyno, M.; Zhang, S.; Safrina, O.; Kozak, J.A.; Wagner, S.L.; Cahalan, M.D.; et al. STIM1, an essential and conserved component of store-operated Ca^{2+} channel function. *J. Cell Biol.* **2005**, *169*, 435–445. [CrossRef]

35. Zhang, S.L.; Yu, Y.; Roos, J.; Kozak, J.A.; Deerinck, T.J.; Ellisman, M.H.; Stauderman, K.A.; Cahalan, M.D. STIM1 is a Ca^{2+} sensor that activates CRAC channels and migrates from the Ca^{2+} store to the plasma membrane. *Nature* **2005**, *437*, 902–905. [CrossRef] [PubMed]

36. Liou, J.; Kim, M.L.; Heo, W.D.; Jones, J.T.; Myers, J.W.; Ferrell, J.E.; Meyer, T. STIM is a Ca^{2+} sensor essential for Ca^{2+}-store-depletion-triggered Ca^{2+} influx. *Curr. Biol.* **2005**, *15*, 1235–1241. [CrossRef]

37. Feske, S.; Gwack, Y.; Prakriya, M.; Srikanth, S.; Puppel, S.-H.; Tanasa, B.; Hogan, P.G.; Lewis, R.S.; Daly, M.J.; Rao, A. A mutation in Orai1 causes immune deficiency by abrogating CRAC channel function. *Nature* **2006**, *441*, 179–185. [CrossRef]

38. Sun, S.; Zhang, H.; Liu, J.; Popugaeva, E.; Xu, N.-J.; Feske, S.; White, C.L.; Bezprozvanny, I. Reduced synaptic STIM2 expression and impaired store-operated calcium entry cause destabilization of mature spines in mutant presenilin mice. *Neuron* **2014**, *82*, 79–93. [CrossRef]

39. Popugaeva, E.; Pchitskaya, E.; Speshilova, A.B.; Alexandrov, S.; Zhang, H.; Vlasova, O.L.; Bezprozvanny, I. STIM2 protects hippocampal mushroom spines from amyloid synaptotoxicity. *Mol. Neurodegener.* **2015**, *10*, 37. [CrossRef]

40. Qu, Z.; Wang, Y.; Li, X.; Wu, L.; Wang, Y. TRPC6 expression in neurons is differentially regulated by NR2A- and NR2B-containing NMDA receptors. *J. Neurochem.* **2017**, *143*, 282–293. [CrossRef]

41. Tu, H.; Nelson, O.; Bezprozvanny, A.; Wang, Z.; Lee, S.; Hao, Y.; Serneels, L.; de Strooper, B.; Yu, G.; Bezprozvanny, I. Presenilins form ER calcium leak channels, a function disrupted by mutations linked to familial Alzheimer's disease. *Cell* **2006**, *126*, 981–993. [CrossRef] [PubMed]

42. Zatti, G.; Burgo, A.; Giacomello, M.; Barbiero, L.; Ghidoni, R.; Sinigaglia, G.; Florean, C.; Bagnoli, S.; Binetti, G.; Sorbi, S.; et al. Presenilin mutations linked to familial Alzheimer's disease reduce endoplasmic reticulum and Golgi apparatus calcium levels. *Cell Calcium* **2006**, *39*, 539–550. [CrossRef] [PubMed]

43. Zhang, H.; Sun, S.; Herreman, A.; De Strooper, B.; Bezprozvanny, I. Role of presenilins in neuronal calcium homeostasis. *J. Neurosci.* **2010**, *30*, 8566–8580. [CrossRef] [PubMed]

44. Zhang, H.; Liu, J.; Sun, S.; Pchitskaya, E.; Popugaeva, E.; Bezprozvanny, I. Calcium signaling, excitability, and synaptic plasticity defects in a mouse model of Alzheimer's disease. *J. Alzheimers Dis.* **2015**, *45*, 561–580. [CrossRef]

45. Tong, B.C.; Lee, C.S.; Cheng, W.; Lai, K.; Foskett, J.K.; Cheung, K. Familial Alzheimer's disease-associated presenilin 1 mutants promote gamma-secretase cleavage of STIM1 to impair store-operated Ca^{2+} entry. *Sci. Signal.* **2016**, *9*, ra89. [CrossRef]

46. Bojarski, L.; Pomorski, P.; Szybinska, A.; Drab, M.; Skibinska-Kijek, A.; Gruszczynska-Biegala, J.; Kuznicki, J. Presenilin-dependent expression of STIM proteins and dysregulation of capacitative Ca^{2+} entry in familial Alzheimer's disease. *Biochim. Biophys. Acta* **2009**, *1793*, 1050–1057. [CrossRef]

47. Popugaeva, E.; Pchitskaya, E.; Bezprozvanny, I. Dysregulation of neuronal calcium homeostasis in Alzheimer's disease—A therapeutic opportunity? *Biochem. Biophys. Res. Commun.* **2017**, *483*, 998–1004. [CrossRef]

48. Pchitskaya, E.; Popugaeva, E.; Bezprozvanny, I. Calcium signaling and molecular mechanisms underlying neurodegenerative diseases. *Cell Calcium* **2018**, *70*, 87–94. [CrossRef]

49. Popugaeva, E.; Chernyuk, D.; Bezprozvanny, I. Correction of calcium dysregulation as potential approach for treating Alzheimer's disease. *Curr. Alzheimer Res.* **2019**, in press.

50. Ryazantseva, M.A.; Goncharova, A.; Skobeleva, K.; Erokhin, M.; Methner, A.; Georgiev, P.; Kaznacheyeva, E.V. Presenilin-1 Delta E9 Mutant Induces STIM1-Driven Store-Operated Calcium Channel Hyperactivation in Hippocampal Neurons. *Mol. Neurobiol.* **2017**, *55*, 4667–4680. [CrossRef]

51. Yoo, A.S.; Cheng, I.; Chung, S.; Grenfell, T.Z.; Lee, H.; Pack-Chung, E.; Handler, M.; Shen, J.; Xia, W.; Tesco, G.; et al. Presenilin-mediated modulation of capacitative calcium entry. *Neuron* **2000**, *27*, 561–572. [CrossRef]

52. Zhang, X.; Le, W. Pathological role of hypoxia in Alzheimer's disease. *Exp. Neurol.* **2010**, *223*, 299–303. [CrossRef]

53. Vijayan, M.; Reddy, P.H. Stroke, Vascular Dementia, and Alzheimer's Disease: Molecular Links. *J. Alzheimers Dis.* **2016**, *54*, 427–443. [CrossRef] [PubMed]

54. Olichney, J.M.; Hansen, L.A.; Galasko, D.; Saitoh, T.; Hofstetter, C.R.; Katzman, R.; Thal, L.J. The apolipoprotein E epsilon 4 allele is associated with increased neuritic plaques and cerebral amyloid angiopathy in Alzheimer's disease and Lewy body variant. *Neurology* **1996**, *47*, 190–196. [CrossRef] [PubMed]

55. Premkumar, D.R.; Cohen, D.L.; Hedera, P.; Friedland, R.P.; Kalaria, R.N. Apolipoprotein E-epsilon4 alleles in cerebral amyloid angiopathy and cerebrovascular pathology associated with Alzheimer's disease. *Am. J. Pathol.* **1996**, *148*, 2083–2095.

56. Xing, C.; Arai, K.; Lo, E.H.; Hommel, M. Pathophysiologic cascades in ischemic stroke. *Int. J. Stroke* **2012**, *7*, 378–385. [CrossRef] [PubMed]

57. Szydlowska, K.; Tymianski, M. Calcium, ischemia and excitotoxicity. *Cell Calcium* **2010**, *47*, 122–129. [CrossRef]

58. Wang, X.; Teng, L.; Li, A.; Ge, J.; Laties, A.M.; Zhang, X. TRPC6 channel protects retinal ganglion cells in a rat model of retinal ischemia/reperfusion-induced cell death. *Investig. Opthalmol. Vis. Sci.* **2010**, *51*, 5751–5758. [CrossRef]

59. Nakuluri, K.; Nishad, R.; Mukhi, D.; Kumar, S.; Nakka, V.P.; Kolligundla, L.P.; Narne, P.; Natuva, S.S.K.; Phanithi, P.B.; Pasupulati, A.K. Cerebral ischemia induces TRPC6 via HIF1alpha/ZEB2 axis in the glomerular podocytes and contributes to proteinuria. *Sci. Rep.* **2019**, *9*, 17897. [CrossRef]

60. Jia, Y.; Zhou, J.; Tai, Y.; Wang, Y. TRPC channels promote cerebellar granule neuron survival. *Nat. Neurosci.* **2007**, *10*, 559–567. [CrossRef]

61. Li, Y.; Jia, Y.; Cui, K.; Li, N.; Zheng, Z.-Y.; Wang, Y.; Yuan, X.-B. Essential role of TRPC channels in the guidance of nerve growth cones by brain-derived neurotrophic factor. *Nature* **2005**, *434*, 894–898. [CrossRef] [PubMed]

62. Caracciolo, L.; Marosi, M.; Mazzitelli, J.; Latifi, S.; Sano, Y.; Galvan, L.; Kawaguchi, R.; Holley, S.; Levine, M.S.; Coppola, G.; et al. CREB controls cortical circuit plasticity and functional recovery after stroke. *Nat. Commun.* **2018**, *9*, 2250. [CrossRef] [PubMed]

63. McCullough, L.D.; Tarabishy, S.; Liu, L.; Benashski, S.; Xu, Y.; Ribar, T.; Means, A.; Li, J. Inhibition of calcium/calmodulin-dependent protein kinase kinase beta and calcium/calmodulin-dependent protein kinase IV is detrimental in cerebral ischemia. *Stroke* **2013**, *44*, 2559–2566. [CrossRef] [PubMed]

64. Du, W.; Huang, J.; Yao, H.; Zhou, K.; Duan, B.; Wang, Y. Inhibition of TRPC6 degradation suppresses ischemic brain damage in rats. *J. Clin. Investig.* **2010**, *120*, 3480–3492. [CrossRef]

65. Fu, J.; Xue, R.; Gu, J.; Xiao, Y.; Zhong, H.; Pan, X.; Ran, R. Neuroprotective effect of calcitriol on ischemic/reperfusion injury through the NR3A/CREB pathways in the rat hippocampus. *Mol. Med. Rep.* **2013**, *8*, 1708–1714. [CrossRef]

66. Lin, Y.; Chen, F.; Zhang, J.; Wang, T.; Wei, X.; Wu, J.; Feng, Y.; Dai, Z.; Wu, Q. Neuroprotective effect of resveratrol on ischemia/reperfusion injury in rats through TRPC6/CREB pathways. *J. Mol. Neurosci.* **2013**, *50*, 504–513. [CrossRef]

67. Omura, T.; Tanaka, Y.; Miyata, N.; Koizumi, C.; Sakurai, T.; Fukasawa, M.; Hachiuma, K.; Minagawa, T.; Susumu, T.; Yoshida, S.; et al. Effect of a new inhibitor of the synthesis of 20-HETE on cerebral ischemia reperfusion injury. *Stroke* **2006**, *37*, 1307–1313. [CrossRef]

68. Renic, M.; Klaus, J.A.; Omura, T.; Kawashima, N.; Onishi, M.; Miyata, N.; Koehler, R.C.; Harder, D.R.; Roman, R.J. Effect of 20-HETE inhibition on infarct volume and cerebral blood flow after transient middle cerebral artery occlusion. *Br. J. Pharmacol.* **2008**, *29*, 629–639. [CrossRef]

69. Shaik, J.S.B.; Poloyac, S.M.; Kochanek, P.M.; Alexander, H.; Tudorascu, D.L.; Clark, R.S.; Manole, M.D. 20-Hydroxyeicosatetraenoic Acid Inhibition by HET0016 Offers Neuroprotection, Decreases Edema, and Increases Cortical Cerebral Blood Flow in a Pediatric Asphyxial Cardiac Arrest Model in Rats. *Br. J. Pharmacol.* **2015**, *35*, 1757–1763. [CrossRef]

70. Urban, N.; Hill, K.; Wang, L.; Kuebler, W.M.; Schaefer, M. Novel pharmacological TRPC inhibitors block hypoxia-induced vasoconstriction. *Cell Calcium* **2012**, *51*, 194–206. [CrossRef]

71. Maier, T.; Follmann, M.; Hessler, G.; Kleemann, H.-W.; Hachtel, S.; Fuchs, B.; Weissmann, N.; Linz, W.; Schmidt, T.; Löhn, M.; et al. Discovery and pharmacological characterization of a novel potent inhibitor of diacylglycerol-sensitive TRPC cation channels. *Br. J. Pharmacol.* **2015**, *172*, 3650–3660. [CrossRef] [PubMed]

72. Urban, N.; Wang, L.; Kwiek, S.; Rademann, J.; Kuebler, W.M.; Schaefer, M. Identification and Validation of Larixyl Acetate as a Potent TRPC6 Inhibitor. *Mol. Pharmacol.* **2015**, *89*, 197–213. [CrossRef] [PubMed]

73. Hou, X.; Xiao, H.; Zhang, Y.; Zeng, X.; Huang, M.; Chen, X.; Birnbaumer, L.; Liao, Y. Transient receptor potential channel 6 knockdown prevents apoptosis of renal tubular epithelial cells upon oxidative stress via autophagy activation. *Cell Death Dis.* **2018**, *9*, 1015. [CrossRef] [PubMed]

74. Chen, X.; Taylor-Nguyen, N.N.; Riley, A.M.; Herring, B.P.; White, F.A.; Obukhov, A.G. The TRPC6 inhibitor, larixyl acetate, is effective in protecting against traumatic brain injury-induced systemic endothelial dysfunction. *J. Neuroinflamm.* **2019**, *16*, 21. [CrossRef]

75. Cross, J.L.; Meloni, B.P.; Bakker, A.J.; Lee, S.; Knuckey, N.W. Modes of Neuronal Calcium Entry and Homeostasis following Cerebral Ischemia. *Stroke Res. Treat.* **2010**, *2010*, 316862. [CrossRef]

76. Wu, Q.J.; Tymianski, M. Targeting NMDA receptors in stroke: New hope in neuroprotection. *Mol. Brain* **2018**, *11*, 15. [CrossRef]

77. Basora, N.; Boulay, G.; Bilodeau, L.; Payet, M.D.; Rousseau, E. 20-hydroxyeicosatetraenoic acid (20-HETE) activates mouse TRPC6 channels expressed in HEK293 cells. *J. Biol. Chem.* **2003**, *278*, 31709–31716. [CrossRef]

78. Aires, V.; Hichami, A.; Boulay, G.; Khan, N.A. Activation of TRPC6 calcium channels by diacylglycerol (DAG)-containing arachidonic acid: A comparative study with DAG-containing docosahexaenoic acid. *Biochim.* **2007**, *89*, 926–937. [CrossRef]

79. Belayev, L.; Khoutorova, L.; Atkins, K.D.; Eady, T.N.; Hong, S.; Lu, Y.; Obenaus, A.; Bazan, N.G. Docosahexaenoic Acid therapy of experimental ischemic stroke. *Transl. Stroke Res.* **2010**, *2*, 33–41. [CrossRef]

80. Yao, C.; Zhang, J.; Chen, F.; Lin, Y. Neuroprotectin D1 attenuates brain damage induced by transient middle cerebral artery occlusion in rats through TRPC6/CREB pathways. *Mol. Med. Rep.* **2013**, *8*, 543–550. [CrossRef]

81. Guinamard, R.; Simard, C.; Del Negro, C. Flufenamic acid as an ion channel modulator. *Pharmacol. Ther.* **2013**, *138*, 272–284. [CrossRef] [PubMed]

82. Qu, C.; Ding, M.; Zhu, Y.; Lu, Y.; Du, J.; Miller, M.; Tian, J.; Zhu, J.; Xu, J.; Wen, M.; et al. Pyrazolopyrimidines as Potent Stimulators for Transient Receptor Potential Canonical 3/6/7 Channels. *J. Med. Chem.* **2017**, *60*, 4680–4692. [CrossRef] [PubMed]

83. Sawamura, S.; Hatano, M.; Takada, Y.; Hino, K.; Kawamura, T.; Tanikawa, J.; Nakagawa, H.; Hase, H.; Nakao, A.; Hirano, M.; et al. Screening of Transient Receptor Potential Canonical Channel Activators Identifies Novel Neurotrophic Piperazine Compounds. *Mol. Pharmacol.* **2016**, *89*, 348–363. [CrossRef] [PubMed]

84. Tiapko, O.; Shrestha, N.; Lindinger, S.; De La Cruz, G.G.; Graziani, A.; Klec, C.; Butorac, C.; Graier, W.F.; Kubista, H.; Freichel, M.; et al. Lipid-independent control of endothelial and neuronal TRPC3 channels by light. *Chem. Sci.* **2019**, *10*, 2837–2842. [CrossRef]

85. Häfner, S.; Urban, N.; Schaefer, M. Discovery and characterization of a positive allosteric modulator of transient receptor potential canonical 6 (TRPC6) channels. *Cell Calcium* **2019**, *78*, 26–34. [CrossRef]

86. Guo, C.; Tong, L.; Xi, M.; Yang, H.; Dong, H.; Wen, A. Neuroprotective effect of calycosin on cerebral ischemia and reperfusion injury in rats. *J. Ethnopharmacol.* **2012**, *144*, 768–774. [CrossRef]

87. Yao, C.; Zhang, J.; Liu, G.; Chen, F.; Lin, Y. Neuroprotection by (−)-epigallocatechin-3-gallate in a rat model of stroke is mediated through inhibition of endoplasmic reticulum stress. *Mol. Med. Rep.* **2014**, *9*, 69–76. [CrossRef]

88. Sysoev, Y.I.; Uzuegbunam, B.C.; Okovityi, S.V. Attenuation of neurological deficit by a novel ethanolamine derivative in rats after brain trauma. *J. Exp. Pharmacol.* **2019**, *11*, 53–63. [CrossRef]

89. Sysoev, Y.I.; Popugaeva, E.A.; Chernyuk, D.P.; Titovich, I.A.; Zagladkina, E.V.; Bolotova, V.C.; Bezprozvanny, I.; Okovityi, S.V. Mechanism of action of the new ethanolamine derivative bis{2-[(2E)-4-hydroxy-4-oxobut-2-enoyloxy]-N,N-diethylethanaminium}butanedioate. *Eksperimental'naya Klin. Farmakol.* **2019**, *82*, 3–10.

90. Harteneck, C.; Klose, C.; Krautwurst, D. Synthetic modulators of TRP channel activity. *Adv. Exp. Med. Biol.* **2011**, *704*, 87–106.

91. Tu, P.; Kunert-Keil, C.; Lucke, S.; Brinkmeier, H.; Bouron, A. Diacylglycerol analogues activate second messenger-operated calcium channels exhibiting TRPC-like properties in cortical neurons. *J. Neurochem.* **2009**, *108*, 126–138. [CrossRef] [PubMed]

92. Montecinos-Oliva, C.; Schuller, A.; Parodi, J.; Melo, F.; Inestrosa, N.C. Effects of tetrahydrohyperforin in mouse hippocampal slices: Neuroprotection, long-term potentiation and TRPC channels. *Curr. Med. Chem.* **2014**, *21*, 3494–3506. [CrossRef] [PubMed]

93. Law, S.-H.; Chan, M.-L.; Marathe, G.K.; Parveen, F.; Chen, C.-H.; Ke, L.-Y. An Updated Review of Lysophosphatidylcholine Metabolism in Human Diseases. *Int. J. Mol. Sci.* **2019**, *20*, 1149. [CrossRef] [PubMed]

94. Koizumi, S.; Yamamoto, S.; Hayasaka, T.; Konishi, Y.; Yamaguchi-Okada, M.; Goto-Inoue, N.; Sugiura, Y.; Setou, M.; Namba, H. Imaging mass spectrometry revealed the production of lyso-phosphatidylcholine in the injured ischemic rat brain. *Neuroscience* **2010**, *168*, 219–225. [CrossRef]

95. Sabogal-Guáqueta, A.M.; Posada-Duque, R.; Cortes, N.C.; Arias-Londoño, J.D.; Cardona-Gómez, G.P. Changes in the hippocampal and peripheral phospholipid profiles are associated with neurodegeneration hallmarks in a long-term global cerebral ischemia model: Attenuation by Linalool. *Neuropharmacology* **2018**, *135*, 555–571. [CrossRef]

96. Jickling, G.C.; Montaner, J. Lysophosphatidylcholine to stratify risk of ischemic stroke in TIA. *Neurology* **2014**, *84*, 17–18. [CrossRef]

97. Chaudhuri, P.; Colles, S.M.; Damron, D.S.; Graham, L.M. Lysophosphatidylcholine inhibits endothelial cell migration by increasing intracellular calcium and activating calpain. *Arter. Thromb. Vasc. Biol.* **2003**, *23*, 218–223. [CrossRef]

98. Chaudhuri, P.; Rosenbaum, M.A.; Birnbaumer, L.; Graham, L.M. Integration of TRPC6 and NADPH oxidase activation in lysophosphatidylcholine-induced TRPC5 externalization. *Am. J. Physiol. Physiol.* **2017**, *313*, C541–C555. [CrossRef]

99. Cloutier, M.; Campbell, S.; Basora, N.; Proteau, S.; Payet, M.D.; Rousseau, E. 20-HETE inotropic effects involve the activation of a nonselective cationic current in airway smooth muscle. *Am. J. Physiol. Cell. Mol. Physiol.* **2003**, *285*, L560–L568. [CrossRef]

100. Lu, L.; Wang, M.; Wei, X.; Li, W. Corrigendum: 20-HETE Inhibition by HET0016 Decreases the Blood-Brain Barrier Permeability and Brain Edema After Traumatic Brain Injury. *Front. Aging Neurosci.* **2018**, *10*, 207. [CrossRef]

101. Wen, H.; Östman, J.; Bubb, K.J.; Panayiotou, C.; Priestley, J.V.; Baker, M.D.; Ahluwalia, A. 20-Hydroxyeicosatetraenoic acid (20-HETE) is a novel activator of transient receptor potential vanilloid 1 (TRPV1) channel. *J. Biol. Chem.* **2012**, *287*, 13868–13876. [CrossRef] [PubMed]

102. Asatryan, A.; Bazan, N.G. Molecular mechanisms of signaling via the docosanoid neuroprotectin D1 for cellular homeostasis and neuroprotection. *J. Biol. Chem.* **2017**, *292*, 12390–12397. [CrossRef] [PubMed]

103. Yao, H.; Zhang, Y.; Shu, H.; Xie, B.; Tao, Y.; Yuan, Y.; Shang, Y.; Yuan, S.; Zhang, J. Hyperforin Promotes Post-stroke Neuroangiogenesis via Astrocytic IL-6-Mediated Negative Immune Regulation in the Ischemic Brain. *Front. Cell. Neurosci.* **2019**, *13*, 201. [CrossRef] [PubMed]

104. Kumar, V.; Mdzinarishvili, A.; Kiewert, C.; Abbruscato, T.; Bickel, U.; Van Der Schyf, C.J.; Klein, J. NMDA receptor-antagonistic properties of hyperforin, a constituent of St. John's Wort. *J. Pharmacol. Sci.* **2006**, *102*, 47–54. [CrossRef]

105. Dinamarca, M.C.; Cerpa, W.; Garrido, J.; Hancke, J.L.; Inestrosa, N.C. Hyperforin prevents beta-amyloid neurotoxicity and spatial memory impairments by disaggregation of Alzheimer's amyloid-beta-deposits. *Mol. Psychiatry* **2006**, *11*, 1032–1048. [CrossRef]

106. Leuner, K.; Li, W.; Amaral, M.D.; Rudolph, S.; Calfa, G.; Schuwald, A.M.; Harteneck, C.; Inoue, T.; Pozzo-Miller, L. Hyperforin modulates dendritic spine morphology in hippocampal pyramidal neurons by activating Ca(2+) -permeable TRPC6 channels. *Hippocampus* **2012**, *23*, 40–52. [CrossRef]

107. Zhang, Y.; Yu, P.; Liu, H.; Yao, H.; Yao, S.; Yuan, S.; Zhang, J.-C. Hyperforin improves post-stroke social isolation-induced exaggeration of PSD and PSA via TGF-β. *Int. J. Mol. Med.* **2018**, *43*, 413–425. [CrossRef]

108. Ma, L.; Pan, X.; Zhou, F.; Liu, K.; Wang, L. Hyperforin protects against acute cerebral ischemic injury through inhibition of interleukin-17A-mediated microglial activation. *Brain Res.* **2018**, *1678*, 254–261. [CrossRef]

109. Zhang, J.; Yao, C.; Chen, J.; Zhang, Y.; Yuan, S.; Lin, Y. Hyperforin promotes post-stroke functional recovery through interleukin (IL)–17A-mediated angiogenesis. *Brain Res.* **2016**, *1646*, 504–513. [CrossRef]

110. Lin, Y.; Zhang, J.-C.; Fu, J.; Chen, F.; Wang, J.; Wu, Z.-L.; Yuan, S.-Y. Hyperforin attenuates brain damage induced by transient middle cerebral artery occlusion (mcao) in rats via inhibition of TRPC6 channels degradation. *Br. J. Pharmacol.* **2012**, *33*, 253–262. [CrossRef]

111. Cerpa, W.; Hancke, J.L.; Morazzoni, P.; Bombardelli, E.; Riva, A.; Marin, P.P.; Inestrosa, N.C. The hyperforin derivative IDN5706 occludes spatial memory impairments and neuropathological changes in a double transgenic Alzheimer's mouse model. *Curr. Alzheimer Res.* **2010**, *7*, 126–133. [CrossRef] [PubMed]

112. Clark, D.; Tuor, U.I.; Thompson, R.; Institoris, A.; Kulynych, A.; Zhang, X.; Kinniburgh, D.W.; Bari, F.; Busija, D.W.; Barber, P.A. Protection against recurrent stroke with resveratrol: Endothelial protection. *PLoS ONE* **2012**, *7*, e47792. [CrossRef] [PubMed]

113. Narayanan, S.V.; Dave, K.R.; Saul, I.; Perez-Pinzon, M.A. Resveratrol Preconditioning Protects Against Cerebral Ischemic Injury via Nuclear Erythroid 2-Related Factor 2. *Stroke* **2015**, *46*, 1626–1632. [CrossRef] [PubMed]

114. Dong, W.; Li, N.; Gao, D.; Zhen, H.; Zhang, X.; Li, F. Resveratrol attenuates ischemic brain damage in the delayed phase after stroke and induces messenger RNA and protein express for angiogenic factors. *J. Vasc. Surg.* **2008**, *48*, 709–714. [CrossRef]

115. Della-Morte, D.; Dave, K.R.; DeFazio, R.A.; Bao, Y.C.; Raval, A.P.; Perez-Pinzon, M.A. Resveratrol pretreatment protects rat brain from cerebral ischemic damage via a sirtuin 1-uncoupling protein 2 pathway. *Neuroscience* **2009**, *159*, 993–1002. [CrossRef]

116. Wang, Q.; Xu, J.; Rottinghaus, G.E.; Simonyi, A.; Lubahn, D.; Sun, G.Y.; Sun, A.Y. Resveratrol protects against global cerebral ischemic injury in gerbils. *Brain Res.* **2002**, *958*, 439–447. [CrossRef]

117. Drygalski, K.; Fereniec, E.; Koryciński, K.; Chomentowski, A.; Kiełczewska, A.; Odrzygóźdź, C.; Modzelewska, B. Resveratrol and Alzheimer's disease. From molecular pathophysiology to clinical trials. *Exp. Gerontol.* **2018**, *113*, 36–47. [CrossRef]

118. Jia, J.; Verma, S.; Nakayama, S.; Quillinan, N.; Grafe, M.R.; Hurn, P.D.; Herson, P.S. Sex differences in neuroprotection provided by inhibition of TRPM2 channels following experimental stroke. *J. Cereb. Blood Flow Metab.* **2011**, *31*, 2160–2168. [CrossRef]

119. Khansari, P.S.; Halliwell, R.F. Mechanisms Underlying Neuroprotection by the NSAID Mefenamic Acid in an Experimental Model of Stroke. *Front. Neurosci.* **2019**, *13*, 64. [CrossRef]

120. Titovich, I.A.; Sysoev, Y.I.; Bolotova, V.C.; Okovityi, S.V. Neurotropic activity of a new aminoethanol derivative under conditions of experimental brain ischemia. *Exp. Clin. Pharmacol. (Rus)* **2017**, *80*, 3–6.

121. Poloyac, S.M.; Zhang, Y.; Bies, R.R.; Kochanek, P.M.; Graham, S.H. Protective effect of the 20-HETE inhibitor het0016 on brain damage after temporary focal ischemia. *J. Cereb. Blood Flow Metab.* **2006**, *26*, 1551–1561. [CrossRef] [PubMed]

122. Zhu, J.; Wang, B.; Lee, J.-H.; Armstrong, J.S.; Kulikowicz, E.; Bhalala, U.S.; Martin, L.J.; Koehler, R.C.; Yang, Z.-J. Additive Neuroprotection of a 20-HETE Inhibitor with Delayed Therapeutic Hypothermia after Hypoxia-Ischemia in Neonatal Piglets. *Dev. Neurosci.* **2015**, *37*, 376–389. [CrossRef] [PubMed]

123. Yang, Z.-J.; Carter, E.L.; Kibler, K.K.; Kwansa, H.; Crafa, D.; Martin, L.J.; Roman, R.J.; Harder, D.R.; Koehler, R.C. Attenuation of neonatal ischemic brain damage using a 20-HETE synthesis inhibitor. *J. Neurochem.* **2012**, *121*, 168–179. [CrossRef] [PubMed]

124. Khansari, P.S.; Halliwell, R.F. Evidence for neuroprotection by the fenamate NSAID, mefenamic acid. *Neurochem. Int.* **2009**, *55*, 683–688. [CrossRef] [PubMed]

125. Daniels, M.J.D.; Rivers-Auty, J.; Schilling, T.; Spencer, N.G.; Watremez, W.; Fasolino, V.; Booth, S.J.; White, C.S.; Baldwin, A.G.; Freeman, S.; et al. Fenamate NSAIDs inhibit the NLRP3 inflammasome and protect against Alzheimer's disease in rodent models. *Nat. Commun.* **2016**, *7*, 12504. [CrossRef]

126. Chauvet, S.; Jarvis, L.; Chevallet, M.; Shrestha, N.; Groschner, K.; Bouron, A. Pharmacological Characterization of the Native Store-Operated Calcium Channels of Cortical Neurons from Embryonic Mouse Brain. *Front. Pharmacol.* **2016**, *7*, 486. [CrossRef]

127. Leuner, K.; Kazanski, V.; Muller, M.; Essin, K.; Henke, B.; Gollasch, M.; Harteneck, C.; Müller, W.E. Hyperforin—A key constituent of St. John's wort specifically activates TRPC6 channels. *FASEB J.* **2007**, *21*, 4101–4111. [CrossRef]

128. Leuner, K.; Heiser, J.H.; Derksen, S.; Mladenov, M.I.; Fehske, C.J.; Schubert, R.; Gollasch, M.; Schneider, G.; Harteneck, C.; Chatterjee, S.S.; et al. Simple 2,4-diacylphloroglucinols as classic transient receptor potential-6 activators—Identification of a novel pharmacophore. *Mol. Pharmacol.* **2009**, *77*, 368–377. [CrossRef]

129. Singer, A.; Wonnemann, M.; Müller, W.E. Hyperforin, a major antidepressant constituent of St. John's Wort, inhibits serotonin uptake by elevating free intracellular Na+1. *J. Pharmacol. Exp. Ther.* **1999**, *290*, 1363–1368.

130. Pochwat, B.; Szewczyk, B.; Kotarska, K.; Rafało-Ulińska, A.; Siwiec, M.; Sowa, J.E.; Tokarski, K.; Siwek, A.; Bouron, A.; Friedland, K.; et al. Hyperforin Potentiates Antidepressant-Like Activity of Lanicemine in Mice. *Front. Mol. Neurosci.* **2018**, *11*, 456. [CrossRef]

131. Gibon, J.; Deloulme, J.C.; Chevallier, T.; Ladevèze, E.; Abrous, D.N.; Bouron, A. The antidepressant hyperforin increases the phosphorylation of CREB and the expression of TrkB in a tissue-specific manner. *Int. J. Neuropsychopharmacol.* **2013**, *16*, 189–198. [CrossRef]

132. Inestrosa, N.C.; Tapia-Rojas, C.; Griffith, T.N.; Carvajal, F.J.; Benito, M.J.; Rivera-Dictter, A.; Alvarez, A.R.; Serrano, F.G.; Hancke, J.L.; Burgos, P.V.; et al. Tetrahydrohyperforin prevents cognitive deficit, Abeta deposition, tau phosphorylation and synaptotoxicity in the APPswe/PSEN1DeltaE9 model of Alzheimer's disease: A possible effect on APP processing. *Transl. Psychiatry* **2011**, *1*, e20. [CrossRef] [PubMed]

133. Müller, M.; Essin, K.; Hill, K.; Beschmann, H.; Rubant, S.; Schempp, C.M.; Gollasch, M.; Boehncke, W.H.; Harteneck, C.; Müller, W.E.; et al. Specific TRPC6 channel activation, a novel approach to stimulate keratinocyte differentiation. *J. Biol. Chem.* **2008**, *283*, 33942–33954. [CrossRef] [PubMed]

134. Moore, L.B.; Goodwin, B.; Jones, S.A.; Wisely, G.B.; Serabjit-Singh, C.J.; Willson, T.M.; Collins, J.L.; Kliewer, S.A. St. John's wort induces hepatic drug metabolism through activation of the pregnane X receptor. *Proc. Natl. Acad. Sci. USA* **2000**, *97*, 7500–7502. [CrossRef] [PubMed]

135. Ting, C.P.; Maimone, T.J. Total Synthesis of Hyperforin. *J. Am. Chem. Soc.* **2015**, *137*, 10516–10519. [CrossRef] [PubMed]

136. Woelk, H.; Burkard, G.; Grünwald, J. Benefits and risks of the hypericum extract LI 160: Drug monitoring study with 3250 patients. *J. Geriatr. Psychiatry Neurol.* **1994**, *7*, 34–38. [CrossRef]

137. Da Silva, D.D.; Arbo, M.D.; Valente, M.J.; Bastos, M.D.L.; Carmo, H. Hepatotoxicity of piperazine designer drugs: Comparison of different in vitro models. *Toxicol. Vitr.* **2015**, *29*, 987–996. [CrossRef]

138. Singh, N.; Agrawal, M.; Doré, S. Neuroprotective properties and mechanisms of resveratrol in in vitro and in vivo experimental cerebral stroke models. *ACS Chem. Neurosci.* **2013**, *4*, 1151–1162. [CrossRef]

139. Sun, A.Y.; Wang, Q.; Simonyi, A.; Sun, G.Y. Resveratrol as a therapeutic agent for neurodegenerative diseases. *Mol. Neurobiol.* **2010**, *41*, 375–383. [CrossRef]

140. Albani, D.; Polito, L.; Forloni, G. Sirtuins as novel targets for Alzheimer's disease and other neurodegenerative disorders: Experimental and genetic evidence. *J. Alzheimers Dis.* **2010**, *19*, 11–26. [CrossRef]

141. Anekonda, T.S. Resveratrol—A boon for treating Alzheimer's disease? *Brain Res. Rev.* **2006**, *52*, 316–326. [CrossRef]

142. Rasouri, S.; Lagouge, M.; Auwerx, J. SIRT1/PGC-1: A neuroprotective axis? *Med. Sci (Paris)* **2007**, *23*, 840–844. [CrossRef] [PubMed]

143. Turner, R.S.; Thomas, R.G.; Craft, S.; Van Dyck, C.H.; Mintzer, J.; Reynolds, B.A.; Brewer, J.B.; Rissman, R.A.; Raman, R.; Aisen, P.S.; et al. A randomized, double-blind, placebo-controlled trial of resveratrol for Alzheimer disease. *Neurology* **2015**, *85*, 1383–1391. [CrossRef] [PubMed]

144. Sawda, C.; Moussa, C.; Turner, R.S. Resveratrol for Alzheimer's disease. *Ann. N. Y. Acad. Sci.* **2017**, *1403*, 142–149. [CrossRef] [PubMed]

145. Piver, B.; Berthou, F.; Dreano, Y.; Lucas, D. Inhibition of CYP3A, CYP1A and CYP2E1 activities by resveratrol and other non volatile red wine components. *Toxicol. Lett.* **2001**, *125*, 83–91. [CrossRef]

146. Chow, H.-H.S.; Garland, L.L.; Hsu, C.-H.; Vining, D.R.; Chew, W.M.; Miller, J.A.; Perloff, M.; Crowell, J.A.; Alberts, D.S. Resveratrol modulates drug- and carcinogen-metabolizing enzymes in a healthy volunteer study. *Cancer Prev. Res.* **2010**, *3*, 1168–1175. [CrossRef]

147. Salehi, B.; Mishra, A.P.; Nigam, M.; Sener, B.; Kilic, M.; Sharifi-Rad, M.; Fokou, P.V.T.; Martins, N.; Sharifi-Rad, J. Resveratrol: A Double-Edged Sword in Health Benefits. *Biomedicines* **2018**, *6*, 91. [CrossRef] [PubMed]

148. Inoue, R.; Okada, T.; Onoue, H.; Hara, Y.; Shimizu, S.; Naitoh, S.; Ito, Y.; Mori, Y. The transient receptor potential protein homologue TRP6 is the essential component of vascular α 1-adrenoceptor-activated ca 2+-permeable cation channel. *Circ. Res.* **2001**, *88*, 325–332. [CrossRef]

149. Foster, R.R.; Zadeh, M.A.; Welsh, G.I.; Satchell, S.C.; Ye, Y.; Mathieson, P.W.; Bates, D.O.; Saleem, M.A. Flufenamic acid is a tool for investigating TRPC6-mediated calcium signalling in human conditionally immortalised podocytes and HEK293 cells. *Cell Calcium* **2009**, *45*, 384–390. [CrossRef]

150. Macianskiene, R.; Gwanyanya, A.; Sipido, K.; Vereecke, J.; Mubagwa, D.K. Induction of a novel cation current in cardiac ventricular myocytes by flufenamic acid and related drugs. *Br. J. Pharmacol.* **2010**, *161*, 416–429. [CrossRef]

151. Klose, C.; Straub, I.; Riehle, M.; Ranta, F.; Krautwurst, D.; Ullrich, S.; Meyerhof, W.; Harteneck, C. Fenamates as TRP channel blockers: Mefenamic acid selectively blocks TRPM3. *Br. J. Pharmacol.* **2011**, *162*, 1757–1769. [CrossRef] [PubMed]

152. Hu, H.; Tian, J.; Zhu, Y.; Wang, C.; Xiao, R.; Herz, J.M.; Wood, J.D.; Zhu, M.X. Activation of TRPA1 channels by fenamate nonsteroidal anti-inflammatory drugs. *Pflug. Arch.* **2009**, *459*, 579–592. [CrossRef]

153. Kochetkov, K.V.; Kazachenko, V.N.; Marinov, B.S. Dose-dependent potentiation and inhibition of single Ca^{2+}-activated K+ channels by flufenamic acid. *Membr. cell Boil.* **2000**, *14*, 285–298.

154. Barrier, M.; Burlet, S.; Estrella, C.; Melnyk, P.; Sergeant, N.; Buee, L.; Verwaerde, P. Sulfate Salts of N-(3-(4-(3-(diisobutylamino)propyl)piperazin-1-yl)propyl)-1H-benzo [d]imidazol-2 Amine, Preparation Thereof and Use of the Same. U.S. Patent 9562018B2, 7 February 2017.

155. U.S. National Library of Medicine. A Study to Assess Tolerability, Safety, Pharmacokinetics and Effect of AZP2006 in Patients with PS. 2019. Available online: https://clinicaltrials.gov/ct2/show/study/NCT04008355?term=azp2006&draw=2&rank=1 (accessed on 1 August 2020).

156. Titovich, I.A.; Radko, S.V.; Lisitsky, D.S.; Okovity, S.V.; Bolotov, V.T.; Belsky, A.V.; Mikhailova, M.V.; Sysoev, I.Y. The study of a novel diethylaminoethanol derivative cognitive function in laboratory animals. *J. Biomed (Ru)* **2017**, *3*, 102–110.

157. Xu, X.; Lozinskaya, I.; Costell, M.; Lin, Z.; Ball, J.A.; Bernard, R.; Behm, D.J.; Marino, J.P.; Schnackenberg, C.G. Characterization of small molecule TRPC3 and TRPC6 agonist and antagonists. *Biophys. J.* **2013**, *104*, 454a. [CrossRef]

158. Inoue, R.; Jensen, L.J.; Jian, Z.; Shi, J.; Hai, L.; Lurie, A.I.; Henriksen, F.H.; Salomonsson, M.; Morita, H.; Kawarabayashi, Y.; et al. Synergistic activation of vascular TRPC6 channel by receptor and mechanical stimulation via phospholipase C/diacylglycerol and phospholipase A2/omega-hydroxylase/20-HETE pathways. *Circ. Res.* **2009**, *104*, 1399–1409. [CrossRef]

159. Butterfield, D.A.; Reed, T.; Sultana, R. Roles of 3-nitrotyrosine- and 4-hydroxynonenal-modified brain proteins in the progression and pathogenesis of Alzheimer's disease. *Free Radic Res.* **2011**, *45*, 59–72. [CrossRef]

160. Jin, Z.; Berthiaume, J.M.; Li, Q.; Henry, F.; Huang, Z.; Sadhukhan, S.; Gao, P.; Tochtrop, G.P.; Puchowicz, M.A.; Zhang, G. Catabolism of (2E)-4-hydroxy-2-nonenal via omega- and omega-1-oxidation stimulated by ketogenic diet. *J. Biol. Chem.* **2014**, *289*, 32327–32338. [CrossRef]

161. van Ierssel, S.H.; Conraads, V.M.; van Craenenbroeck, E.M.; Liu, Y.; Maas, A.I.; Parizel, P.M.; Hoymans, V.Y.; Vrints, C.J.; Jorens, P.G. Endothelial dysfunction in acute brain injury and the development of cerebral ischemia. *J. Neurosci. Res.* **2015**, *93*, 866–872. [CrossRef]

162. Clementi, E.; Meldolesi, J. Pharmacological and functional properties of voltagemi independent Ca^{2+} channels. *Cell Calcium* **1996**, *19*, 269–279. [CrossRef]

163. Tesfai, Y.; Brereton, H.M.; Barritt, G.J. A diacylglycerol-activated Ca^{2+} channel in PC12 cells (an adrenal chromaffin cell line) correlates with expression of the TRP-6 (transient receptor potential) protein. *Biochem. J.* **2001**, *358*, 717. [CrossRef]

164. Zhang, S.L.; Kozak, J.A.; Jiang, W.; Yeromin, A.V.; Chen, J.; Yu, Y.; Penna, A.; Shen, W.; Chi, V.; Cahalan, M.D. Store-dependent and -independent modes regulating Ca^{2+} release-activated Ca^{2+} channel activity of human Orai1 and Orai3. *J. Biol. Chem.* **2008**, *283*, 17662–17671. [CrossRef] [PubMed]

165. Nilius, B.; Flockerzi, V. Mammalian transient receptor potential (TRP) cation channels. In *Handbook of Experimental Pharmacology*; Springer: Heidelberg, Germany, 2014; Volume 223.

166. Wu, J.; Ryskamp, D.A.; Liang, X.; Egorova, P.; Zakharova, O.; Hung, G.; Bezprozvanny, I. Enhanced Store-Operated Calcium Entry Leads to Striatal Synaptic Loss in a Huntington's Disease Mouse Model. *J. Neurosci.* **2016**, *36*, 125–141. [CrossRef] [PubMed]

167. Tobe, M.; Yoshiaki Isobe, H.T.; Nagasaki, T.; Takahashi, H.; Fukazawa, T.; Hayashi, H. Discovery of quinazolines as a novel structural class of potent inhibitors of NF-kappa B activation. *Bioorg. Med. Chem.* **2003**, *11*, 383–391. [CrossRef]

168. Ma, R.; Wang, Y.X.; Ding, M. Nuclear transcription factor kappa B (NF-kB) mediates ROS and PKC-induced decrease in TRPC6 protein expression in human glomerular mesangial cells (HMCs). *Faseb. J.* **2012**, *26*, 687.

169. Tang, Q.; Guo, W.; Zheng, L.; Wu, J.-X.; Liu, M.; Zhou, X.; Zhang, X.; Chen, L. Structure of the receptor-activated human TRPC6 and TRPC3 ion channels. *Cell Res.* **2018**, *28*, 746–755. [CrossRef] [PubMed]

Publisher's Note: MDPI stays neutral with regard to jurisdictional claims in published maps and institutional affiliations.

Review

Therapeutic Strategies to Target Calcium Dysregulation in Alzheimer's Disease

Maria Calvo-Rodriguez, Elizabeth K. Kharitonova and Brian J. Bacskai *

Alzheimer Research Unit, Department of Neurology, Massachusetts General Hospital and Harvard Medical School, Charlestown, MA 02129, USA; mcalvorodriguez@mgh.harvard.edu (M.C.-R.); ekharitonova@mgh.harvard.edu (E.K.K.)
* Correspondence: bbacskai@mgh.harvard.edu; Tel.: +1-617-724-5306

Received: 27 October 2020; Accepted: 18 November 2020; Published: 20 November 2020

Abstract: Alzheimer's disease (AD) is the most common form of dementia, affecting millions of people worldwide. Unfortunately, none of the current treatments are effective at improving cognitive function in AD patients and, therefore, there is an urgent need for the development of new therapies that target the early cause(s) of AD. Intracellular calcium (Ca^{2+}) regulation is critical for proper cellular and neuronal function. It has been suggested that Ca^{2+} dyshomeostasis is an upstream factor of many neurodegenerative diseases, including AD. For this reason, chemical agents or small molecules aimed at targeting or correcting this Ca^{2+} dysregulation might serve as therapeutic strategies to prevent the development of AD. Moreover, neurons are not alone in exhibiting Ca^{2+} dyshomeostasis, since Ca^{2+} disruption is observed in other cell types in the brain in AD. In this review, we examine the distinct Ca^{2+} channels and compartments involved in the disease mechanisms that could be potential targets in AD.

Keywords: calcium homeostasis; Alzheimer's disease; therapeutics; amyloid; tau; endoplasmic reticulum; mitochondria; lysosomes

---

1. Calcium Dysregulation Is a Hallmark of Alzheimer's Disease

Alzheimer's disease (AD) is the most common form of dementia, affecting more than 30 million people worldwide. It is characterized by accumulation of extracellular amyloid β (Aβ) plaques—or senile plaques—composed of Aβ peptide, intraneuronal fibrillary tangles (NFTs) comprising hyperphosphorylated and misfolded microtubule-associated protein tau, and selective neuronal loss, particularly in brain regions like the neocortex and hippocampus, eventually leading to memory loss and a decline in cognitive function. Most AD cases are sporadic (SAD), with less than 1% due to genetic mutations. Risk factors, such as aging, lifestyle, obesity, or diabetes, or genetic factors such as carrying the allele ε4 in the apolipoprotein E (ApoE) gene predispose individuals to SAD development [1]. Genetically inherited forms of AD (familial AD, FAD) show early onset and are caused by mutations in genes coding for presenilin (PS) 1, PS2, or amyloid precursor protein (APP), all involved in the Aβ generation pathway. Other than the onset, there are no clear differences regarding symptoms or histopathological features between SAD and FAD. Different hypotheses have been proposed to explain the origin of AD. The relation to genetics in FAD supported the "amyloid cascade hypothesis", which suggests that AD pathogenesis is initiated by overproduction of Aβ and/or failure of its clearance mechanisms, upstream of tau dysregulation [2]. However, other hypotheses that explain the etiology of AD are being considered. The "cholinergic hypothesis" [3], "tau propagation hypothesis" [4], "inflammatory hypothesis" [5], or "glymphatic system hypothesis" [6] are among the most relevant.

Intracellular calcium (Ca^{2+}) is an important second messenger that regulates multiple cellular functions, such as synaptic plasticity, action potentials, and learning and memory. Ca^{2+} dyshomeostasis, on the other hand, contributes to detrimental mechanisms such as necrosis, apoptosis, autophagy deficits, and neurodegeneration. Perturbations in intracellular Ca^{2+} are involved in many neurodegenerative diseases including AD, Parkinson's disease, and Huntington's disease [7]. Back in the mid-1980s, Khachaturian proposed that Ca^{2+} dysregulation led to neurodegeneration, suggesting that a sustained imbalance of cellular Ca^{2+} could disrupt normal neuronal functions and lead to neurodegenerative diseases such as AD [8]. Since then, many reports have shown Ca^{2+} dysregulation in AD (both in SAD [9,10] and in FAD [11]), animal models of the disorder [12–19], and cells from human AD patients [20]. The "Ca^{2+} hypothesis of Alzheimer's disease" [21] postulates that activation of the amyloidogenic pathway causes a remodeling of normal neuronal Ca^{2+} signaling pathways, which then alters Ca^{2+} homeostasis and leads to the disruption of the mechanisms involved in learning and memory. Neuronal Ca^{2+} dyshomeostasis seems to manifest early in AD progression prior to the development of histopathological markers or clinical symptoms [22]. Similarly, AD is also marked by Ca^{2+} disruption in other cells in the brain such as astrocytes and microglia. Whether disruption of Ca^{2+} homeostasis is cause or consequence of AD pathology is still a matter of debate.

Up to date, there are only two types of Food and Drug Administration (FDA)-approved therapies for AD treatment (www.alzforum.org)—acetylcholinesterase inhibitors and *N*-methyl-D-aspartate receptor (NMDAR) antagonists—and neither can cure or reverse the disease, but can, at least, transiently relieve patients' symptoms [23]. Unfortunately, drugs targeting Aβ have been mostly unsuccessful. Although these therapies have shown some success in clearing Aβ plaques from the AD brain, they have failed to relieve the cognitive decline of AD patients in clinical trials [24], with the exception of aducanumab, which demonstrated both clearance of plaques and modest gains in cognitive function [25]. In addition, the well-known lack of correlation between cognitive symptoms and Aβ deposition further supports the idea of the need for different approaches [26]. Ca^{2+} dyshomeostasis is an early molecular defect in AD and might precede Aβ and tau deposition [22]. Therefore, therapeutics that stabilize Ca^{2+} signals may represent an alternative strategy for treating AD. In the remaining sections, we review human data and those generated from experimental models, and we discuss the different strategies for targeting Ca^{2+} dysregulation—including specific Ca^{2+} channels and different cell types—that could be used as therapeutics in AD.

2. Neuronal Ca^{2+} as a Therapeutic Target in AD

Ca^{2+} is a fundamental regulator of neuronal fate; thus, intracellular Ca^{2+} homeostasis must be finely tuned in physiological conditions. In the extracellular space, Ca^{2+} concentration is maintained between 1.1 and 1.4 mM, whereas resting cytosolic levels within neurons are maintained in the nM range (50–300 nM) [27]. After cell activation, intracellular Ca^{2+} concentrations increase rapidly to the µM range. This Ca^{2+} gradient allows the initiation of different signaling cascades. Ca^{2+} levels in the endoplasmic reticulum (ER) are nearly a thousand times higher than those of the cytoplasm [28]. Ca^{2+} signals are generated by the influx of Ca^{2+} from the extracellular space or by Ca^{2+} release from intracellular stores. Ca^{2+} enters neurons mainly through plasma membrane channels and is then buffered by Ca^{2+}-binding proteins and organelles such as mitochondria. Even though the mechanisms responsible for neuronal Ca^{2+} dysregulation in AD are not completely understood, as discussed in the sections below, evidence shows that different compartments and/or organelles are involved (Figure 1).

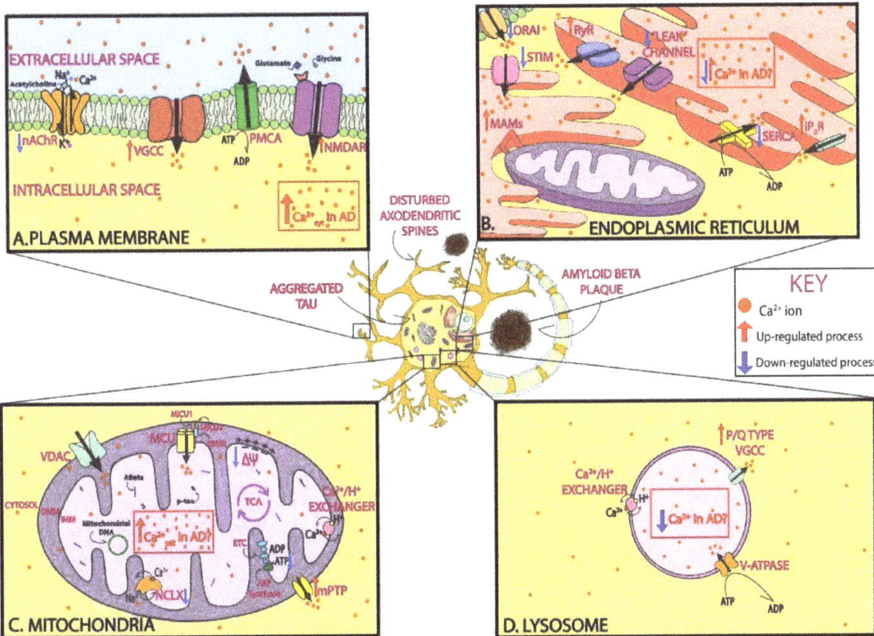

Figure 1. Neuronal Ca^{2+} as a therapeutic target in Alzheimer's disease (AD). Schematic of Ca^{2+} dysregulation in neurons in AD that could be used as potential targets. In AD, Ca^{2+} dysregulation is present in many of the different compartments within neurons. In the plasma membrane, voltage-gated Ca^{2+} channels (VGCCs) and receptor operated Ca^{2+} channels, including N-methyl-D-aspartate receptors (NMDARs) and nicotinic acetylcholine receptors (nAChRs), allow for the influx of Ca^{2+} ions into the neuron after depolarization or ligand binding, respectively. Both Aβ and tau overactivate these channels and increase their function (**A**). In the endoplasmic reticulum (ER), Ca^{2+} is released via ryanodine receptors (RyRs) and inositol 1,4,5-trisphosphate receptors (IP$_3$Rs) to the cytosol after stimulation. Ca^{2+} is then extruded by the sarco-endoplasmic reticulum ATPase (SERCA) pump, which actively consumes ATP while bringing Ca^{2+} into the lumen. AD-associated presenilin (PS) mutations impair IP$_3$R and RyR signaling, increasing Ca^{2+} release into the cytosol, and diminish SERCA activity, increasing cytosolic Ca^{2+} concentration. Following ER Ca^{2+} depletion, the stromal-interacting molecule (STIM) interacts with the Orai channel in the plasma membrane to activate the store-operated Ca^{2+} entry (SOCE) pathway. SOCE is decreased by diverse familial AD (FAD) PS mutations and by soluble Aβ. Lastly, in order to facilitate the communication between mitochondria and ER, contact points known as mitochondrial-associated membranes (MAMs) are established. Increased association between the ER and mitochondria and enhanced Ca^{2+} transfer have been observed in AD (**B**). In the mitochondria, the voltage-dependent anion-selective channel protein (VDAC) lets Ca^{2+} across the outer mitochondrial membrane (OMM), and the mitochondrial Ca^{2+} uniporter (MCU) complex allows the influx of Ca^{2+} across the inner mitochondrial membrane (IMM). Ca^{2+} efflux is partially managed by the Na^+/Ca^{2+} exchanger (NCLX). Both Aβ and tau (phospho-tau, p-tau) have been found in mitochondria. Elevated mitochondrial Ca^{2+} levels and decreased NCLX activity have been observed in AD (**C**). In the lysosome, the P/Q type VGCCs in their membrane regulate Ca^{2+} efflux into the cytosol, while the V-ATPase and Ca^{2+}/H^+ exchanger are in charge of lysosomal Ca^{2+} refilling (**D**). Additionally, Aβ and tau accumulate extracellularly and intracellularly, respectively, and lead to loss of dendritic spine density and synaptic function.

2.1. Targeting Plasma Membrane Receptors and Cytosolic Ca^{2+}

Proper intracellular Ca^{2+} homeostasis is crucial for many neuronal functions. Disruption of this homeostasis might be one of the main mechanisms via which Aβ and tau exert their neurotoxicity. The main plasma membrane channels involved in neuronal Ca^{2+} influx from the extracellular space are voltage-gated Ca^{2+} channels (VGCCs), which allow Ca^{2+} influx following neuronal depolarization, and receptor-operated Ca^{2+} channels (ROCs), which open upon specific binding of the agonist, with NMDARs among the most important examples.

Toxic Aβ increases cytosolic Ca^{2+}, which may affect a variety of enzymes (such as proteases or phosphatases), promote cytoskeletal modifications, cause the generation of free radicals, or trigger neuronal apoptosis [29]. It has been proposed that Aβ can overactivate channels and/or form pores in the cytosolic plasma membrane, allowing massive influx of Ca^{2+} from the extracellular space and increasing the overall Ca^{2+} levels in the cytosol, severely limiting normal cellular function [30,31]. Aβ potentiates Ca^{2+} influx through VGCCs, particularly L-type VGCCs. Excessive Ca^{2+} influx through these channels has been observed in cultured neurons following Aβ exposure and was shown to be blocked by the L-type VGCC inhibitor nimodipine [32]. This phenomenon, however, was not observed in brain slices from AD mouse models [33]. In addition, AD patients taking L-type VGCCs inhibitors such as nilvadipine (NILVAD multicenter trial) showed reduced Aβ levels but no improvement in cognitive decline [34,35]. Acute application of tau aggregates has also been observed to increase cytosolic Ca^{2+} and elevate reactive oxygen species (ROS) production via nicotinamide adenine dinucleotide phosphate (NADPH), an effect that can be prevented by nifedipine and verapamil, both L-type VGCC inhibitors [36]. This suggests that tau fibrils could also incorporate into the cell membrane to activate VGCCs and lead to neuronal dysfunction [36].

CALHM1 (Ca^{2+} homeostasis modulator protein 1) is a Ca^{2+} channel highly expressed in neurons in the hippocampus that allows cytosolic Ca^{2+} influx in response to decreases in extracellular Ca^{2+} [37]. Its activation triggers different kinase signaling cascades in neurons. The *CALMH1* polymorphism P86L has been proposed as a risk factor for late-onset SAD [9,37], an argument that has been challenged by other groups [38,39]. Nevertheless, increased levels of Aβ have been observed in transfected cells expressing the P86L polymorphism, suggesting a role for CALMH1 in AD [37]. Additionally, the P86L polymorphism alters the channel permeability to Ca^{2+} [37]. A recent study demonstrated that CALHM1 deficiency in mice leads to cognitive and neuronal deficits, which manifest memory impairment and hippocampal long-term potentiation (LTP) [40], pointing to CALHM1 as a potential treatment target in AD.

NMDARs are a subfamily of ionotropic glutamate receptors involved in the excitatory synaptic transmission and synaptic plasticity of the brain. Specific types of NMDARs are much more permeable to Ca^{2+} than other ionotropic glutamate receptors and are often implicated in neuronal pathophysiology. NMDARs are mainly composed of GluN2A and GluN2B in the brain areas most affected in AD [41]. Extra-synaptic GluN2B-containing NMDARs have been associated with excitotoxicity (the excessive neuronal death induced by cellular Ca^{2+} overload due to excessive stimulation of glutamate receptors) and the toxic effect of Aβ oligomers in AD [42,43]. For this reason, selective GluN2B subunit antagonists may be a strategy to prevent synaptic dysfunction in AD. Aβ$_{42}$ peptides interact with NMDARs, potentiating their activity and leading to increased Ca^{2+} influx, thus contributing to the synapse loss observed in AD [44,45]. Additionally, as demonstrated in mouse models of AD, glutamate-induced excitotoxicity is inhibited by tau reduction [46] and exacerbated by tau overexpression [47,48]. In turn, glutamate-induced excitotoxicity increases tau expression [49] and phosphorylation [50], while activation of extra-synaptic NMDAR leads to tau overexpression, neuronal degeneration, and cell loss [51]. Memantine—a weak NMDAR antagonist—is one of the two FDA-approved drugs to treat AD patients and the only NMDAR antagonist [23]. It provides modest improvements to memory and cognitive performance in moderate to severe AD patients [52,53]. Memantine restricts excessive Ca^{2+} influx, thus reducing neuronal excitotoxicity, and, due to its low activity, the basal NMDAR function is preserved. Memantine has also shown neuroprotective effects against oxidative stress, neuroinflammation, and tau phosphorylation [54,55].

Ionotropic neuronal nicotinic acetylcholine receptors (nAChRs) respond to the neurotransmitter acetylcholine (ACh) and to drugs such as the agonist nicotine. They are permeable to Na^+, K^+, and Ca^{2+}. The nAChRs expressing the $\alpha7$ subunit have the highest conductance for Ca^{2+} and are found in brain regions most susceptible to AD [56]. In the basal forebrain, cholinergic neuronal loss and decreased levels of ACh mediate cholinergic impairment, which eventually leads to short-term memory loss [57–59]. The loss of cholinergic innervation in early AD led to the "cholinergic hypothesis of AD" [3]. Galantamine and rivastigmine (for use in mild to moderate AD) and donepezil (in mild to severe AD) are the cholinesterase inhibitors FDA-approved to treat AD [23]. These drugs act by increasing ACh levels, which delay the progression of AD through Ca^{2+}-dependent mechanisms. Furthermore, supplemented with memantine, it has been proposed that this combination could provide greater benefits on behavior, cognition, and global outcomes in AD [60].

Exposure of hippocampal and cortical neurons to tau also increased intracellular Ca^{2+} levels through muscarinic receptors [61]. Interestingly, Ca^{2+} activates many kinases, including those responsible for tau phosphorylation—such as glycogen synthase kinase 3β (GSK3β)—and, therefore, Ca^{2+} dyshomeostasis may increase tau phosphorylation and NFT formation [62]. Given that tau pathology correlates better with cognitive impairments than Aβ deposition, tau targeting is expected to be more effective once clinical symptoms emerge [63]. It has long been known that tau-expressing cells secrete normal and pathological tau [64], which can be taken up by other cells, seeding and spreading tau pathology [4,65–67]. Led by immunotherapy approaches, the efforts to target tau with therapeutics focus on reducing tau pathology by limiting the spread of extracellular tau across brain regions [68]. Anti-tau immunotherapy has shown potential in numerous clinical studies. Both active and passive tau immunization seem to offer a promising option by reducing tau pathology [69]. Active tau immunization, however, seems to elicit a risk of adverse immune reactions from targeting the normal protein. Other tested approaches involve reducing tau expression (with small interfering RNA or antisense oligonucleotides; siRNA and ASOs, respectively), targeting tau modifications, reducing tau aggregation, and stabilizing microtubules. Preventing or reducing pathologic tau has been shown to improve cognitive and motor impairments in animal models with neurofibrillary pathology, and several tau antibodies and vaccines have been tested in preclinical studies in the last years. Immunotherapy is currently at the stage of drug development (recently reviewed in [68–70]), and, as of today, eight humanized tau antibodies and two tau vaccines are under clinical trial for AD or frontotemporal dementia [71] (www.alzforum.org).

The use of intravital imaging and transgenic mouse models of AD have allowed for direct observation of cytosolic Ca^{2+} dysregulation. In vivo, neuronal cytosolic Ca^{2+} dyshomeostasis is more likely to be observed in the vicinity of amyloid β plaques, but is detectable in neurons throughout the cortex [15]. Higher Ca^{2+} levels were observed in neurons close to amyloid plaques in a commonly used mouse model of cerebral amyloidosis (APP$_{Swe}$xPS1ΔE9, APP/PS1) [15], but only after plaque deposition and not before. Cytosolic Ca^{2+} overload was absent in mice harboring only the PS1 mutation (typically lacking plaque deposition). The mechanisms of Ca^{2+} dysregulation involved activation of calcineurin (CaN), a Ca^{2+}/calmodulin-dependent protein phosphatase sensitive to subtle rises in intracellular Ca^{2+} levels, and whose activation induces long-term depression (LTD). Ca^{2+} dysregulation in neurites was linked to neurodegeneration (neuritic blebbing and beading), which can be partially prevented by inhibiting CaN [15]. Elevated Ca^{2+} levels in the neurites impair synapses, by increasing the frequency of spontaneous synaptic potentials and reducing plasticity. In addition, pathological increases in neuronal network activity—observed as increased frequencies of somatic Ca^{2+} transients—potentiate Aβ release into the extracellular space [72,73]. In the APP23/PS45 mouse model of AD (overexpressing mutant APP$_{Swe}$ and mutant PS1$_{G348A}$), neuronal hyperactivity was observed around amyloid plaques in the cortex, only after plaque deposition [14]. Hyperactive neurons, however, were found in the CA1 region of the hippocampus in pre-depositing animals [13]. Direct application of soluble Aβ onto the wild-type (Wt) naïve brain increased cytosolic Ca^{2+} levels [12] and induced neuronal hyperactivity [13]. Acute treatment with the γ-secretase inhibitor LY-411575, which reduces soluble Aβ levels, normalized the frequency of Ca^{2+} transients prior to plaque deposition [13].

Interestingly, AD patients are more prone to developing epileptic seizures [74]. Blocking network hyperactivity with the antiepileptic drug levetiracetam improves learning and memory, reverses behavioral abnormalities, and reverts synaptic deficits in the hippocampus in an AD mouse model [75]. In the same way, it has been observed that tau is implicated in neuronal circuit deficits in mouse models of AD expressing both Aβ and tau. Tau effects dominate those of Aβ and are mostly dependent on the presence of soluble tau [16]. According to the authors, this dramatic effect could suggest a possible cellular explanation contributing to disappointing results of anti-Aβ therapeutic trials. This abnormal network activity and its resultant AD-related cognitive deficits in mice point to neuronal hyperactivity as a promising therapeutic target in AD.

Aducanumab is a high-affinity, fully human immunoglobulin G1 (IgG1) monoclonal antibody that selectively binds to aggregated Aβ fibrils and soluble oligomers (and not monomers) in the brain parenchyma [25]. It was shown that it could ameliorate Ca^{2+} dysregulation in AD. Using multiphoton microscopy and a Ca^{2+} reporter, it was observed that a single topical application of the antibody onto the brain surface of mice depositing amyloid plaques (Tg2576 AD model) led to a reduction in existing amyloid deposits [76]. Peripheral administration of the antibody over a period of 6 months rescued Ca^{2+} overload in transgenic neurites, restoring them to control levels within 2 weeks. The authors suggested that aducanumab exerted its function by targeting amyloid deposits, including soluble oligomeric Aβ [76]. In March 2019, the termination of all aducanumab clinical trials was announced after an interim analysis of EMERGE and ENGAGE trials predicted the phase III placebo-controlled studies would not meet their primary end points. However, in a subsequent analysis of a larger dataset from the EMERGE trial, aducanumab met the primary end point, and the FDA accepted the aducanumab application for review [77]. If the case is approved, aducanumab would be the first drug to combat the root causes of AD.

2.2. Targeting ER Ca^{2+} and SOCE

The ER is an important subcellular organelle involved in protein synthesis, modification, and folding. Additionally, it is a dominant Ca^{2+} reservoir in the cell, critical for maintaining intracellular Ca^{2+} levels [27]. Ca^{2+} is released from the ER after activation of either inositol 1,4,5-trisphosphate receptors (IP$_3$Rs) or ryanodine receptors (RyRs). Ca^{2+} efflux from the ER modulates a range of neuronal processes, including regulation of axodendritic growth and morphology or synaptic vesicle release [78]. The sarco-endoplasmic reticulum ATPase (SERCA) pump, which actively consumes ATP, is important for extruding Ca^{2+} into the ER lumen, where it is sequestered by binding to proteins such as calsequestrin and calretinin, priming this organelle as a critical component of Ca^{2+} buffering.

Impaired IP$_3$R signaling in the ER was an early discovery in AD. It was shown that human cells from FAD patients exhibited enhanced Ca^{2+} release in response to IP$_3$R-generating stimuli [79]. Fibroblasts from asymptomatic members of AD families [80], as well as PS1 knock-in mice and other presymptomatic AD mouse models, showed the same enhancement [81]. These observations suggested that FAD mutations contribute to Ca^{2+} dysregulation, even before pathology deposition or cognitive impairments were evident. A reduction in IP$_3$R expression can normalize Ca^{2+} homeostasis and restore hippocampal LTP in mouse models of AD [82]. PSs are transmembrane proteins found in the ER membranes and form the catalytic core of the γ-secretase complex that processes APP and other type 1 transmembrane proteins, such as Notch [83,84]. PSs are essential for learning and memory, as well as neuronal survival during aging in the murine cerebral cortex [85,86]. Mutations in PSs have been shown to affect APP processing, leading to increased production of the more hydrophobic neurotoxic form Aβ_{42} [87–89] or increasing the Aβ42/40 production ratio [90]. It has also been proposed that *PSEN* mutations cause a loss of presenilin function in the brain, triggering neurodegeneration and dementia in FAD [91]. Mutations in PS1 and PS2 might stimulate IP$_3$Rs, leading to exaggerated Ca^{2+}

release through these channels [79–81]. An alternative hypothesis suggested that PSs function as passive low-conductance leak channels in the ER membrane. AD-associated PS mutations might impair this leak function, resulting in ER Ca^{2+} overload [92] and leading to exaggerated increases in cytosolic Ca^{2+} upon stimulation of Ca^{2+} release. However, these observations have not been supported by other groups [93,94], and, despite extensive research, this subject is still a matter of controversy. Recently, it was proposed that the ER-based transmembrane and coiled-coil domain TMCO1 could be responsible for the ER Ca^{2+} leak [95].

RyR Ca^{2+} dysregulation was also observed before the histopathology and cognitive decline in AD. Both human brain tissue from AD patients and AD mouse models have shown increased expression of RyR (particularly RyR_2) in affected brain regions in AD [96]. Exaggerated Ca^{2+} release from RyR has been related to impaired neurophysiology and synaptic signaling events, contributing to memory impairment in AD [33,97,98]. The FAD PS mutations also exaggerate Ca^{2+} release through the RyR, as a result of either increased expression of RyRs or sensitization of the channel activity [99,100]. Furthermore, RyR-mediated Ca^{2+} release upregulates secretases, increasing APP cleavage, Aβ fragments, and plaque deposition, and its blockage leads to Aβ reduction and improved memory impairment [101]. RyRs can also themselves be activated by Ca^{2+}, which amplify IP_3R activity via Ca^{2+}-induced Ca^{2+} release mechanisms [102]. Additionally, Aβ aggregates themselves trigger ER Ca^{2+} release through IP_3Rs and RyRs [103,104]. Recently, stabilization of RyR_2 macromolecular complex by S107 (Rycal)—a benzothiazepine that prevents the dissociation calstabin2 from the RyR_2 complex—showed therapeutic potential in vitro and in mouse models of AD in vivo. Application or administration of S107 reversed ER Ca^{2+} leak, reduced APP cleavage and Aβ production, and restored synaptic plasticity and cognitive deficits [105,106].

It has been found that, in cells lacking PS1, PS2, or PS1/2, or cells expressing either PS2 or FAD-PS2, SERCA activity is diminished, resulting in increased cytosolic Ca^{2+} [107]. Conversely, other studies have shown that mutations in PS influence SERCA by accelerating Ca^{2+} sequestration via ATPase [108], leading to an overfilled ER. In any case, these data suggest that normal PSs are required for normal SERCA functioning and suggest that PSs are a candidate target for development of therapeutics, independent of their role in APP processing.

Increased Ca^{2+} release from intracellular ER Ca^{2+} stores might exacerbate disease-mediated pathology. Accordingly, dantrolene, a negative allosteric modulator of RyR—and its central nervous system (CNS)-penetrant version Ryanodex—has been shown to reduce amyloid pathology, normalize ER Ca^{2+} homeostasis, restore synaptic structure and density, normalize synaptic plasticity, and improve behavioral performance in mouse models of AD [101,109,110]. This builds on RyR as a therapeutic target for AD, and further emphasizes the role of dysregulated ER Ca^{2+} as a key component in the AD pathogenesis.

ER Ca^{2+} depletion triggers a sustained extracellular Ca^{2+} influx to the cytosol through the store-operated Ca^{2+} entry (SOCE) pathway by activating STIM (stromal-interacting molecule) protein—which senses low Ca^{2+} concentration upon depletion of the ER stores—and plasma membrane channels Orai and TRPC (transient receptor potential canonical) [111]. Two forms of STIM are expressed in the brain (STIM1, predominantly in the cerebellum, and STIM2 in the hippocampus and cortex) [112]. SOCE refills the ER, keeping it ready for the next ER Ca^{2+} signal [113]. Disrupted SOCE has been observed in AD. SOCE is decreased by diverse FAD PS mutations [114,115] and in the presence of soluble Aβ [116]. It has also been proposed that SOCE deficits may be due to the decreased expression of STIM1 and/or STIM2 in FAD-linked PS1 mutations [117]. Related to this, overexpression of the dominant negative PS1 variant potentiates SOCE [118]. It has also been proposed that SOCE deficits might result from overfilled ER Ca^{2+} stores [114]. These findings, however, are inconsistent, as other groups have observed no differences or decreased ER Ca^{2+} concentration in mutant PS expressing cells [11,93,107]. Recently, it has been proposed that neuronal SOCE is required for maintaining the morphology of mushroom spines, modulating Aβ production and promoting memory functions [117,119]. Ca^{2+} entry via SOCE activates Ca^{2+}/CaM-dependent kinase II (CamKII), which is upstream of gene transcription for maintenance of mature spines. Attenuated SOCE-mediated Ca^{2+} influx might reduce CaMKII activity while inducing destabilization of mushroom spines. This can reduce LTP-mediated memory formation [117]. Attenuated SOCE may also lead to inadequate ER refill, which might induce neuronal cell death via apoptosis [120,121]. STIM2 overexpression in AD models restores spine morphology, implicating SOCE in AD [122] and suggesting that targeting SOCE in AD may avoid or restore dendritic spine loss. Additionally, it was recently found that expression of TRPC1, a subfamily of TRPCs, is decreased in AD cells and mouse models. While deletion of TRPC1 did not impair cognitive function or lead to cell death in physiological conditions, it did exacerbate memory deficits and increase neuronal apoptosis induced by Aβ. On the contrary, overexpression of TRPC1 inhibited Aβ production and decreased apoptosis [123]. Together, these studies suggest another mechanistic target for therapeutic development within the Ca^{2+} hypothesis of AD.

2.3. Targeting Mitochondrial Ca^{2+}

Mitochondria are crucial organelles that provide energy to the cell in the form of adenosine triphosphate (ATP) via the process of oxidative phosphorylation. Mitochondria form a dynamic tubular network that extends throughout the cytosol, undergoing fusion and fission, which regulates the morphology and structure of the mitochondrial network [124]. Neurons rely strictly on mitochondria to produce ATP, with mitochondria being recruited in areas like synapses, where high energy is required. Mitochondria also buffer Ca^{2+} and shape its signal [125], which is involved in neurotransmission and maintenance of the membrane potential along the axon. At the synaptic level, mitochondria regulate the Ca^{2+} levels necessary for synaptic functions [126]. Mitochondrial Ca^{2+} uptake activates some dehydrogenases at the electron transport chain (ETC), activating mitochondrial respiration and ATP production [127]. The electrochemical gradient created by the ETC allows mitochondria to take up Ca^{2+}. This mitochondrial Ca^{2+} participates in signal transduction and the production of energy. Mitochondria contain two major membranes, the outer mitochondrial membrane (OMM), which contains voltage-dependent anion-selective channel protein (VDAC), permeable to most molecules, and the inner mitochondrial membrane (IMM), which is impermeable to molecules and ions, unless they contain specific channels or transporters.

Ca^{2+} is taken up into the mitochondrial matrix through the mitochondrial Ca^{2+} uniporter (MCU) complex, a highly Ca^{2+}-sensitive ion conductance channel [128,129]. The MCU is a macromolecular complex of proteins, which includes the pore and several regulatory subunits. It is ubiquitously expressed among organisms and defines the pore domain of the complex [128]. Two other proteins participate in the Ca^{2+} permeant pore: MCUb, whose expression is restricted to most vertebrates [130,131], and the essential MCU regulator (EMRE) [132]. The response of the MCU to extramitochondrial Ca^{2+} is regulated by the mitochondrial Ca^{2+} uptake (MICU) family of proteins, which are in the intermembrane space. MICU1 and MICU2 act as Ca^{2+} sensors, each with two Ca^{2+}-binding EF-hand motifs that confer sensitivity to Ca^{2+} [133]. MICU1 and MICU2 also act as gatekeepers of MCU [134], with MICU1 getting involved when the extramitochondrial Ca^{2+} concentration is high, activating the channel open state. At low concentrations, the main player seems to be MICU2, leading to minimal accumulation of Ca^{2+} within mitochondria [135,136], thus preventing mitochondrial Ca^{2+} overload at resting conditions. MICU3, a paralog of MICU1 and MICU2, is mainly expressed in the CNS [137], and has been proposed to enhance mitochondrial Ca^{2+} uptake in neurons [138]. In regulating the MCU pore, the mitochondrial Ca^{2+} uniporter regulator 1 (MCUR1) also plays a role [139]. It has been suggested as a necessary player in MCU-mediated mitochondrial Ca^{2+} uptake. The small Ca^{2+}-binding mitochondrial carrier protein (SCaMC, also known as SLC25A23) [140] seems to also participate in the mitochondrial Ca^{2+} uptake by interacting with MCU and M1CU1.

Mitochondrial Ca^{2+} efflux occurs via the Na^+/Ca^{2+} exchanger (NCLX) [141] and leucine zipper- and EF hand-containing transmembrane protein 1 (Letm1), located at the IMM [142]. Excessive Ca^{2+} in the mitochondrial matrix induces the activation of the mitochondrial permeability transition pore (mPTP) and allows the release of Ca^{2+} ions and small molecules such as cytochrome c [143].

Mitochondrial Ca^{2+} levels are tightly regulated since excessive levels of Ca^{2+} within mitochondria, i.e., mitochondrial Ca^{2+} overload, result in the impairment of mitochondrial function, suppression of ATP production, increase in reactive oxygen species (ROS) production, and mPTP opening. This can lead to caspase activation and cell death via apoptosis [144].

Mitochondrial function has long been considered one of the intracellular processes compromised at the early stages in AD and likely in other neurodegenerative diseases. Moreover, the "mitochondrial cascade hypothesis" was proposed to explain the onset of SAD [145], which posits that mitochondrial dysfunction is the primary process to trigger the cascade of events that lead to late-onset AD. Even though the validity of this hypothesis has yet to be demonstrated, numerous mitochondrial functions are disrupted in AD [146], including mitochondrial morphology and number [147], oxidative phosphorylation, mitochondrial membrane potential, ROS production [148], mitochondrial DNA (mtDNA) oxidation and mutation [149], mitochondrial–ER contacts [150], and mitochondrial dynamics, including mitochondrial transport along the axon and mitophagy [151]. Additionally, both Aβ and tau have been found in mitochondria. Aβ is imported to mitochondrial matrix via translocase of the outer membrane (TOM) [152], and a fraction of intracellular tau has been found within the inner mitochondrial space [153]. Once in mitochondria, they interact with specific intramitochondrial targets, leading to the dysfunction of the organelle. Furthermore, tau accumulation in mitochondrial synaptosomes has been proposed to correlate with synaptic loss in AD brains [154].

Mitochondrial Ca^{2+} dysregulation is considered a fingerprint of AD. Mitochondrial Ca^{2+} overload can be a result of three different processes: (i) increased mitochondrial Ca^{2+} influx (following Ca^{2+} influx from extracellular space or Ca^{2+} transfer from ER), (ii) decreased mitochondrial Ca^{2+} efflux through NCLX, or (iii) reduced mitochondrial Ca^{2+} buffering. Neurotoxic Aβ can lead to mitochondrial Ca^{2+} overload, as shown in in vitro and in vivo models [155–157]. Primary neurons in culture exposed to Aβ oligomers triggered mitochondrial Ca^{2+} overload, leading to mPTP opening, release of cytochrome c, and cell death via mitochondrial-mediated apoptosis [156]. Additionally, studies in mouse neuroblastoma N2a cells co-transfected with the Swedish mutant APP and Δ9 deleted PS1 showed similar mitochondrial impairment, evidenced by the increased mitochondrial apoptotic pathway and caspase-3 activity [158]. Furthermore, Aβ can interact with cyclophilin D—a regulator of mPTP—and promote the release of cytochrome c through the opening of mPTP [159]. This causes neuronal injury and decline of cognitive functions, as shown in a mouse model of AD. Genetic deletion of CypD in Tg AD mice rescues mitochondrial impairment and improves learning and memory [160], suggesting that CypD could represent a potential therapeutic target in AD.

Recently, we showed mitochondrial Ca^{2+} overload in a mouse model of cerebral amyloidosis (APP/PS1). Using in vivo multiphoton imaging and a ratiometric Ca^{2+} reporter, we demonstrated increased levels of mitochondrial Ca^{2+} following Aβ deposition, which preceded neuronal cell death. Moreover, naturally secreted soluble oligomers applied to the healthy brain of Wt mice also increased mitochondrial Ca^{2+} levels, a process that could be prevented by MCU inhibition with the specific channel blocker Ru360 [157]. We also showed, for the first time, that the expression of mitochondrial Ca^{2+} transport-related genes in brain tissue from AD patients was impaired compared to control cases. In particular, genes involved in mitochondrial Ca^{2+} uptake (MCU complex) were downregulated, whereas the only one encoding for Ca^{2+} efflux (NCLX) was upregulated, suggesting a compensatory response to prevent mitochondrial Ca^{2+} overload [157]. However, others reported that different techniques used for evaluating expression showed conflicting results [161]. Another mechanism proposed for mitochondrial Ca^{2+} overload in AD is impairment of mitochondrial Ca^{2+} efflux. Loss of NCLX expression and functionality has also been suggested in AD, whereas genetic rescue of NCLX expression in neurons restored cognitive decline and cellular impairment in transgenic mouse models of AD [161]. Additionally, the more general cytosolic Ca^{2+} overload observed in vivo (as previously cited) may contribute to the observed mitochondrial Ca^{2+} overload. These observations suggest that restoring mitochondrial Ca^{2+} levels in AD could be a promising new therapeutic target against AD.

It has been previously proposed that nonsteroidal anti-inflammatory drugs (NSAIDs) may help in preventing the cognitive decline associated with aging [162]. Unfortunately, results from several clinical trials have given rather pessimistic results [163–165], partly due to inadequate CNS drug penetration of existing NSAIDs, suboptimal doses, unknown molecular targets (and, therefore, unknown pharmacodynamics), and toxicities. Nevertheless, in vitro studies have shown that NSAIDs such as salicylate and the enantiomer (*R*)-Flurbiprofen lacking anti-inflammatory activity, at low concentrations, are able to depolarize mitochondria and inhibit the driving force for mitochondrial Ca^{2+} uptake [166,167]. They act as mild mitochondrial uncouplers without altering cytosolic Ca^{2+} levels. This mild mitochondrial depolarization was able to prevent NMDA- and Aβ-induced mitochondrial Ca^{2+} uptake and cell death [155,156,168]. These results point to mitochondrial Ca^{2+} as a key player in Aβ-driven neurotoxicity and suggest a new mechanism of neuroprotection by NSAIDs independent of their anti-inflammatory activity. Another compound, TG-2112x, has been recently suggested as neuroprotective and proposed as a new therapeutic opportunity. Tg-2112x partially inhibits mitochondrial Ca^{2+} uptake without affecting the mitochondrial membrane potential or mitochondrial bioenergetics, protecting neurons against glutamate excitotoxicity [169].

Abnormal tau hyperphosphorylation also influences mitochondrial transport along the neuronal axon, which leads to a reduction in and impairment of mitochondria at the presynaptic terminal with detrimental consequences and eventual cell death [170,171]. In vitro and in vivo studies have shown that tau dysregulates Ca^{2+} homeostasis in mitochondria. Mitochondrial Ca^{2+} buffering and homeostasis are disrupted in cells overexpressing tau and those exposed to extracellular tau aggregates [61,172]. Additionally, basal mitochondrial Ca^{2+} levels have been shown to be elevated in patient-derived human induced pluripotent stem cell (iPSC) neurons expressing a tau mutation, likely due to the inhibition of NCLX by tau [173]. Elevation in mitochondrial Ca^{2+} levels by tau also increased the vulnerability to Ca^{2+}-induced cell death [173]. Phosphorylated tau has also been found to interact with VDAC in AD brains, leading to mitochondrial dysfunction [174].

Mitochondria and ER membranes are juxtaposed and establish contact points known as mitochondrial-associated membranes (MAMs). They are dynamic lipid rafts enriched in cholesterol and sphingomyelin, as well as in proteins associated with Ca^{2+} dynamics [175,176]. MAMs allow for communication between ER and mitochondria, including metabolic pathways and Ca^{2+} transfer from ER to mitochondria [177]. Increased contacts between ER and mitochondria have been found in human fibroblast cells derived from FAD patients, human brain tissue, and AD mouse models [178,179]. An increased association between the ER and mitochondria has also been observed in a Tg mouse model of tauopathy [180]. Increased contact promotes mitochondrial bioenergetics, but excessive Ca^{2+} transfer can contribute to mitochondrial Ca^{2+} overload and suppression of normal mitochondrial functions, and Aβ oligomers have been found to induce massive Ca^{2+} transfer from ER to mitochondria [116,181–183].

Mitochondria-targeted protective compounds that prevent or minimize mitochondrial dysfunction could represent potential therapeutic strategies in the prevention or treatment of AD. However, several compounds targeting mitochondrial function have been tested in AD without a favorable outcome [184]. Nevertheless, the idea of AD as a multifactorial disease is widespread, and mitochondria as a therapeutic target combined with other medications is emerging as a valid therapy for AD. The list of pharmacologic approaches that directly target mitochondria includes antioxidants (such as vitamin E and C, coenzyme Q10, mitoQ, and melatonin) and phenylpropanoids (such as resveratrol, quercetin, or curcumin) [185]. Antioxidants are generally used to decrease oxidative stress and slow the progression of symptoms that generally accompany AD. Antioxidants such as coenzyme Q10 and mitoquinone mesylate (MitoQ) are antioxidants that directly target mitochondria [186]. Currently, there is a small clinical trial testing MitoQ on cerebrovascular blood flow in AD [187]. The Szeto-Schiller (SS) tetrapeptides, an alternative type of antioxidants that target mitochondria, are small molecules that can reach the mitochondrial matrix and act as antioxidants [188]. Specifically, SS31 (also known as elamipretide) selectively binds to cardiolipin and promotes electron transport while optimizing mitochondrial ATP synthesis [189].

In addition, SS31 inhibits mitochondria swelling and oxidative cell death. In mouse models of cerebral amyloidosis, it was shown that SS31 reduces Aβ production and mitochondrial dysfunction, and enhances mitochondrial biogenesis and synaptic activity [190]. Recently, SS31 combined with the mitochondrial division inhibitor 1 (Mdivi1) was tested in vitro with a positive outcome, suggesting this combination as a possible type of mitochondria-targeted antioxidant in AD [191]. Ongoing clinical trials regarding mitochondria in AD are reviewed in [187,192] and at www.clinicaltrials.gov.

2.4. Targeting Lysosomal Ca^{2+}

Lysosomes are acidic organelles that participate in the endolysosomal system. They are important for autophagy and intracellular Ca^{2+} storage (with comparable Ca^{2+} levels to those of the ER) [193]. The Ca^{2+} transport in and out of the lysosomal lumen provides signals that modulate the fusion of autophagosomes and lysosomes. In order to maintain lysosomal Ca^{2+} homeostasis, lysosomes contain P/Q type VGCCs expressed in the lysosomal membrane that provide Ca^{2+} to the cytosol. Dysregulation in lysosomal Ca^{2+} release via VGCCs leads to defective autophagic fusion and flux [194]. The vacuolar-type H^+-ATPase (V-ATPase) and Ca^{2+}/H^+ exchanger are in charge of lysosomal Ca^{2+} refilling [195]. It has been suggested that this refilling is largely dependent on ER Ca^{2+} [196]. V-ATPase activity predominantly maintains lysosomal pH; however, other ion channels localized to the lysosomal membrane participate in pH regulation during lysosomal proteolysis, including the chloride channel CLC7 [197] and the Ca^{2+} channel TRPML1 (mucolipin) [198,199]. Additionally, Ca^{2+} microdomains generated at the mouth of these channels have been suggested to take part in the regulation of autophagy [200].

Lysosomal Ca^{2+} efflux has been linked to changes in lysosomal pH. Recent reports suggested that decreased lysosomal Ca^{2+} in AD-linked mutations or PS1 knockout (KO) cells is a consequence of elevated lysosomal pH [201]. Raising lysosomal pH leads to autophagy defects and lysosomal Ca^{2+} efflux. PS1 mutant cells exhibit these defects. PS1 KO cells show deficiencies in lysosomal V-ATPase content and function, defective autophagy, and abnormal Ca^{2+} efflux. [201]. Reversal of lysosomal pH abnormalities in PS1 KO cells, but not Ca^{2+} efflux deficits, was sufficient to rescue these same deficits [201]. These data suggest that lysosomal Ca^{2+} defects are secondary to lysosomal pH elevation, and that lysosomal Ca^{2+} dyshomeostasis contributes significantly to the overall Ca^{2+} dysregulation observed in PS1-deficient cells. However, other studies do not support these arguments, citing that, although the autophagosome and lysosome accumulation was apparent in PS1 or PS2 cells, defective lysosome acidification was not found [202]. In addition, defects in lysosome acidification or Ca^{2+} homeostasis have not been observed in FAD-PS2 models [203]. On the contrary, other studies have shown both reduced cytosolic Ca^{2+} signal and lower ER content in FAD-PS2 models. In particular, it was proposed that FAD-PS2 decreases ER and *cis*–medial Golgi Ca^{2+} levels by reducing SERCA activity, which could lead to defective autophagosome–lysosome fusion [203]. Further studies are necessary to confirm these observations and demonstrate whether or not lysosomal Ca^{2+} or pH could be potential therapeutic targets for AD.

Autophagy is a lysosomal degradative pathway responsible for the recycling of different cellular constituents. Especially important under conditions of metabolic stress, this pathway aids in the cellular turnover of damaged or obsolete organelles in order to eliminate misfolded and aggregated proteins left behind by the ubiquitin-proteasome system [204]. Materials are engulfed within double-membrane vesicles (autophagosomes) and targeted to lysosomes for degradation of molecular components. Disruption of autophagy results in accumulation of autophagic vacuoles within swollen dystrophic neurites of affected neurons [205]. Lysosomal Ca^{2+} has been proposed to trigger transcriptional activation of autophagic proteins [200]. Impairment of the autophagy–lysosomal pathway has been described as a hallmark of AD related to lysosomal Ca^{2+} dyshomeostasis. This dysregulation impacts clearance of Aβ and hyperphosphorylated tau, and contributes to their accumulation in the brain [206,207].

3. Astrocytic Ca^{2+} as a Therapeutic Target in AD

Astrocytes, the most abundant cells in the brain, are key regulators of molecular homeostasis in the nervous system. They provide trophic and metabolic support to neurons, sense and modulate neuronal network excitability, and participate in neurovascular coupling and maintenance of the blood–brain barrier [208–210]. Astrocytes do not generate action potentials, but exhibit Ca^{2+} transients followed by a release of gliotrasmitters—such as ATP, glutamate, or gamma-aminobutyric acid (GABA)—in response to neurotransmitters [211]. It has been proposed that the Ca^{2+} global signals—propagating waves—rely on Ca^{2+} release from the ER (mostly mediated by IP$_3$R). Local Ca^{2+} microdomains, on the other hand, result from Ca^{2+} influx via ionotropic receptors, TRPs, SOCE, mitochondrial Ca^{2+} activity, or reversed Na$^+$/Ca^{2+} exchangers [212].

In AD, astrocytes become activated. Reactive astrogliosis is characterized by the biochemical, functional, and morphological reshaping of astrocytes aimed at neuroprotection [213]. Reactive astrocytes upregulate activation markers such as glial fibrillary acidic protein and vimentin. Using postmortem human tissue, it has been shown that reactive astrocytes associated with plaques express higher levels of the glutamate metabotropic receptor mGluR5, which induces Ca^{2+} release form intracellular stores [214]. In vitro, exposure of astrocytes to Aβ increases basal intracellular Ca^{2+} levels as a result of extracellular Ca^{2+} entry, release from mGluR5 and IP$_3$R, and induced Ca^{2+} oscillations or transients [214,215]. Pharmacological inhibition of ER Ca^{2+} release blocks the Aβ-induced astrogliosis both in cultured astrocytes and in organotypic slices [216]. As observed in co-cultures of neurons and astrocytes, the Aβ-induced astrocytic Ca^{2+} transients are followed by neuronal death, suggesting that aberrant astrocytic Ca^{2+} signal results in neurotoxicity [217]. These results, however, are not universal, and other groups have not replicated these observations [218]. It has also been suggested that *APOE4* dysregulates Ca^{2+} excitability in astrocytes by modifying membrane lipid composition. This phenomenon was observed in hippocampal slices from *APOE3* and *APOE4* mice, specifically in male mice [219], suggesting that the *APOE* genotype modulates Ca^{2+} fluxes in astrocytes in a lipid and sex-dependent manner. As demonstrated in primary cortical co-cultures of neurons and astrocytes, exposure to insoluble aggregates of tau failed to induce a Ca^{2+} response in astrocytes [36]. Unfortunately, little else is known about the effects of tau on astrocytic Ca^{2+}, and further research is clearly warranted.

In the intact brain in vivo under physiological conditions, astrocytes show sporadic Ca^{2+} transients as a hallmark of astrocytic activity [220]. As demonstrated in cortical astrocytes of amyloid-depositing mice (APP/PS1), under pathological conditions, the frequency of spontaneous Ca^{2+} waves increases [18]. These same astrocytes exhibit higher resting Ca^{2+} levels. While overall astrocytic hyperactivity was noticed throughout the cortical tissue and not just in the vicinity of amyloid plaques, the astrocytes initiating the intracellular Ca^{2+} waves were located in plaque vicinity [18]. Further studies are needed to determine whether this was an effect of soluble Aβ oligomers or Aβ fibrils. Ca^{2+} hyperactivity in astrocytes has been associated with abnormal purinergic signaling, suggesting that reactive astrocytes release excessive amounts of ATP. This in turn activates P2Y purinoceptors mediating abnormal cytosolic Ca^{2+} signaling [17]. It has also been suggested that alterations in extracellular Ca^{2+} levels can be involved in astrocytic hyperactivity. During increased neuronal activity, extracellular Ca^{2+} decreases following ionotropic glutamate receptor and VGCC activation. Astrocytes sense the extracellular Ca^{2+} decrease and release ATP in response [221]. Increases in extracellular ATP trigger astrocytic Ca^{2+} transients and could contribute to AD-associated astrocytic hyperactivity.

4. Microglial Ca^{2+} as a Therapeutic Target in AD

Microglia are the major immune cells in the brain. They sense and react to alterations in brain homeostasis. They are also involved in synaptic pruning, which occurs during the first weeks of postnatal development and is critical for the maturation of neuronal networks [222]. Microglial activation is characterized by morphological alterations and production of pro- and anti-inflammatory mediators [223]. Intracellular Ca^{2+} regulates microglial activation from its homeostatic resting state to a neurotoxic-activated state. Some microglial functions, including the production and release of proinflammatory factors, such as nitric oxide (NO) and certain cytokines, are Ca^{2+}-dependent processes [224]. In turn, proinflammatory cytokines, tumor necrosis factor α (TNFα), interleukin 1β, and interferon γ, all increase intracellular Ca^{2+} levels in microglia [225–227], while anti-inflammatory cytokines decrease them [228].

AD has long been linked to microglial activation. Microglia surround amyloid plaques [229] after they get recruited within the first days after plaque formation [230]. Once activated, microglia internalize and break down Aβ. Microglia activation is an early process in AD, and it has been shown to be correlated with cognitive deficits [231]. Released proinflammatory cytokines (such as IL-1β and TNF-α) by microglia might stimulate the release of proinflammatory substances by astrocytes, amplifying the inflammatory signal and its neurotoxicity [232,233]. Therefore, neurons and astrocytes in the vicinity of these plaques are likely subjected to high levels of proinflammatory mediators released by activated microglia. These mediators can cause alterations in the Ca^{2+} homeostasis of these cells [234]. Additionally, microglial cultures exposed to Aβ increase their immune response (i.e., cytokine production) and intracellular Ca^{2+}, a process that can be blocked by the dihydropyridine nifedipine and the non-dihydropyridine L-type VGCC antagonist verapamil or diltiazem [235].

Observations from in vitro data have shown that intracellular Ca^{2+} homeostasis is impaired in activated microglia. Microglia isolated from AD brain tissue have elevated cytosolic Ca^{2+} levels compared to controls and exhibit reduced responsiveness to stimuli in vitro [236]. Additionally, mouse microglia activated by lipopolysaccharide (LPS) display increased basal Ca^{2+} levels and a reduced agonist-induced Ca^{2+} signal [224]. Ramified activated microglia display large intracellular Ca^{2+} transients in response to the damage of individual cells in their vicinity. The use of in vivo Ca^{2+} imaging and multiphoton microscopy has allowed the study of microglial Ca^{2+} dynamics in the intact living brain. Microglia display rare Ca^{2+} transients in their resting state, but respond with larger Ca^{2+} transients when activated [237]. Microglial Ca^{2+} transients are attributed to Ca^{2+} release from intracellular stores and are prevented by the activation of ATP receptors [237]. Blocking SOCE—via knocking down or knocking out STIM1/2 and Orai—reduces immune functioning, including phagocytosis, migration, and cytokine production in primary isolated murine microglia [238,239]. On the other hand, blocking RyR prevents LPS-induced neurotoxicity mediated by microglia [240]. Although they have not been studied in depth, it has been suggested that microglia display Ca^{2+} microdomains. Some observations suggest that global Ca^{2+} elevations in microglia trigger phagocytosis and migration, whereas local Ca^{2+} increases in their processes regulate acute chemotactic migration [241]. Taken together, elevated Ca^{2+} levels seem to be a hallmark of activated microglia and their regulation, a potential therapeutic target for AD therapy.

Microglia express P2X receptors, a subfamily of purinergic ionotropic receptors, located in the plasma membrane which are permeable to Ca^{2+}, Na^+, and K^+ [242]. Overactivation of P2X receptors may lead to cell death via membrane depolarization, mitochondrial stress, and ROS production [243]. As measured in postmortem brain tissue from AD patients, P2X$_7$ expression is upregulated in microglia in AD [244]. This same effect was shown in plaque-associated microglia in an AD mouse model and after intrahippocampal injection of Aβ_{42} [245]. It is believed that these high levels of P2X$_7$ contribute to the enhanced inflammatory responses observed in AD [244,246,247]. Inhibition of P2X$_7$ receptors has been shown to be neuroprotective as it reduces the dendritic spine loss induced by Aβ [248], as well as Aβ production in general [249]. P2X$_7$ KO mice express reduced plaque size and improved behavioral scores [250], suggesting P2X$_7$ as a potential therapeutic target in AD. Additional in vitro studies have shown that P2X$_7$ activation leads to microglial NOD-, LRR- and pyrin domain-containing protein 3 (NLRP3) inflammasome activation, which requires Ca^{2+} mobilization from intracellular stores [251,252]. Aβ also triggers this NLRP3 activation [253], and NLRP3 is highly activated in microglial cells surrounding amyloid plaques [253]. NLRP3 KO mice show reduced amyloid burden in the brain and have reduced memory impairment [254]. This confirms that P2X$_7$ and NLRP3 could be candidate targets for AD therapeutics.

The triggering receptor expressed on myeloid cells 2 (*TREM2*) gene has been recently identified as a risk gene for AD [255]. Its low-frequency variants increase the risk of developing AD similar to the APOE4 allele. TREM2 is a transmembrane protein receptor expressed on microglia. It stimulates phagocytosis and suppresses inflammation [255]. TREM2 overexpression in a mouse model of AD (APP/PS1) decreased AD-related pathology and improved cognitive functions [256], suggesting that modeling microglial functions could be a protective target in AD. Immunotherapy using antibodies to stimulate TREM2 signaling in order to improve AD pathology is currently being developed by different groups. Stimulation with anti-TREM2 antibodies in vitro produced Ca^{2+} influx and extracellular signal-regulated kinase (ERK) signaling activation in human dendritic cells [257]. When to stimulate TREM2 to treat AD, however, is not clear, and it must be kept in mind that the use of these antibodies could alter the binding of other TREM2 ligands. Further studies will be needed to fully understand TREM2 function and its role in AD therapy.

5. Conclusions and Future Directions

AD is a multifactorial complex disease that leads to progressive dementia. Its nature brings upon the equally complex task of developing a treatment strategy. Current medications for Alzheimer's disease only treat the symptoms and cannot stop the damage that AD pathology causes to brain cells. Therefore, an urgent need exists for new target discovery that directly targets AD pathology and alters the course of its progression. On the basis of a wide range of studies, evidence suggests that treatment should be initiated in AD's earliest stages, before the start of deposition of pathology and occurrence of irreversible mental decline. Ca^{2+} dyshomeostasis is an early event in the AD timeline. Ca^{2+} dysregulation in AD comes as a result of hyperactivity of Ca^{2+} channels in the plasma membrane and intracellular compartments. It does not seem to be restricted to neurons, but rather is a global phenomenon that affects many cell types in the brain (Figure 2).

Intracellular Ca^{2+} homeostasis is mediated by several organelles, such as the ER, mitochondria, and lysosomes, which contribute to cell stress regulation. Increased Ca^{2+} concentrations in these compartments disrupt normal homeostasis, eventually leading to accumulated pathogenic proteins, which in turn further impair Ca^{2+} homeostasis, leading to severe alterations in neuronal circuitry. With these underlying data, it is clear that isolating potential therapeutic strategies aimed at normalizing Ca^{2+} levels is important. Current FDA-approved AD treatments target plasma Ca^{2+} channels, but more specific approaches are needed to target other prevalent and disrupted intracellular Ca^{2+} signaling pathways, such as those of the ER or mitochondria. As our knowledge in Ca^{2+} dysregulation in AD grows, it seems more obvious that targeting these other sources of Ca^{2+} dysregulation could be an effective therapeutic strategy.

A better understanding of the onset and progression of neurodegenerative diseases will facilitate rapid diagnosis and target selection, allowing for early treatment. A truly effective method for preventing or treating Alzheimer's disease will likely involve a combination approach for targets, such as Aβ plaque clearance or soluble tau removal. Additionally, reversal of cellular processes that are disrupted by Aβ or tau accumulation (including Ca^{2+} dyshomeostasis), early diagnosis, and/or lifestyle changes would also be necessary for successful therapeutic intervention. Gene therapy is an emerging therapeutic strategy for the treatment of neurodegenerative disorders, including AD, particularly when traditional therapies are not responsive to well-validated genetic targets. Gene therapy has already shown efficacy in preclinical studies, utilizing different routes for gene delivery [258,259]. Recently, different groups proposed gene therapy as a strategy in the battle against AD, as it is designed to focus on one specific target in affected brain regions. Several gene therapy strategies for AD have already been tested. These include acting directly on APP metabolism, neuroprotection, targeting inflammatory pathways, or modulating genes related to lipid metabolism [260]. Unfortunately, they have not provided an encouraging outcome so far, as they sometimes show unexpected or undesirable side effects. One of the ongoing clinical trials is designed to evaluate gene therapy use in AD patients (already clinically diagnosed) that are APOE4 homozygotes (www.clinicaltrials.gov). The study aims to evaluate whether intracisternal administration of APOE2 to APOE4 homozygotes AD patients will lead to conversion of the APOE protein isoforms from APOE4 homozygotes to APOE2–APOE4, which has given positive results in mice and monkeys previously [261,262]. If this therapy slows the illness in people with advanced AD, this could also function as a method for disease prevention, reducing the risk of disease development in healthy people. A combination of these targets and therapies should reduce stress levels and cell death in AD, offering pathological and potentially symptomatic relief.

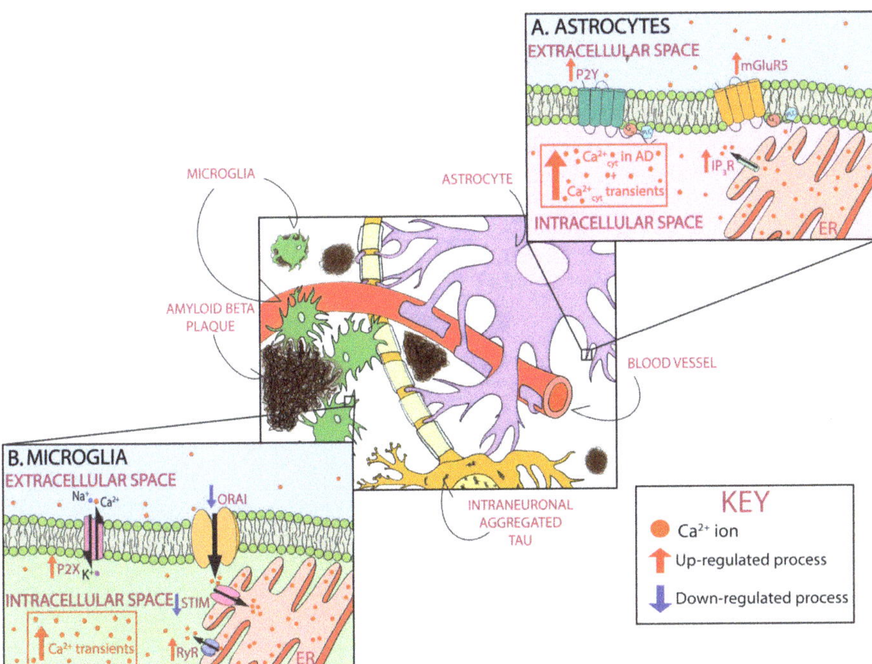

Figure 2. Astrocytic and microglial Ca^{2+} as a therapeutic target in AD. Schematic of glial Ca^{2+} cells dysregulation in the presence of AD pathology. In astrocytes, P2Y purinoceptors and glutamate metabotropic receptors mGluR5, when activated, cause Ca^{2+} increase by releasing Ca^{2+} from intracellular stores. As shown in red, all three receptors are upregulated in AD. In addition, cytosolic Ca^{2+} levels are increased in astrocytes, and they exhibit Ca^{2+} transients (**A**). In microglia, P2X receptors are upregulated in AD, thus leading to Ca^{2+} dysregulation. SOCE, with involves STIM and Orai, is also responsible for Ca^{2+} influx, specifically into the lumen of the endoplasmic reticulum (ER). This pathway is downregulated in AD (shown in blue). RyRs mediate Ca^{2+} efflux from the ER, a process that is upregulated in AD. Microglia also show Ca^{2+} dysregulation by showing cytosolic Ca^{2+} transients (**B**).

Author Contributions: M.C.-R. and E.K.K. wrote the original draft. B.J.B. provided feedback and edited the manuscript. All authors have read and agreed to the published version of the manuscript.

Funding: This work was supported by NIHR01AG0442603, S10 RR025645, and R56AG060974 (B.J.B.), and by the Tosteson & Fund for Medical Discovery and the BrightFocus Foundation A2019488F (M.C.-R.).

Conflicts of Interest: The authors declare no competing interests.

References

1. Corder, E.H.; Saunders, A.M.; Strittmatter, W.J.; E Schmechel, D.; Gaskell, P.C.; Small, G.W.; Roses, A.D.; Haines, J.L.; A Pericak-Vance, M. Gene dose of apolipoprotein E type 4 allele and the risk of Alzheimer's disease in late onset families. *Science* **1993**, *261*, 921–923. [CrossRef] [PubMed]
2. A Hardy, J.; A Higgins, G.; Mayford, M.; Barzilai, A.; Keller, F.; Schacher, S.; Kandel, E. Alzheimer's disease: The amyloid cascade hypothesis. *Science* **1992**, *256*, 184–185. [CrossRef] [PubMed]
3. Davies, P. Selective Loss Of Central Cholinergic Neurons In Alzheimer's Disease. *Lancet* **1976**, *308*, 1403. [CrossRef]
4. Frost, B.; Jacks, R.L.; Diamond, M.I. Propagation of Tau Misfolding from the Outside to the Inside of a Cell. *J. Biol. Chem.* **2009**, *284*, 12845–12852. [CrossRef]
5. McGeer, P.L.; Rogers, J. Anti-inflammatory agents as a therapeutic approach to Alzheimer's disease. *Neurology* **1992**, *42*, 447. [CrossRef] [PubMed]
6. Tarasoff-Conway, J.M.; Carare, R.O.; Osorio, R.S.; Glodzik, L.; Butler, T.; Fieremans, E.; Axel, L.; Rusinek, H.; Nicholson, C.; Zlokovic, B.V.; et al. Clearance systems in the brain—Implications for Alzheimer disease. *Nat. Rev. Neurol.* **2015**, *11*, 457–470. [CrossRef] [PubMed]
7. Mattson, M.P. Calcium and neurodegeneration. *Aging Cell* **2007**, *6*, 337–350. [CrossRef]
8. Khachaturian, Z.S. Hypothesis on the Regulation of Cytosol Calcium Concentration and the Aging Brain. *Neurobiol. Aging* **1987**, *8*, 345–346. [CrossRef]
9. Boada, M.; Antúnez, C.; López-Arrieta, J.; Galán, J.J.; Morón, F.J.; Hernández, I.; Marín, J.; Martínez-Lage, P.; Alegret, M.; Carrasco, J.M.; et al. CALHM1 P86L Polymorphism is Associated with Late-Onset Alzheimer's Disease in a Recessive Model. *J. Alzheimer's Dis.* **2010**, *20*, 247–251. [CrossRef]
10. Tolar, M.; Keller, J.N.; Chan, S.; Mattson, M.P.; Marques, M.A.; Crutcher, K.A. Truncated Apolipoprotein E (ApoE) Causes Increased Intracellular Calcium and May Mediate ApoE Neurotoxicity. *J. Neurosci.* **1999**, *19*, 7100–7110. [CrossRef]
11. Zatti, G.; Ghidoni, R.; Barbiero, L.; Binetti, G.; Pozzan, T.; Fasolato, C.; Pizzo, P. The presenilin 2 M239I mutation associated with familial Alzheimer's disease reduces Ca^{2+} release from intracellular stores. *Neurobiol. Dis.* **2004**, *15*, 269–278. [CrossRef] [PubMed]
12. Arbel-Ornath, M.; Hudry, E.; Boivin, J.R.; Hashimoto, T.; Takeda, S.; Kuchibhotla, K.V.; Hou, S.; Lattarulo, C.R.; Belcher, A.M.; Trujillo, P.B.; et al. Soluble oligomeric amyloid-beta induces calcium dyshomeostasis that precedes synapse loss in the living mouse brain. *Mol. Neurodegener.* **2017**, *12*, 27. [CrossRef] [PubMed]
13. Busche, M.A.; Chen, X.; Henning, H.A.; Reichwald, J.; Staufenbiel, M.; Sakmann, B.; Konnerth, A. Critical role of soluble amyloid-beta for early hippocampal hyperactivity in a mouse model of Alzheimer's disease. *Proc. Natl. Acad. Sci. USA* **2012**, *109*, 8740–8745. [CrossRef] [PubMed]
14. Busche, M.A.; Eichhoff, G.; Adelsberger, H.; Abramowski, D.; Wiederhold, K.-H.; Haass, C.; Staufenbiel, M.; Konnerth, A.; Garaschuk, O. Clusters of Hyperactive Neurons Near Amyloid Plaques in a Mouse Model of Alzheimer's Disease. *Science* **2008**, *321*, 1686–1689. [CrossRef] [PubMed]
15. Kuchibhotla, K.V.; Goldman, S.T.; Lattarulo, C.R.; Wu, H.Y.; Hyman, B.T.; Bacskai, B.J. Abeta plaques lead to aberrant regulation of calcium homeostasis in vivo resulting in structural and functional disruption of neuronal networks. *Neuron* **2008**, *59*, 214–225. [CrossRef] [PubMed]
16. Busche, M.A.; Wegmann, S.; Dujardin, S.; Commins, C.; Schiantarelli, J.; Klickstein, N.; Kamath, T.V.; Carlson, G.A.; Nelken, I.; Hyman, B.T. Tau impairs neural circuits, dominating amyloid-beta effects, in Alzheimer models in vivo. *Nat. Neurosci.* **2019**, *22*, 57–64. [CrossRef]
17. Delekate, A.; Füchtemeier, M.; Schumacher, T.; Ulbrich, C.; Foddis, M.; Petzold, G.C. Metabotropic P2Y1 receptor signalling mediates astrocytic hyperactivity in vivo in an Alzheimer's disease mouse model. *Nat. Commun.* **2014**, *5*, 5422. [CrossRef]
18. Kuchibhotla, K.V.; Lattarulo, C.R.; Hyman, B.T.; Bacskai, B.J. Synchronous Hyperactivity and Intercellular Calcium Waves in Astrocytes in Alzheimer Mice. *Science* **2009**, *323*, 1211–1215. [CrossRef]
19. Wang, X.; Kastanenka, K.V.; Arbel-Ornath, M.; Commins, C.; Kuzuya, A.; Lariviere, A.J.; Krafft, G.A.; Hefti, F.; Jerecic, J.; Bacskai, B.J. An acute functional screen identifies an effective antibody targeting amyloid-beta oligomers based on calcium imaging. *Sci. Rep.* **2018**, *8*, 4634. [CrossRef]
20. Palop, J.J.; Mucke, L. Network abnormalities and interneuron dysfunction in Alzheimer disease. *Nat. Rev. Neurosci.* **2016**, *17*, 777–792. [CrossRef]

21. Khachaturian, Z.S. The role of calcium regulation in brain aging: Reexamination of a hypothesis. *Aging Clin. Exp. Res.* **1989**, *1*, 17–34. [CrossRef] [PubMed]
22. LaFerla, F.M. Calcium dyshomeostasis and intracellular signalling in alzheimer's disease. *Nat. Rev. Neurosci.* **2002**, *3*, 862–872. [CrossRef] [PubMed]
23. Abeysinghe, A.; Deshapriya, R.; Udawatte, C. Alzheimer's disease; a review of the pathophysiological basis and therapeutic interventions. *Life Sci.* **2020**, *256*, 117996. [CrossRef] [PubMed]
24. Panza, F.; Solfrizzi, V.; Imbimbo, B.P.; Logroscino, G. Amyloid-directed monoclonal antibodies for the treatment of Alzheimer's disease: The point of no return? *Expert Opin. Biol. Ther.* **2014**, *14*, 1465–1476. [CrossRef]
25. Sevigny, J.; Chiao, P.; Bussiere, T.; Weinreb, P.H.; Williams, L.; Maier, M.; Dunstan, R.; Salloway, S.; Chen, T.; Ling, Y.; et al. The antibody aducanumab reduces Abeta plaques in Alzheimer's disease. *Nature* **2016**, *537*, 50–56. [CrossRef]
26. Selkoe, D.J.; Hardy, J. The amyloid hypothesis of Alzheimer's disease at 25 years. *EMBO Mol. Med.* **2016**, *8*, 595–608. [CrossRef]
27. Bagur, R.; Hajnóczky, G. Intracellular Ca^{2+} Sensing: Its Role in Calcium Homeostasis and Signaling. *Mol. Cell* **2017**, *66*, 780–788. [CrossRef]
28. Berridge, M.J.; Lipp, P.; Bootman, M.D. The versatility and universality of calcium signalling. *Nat. Rev. Mol. Cell Biol.* **2000**, *1*, 11–21. [CrossRef]
29. Mattson, M.P. Calcium and neuronal injury in Alzheimer's disease. Contributions of beta-amyloid precursor protein mismetabolism, free radicals, and metabolic compromise. *Ann. N. Y. Acad. Sci.* **1994**, *747*, 50–76. [CrossRef]
30. Arispe, N.; Rojas, E.; Pollard, H.B. Alzheimer disease amyloid beta protein forms calcium channels in bilayer membranes: Blockade by tromethamine and aluminum. *Proc. Natl. Acad. Sci. USA* **1993**, *90*, 567–571. [CrossRef]
31. De Felice, F.G.; Velasco, P.T.; Lambert, M.P.; Viola, K.; Fernandez, S.J.; Ferreira, S.T.; Klein, W.L. Abeta oligomers induce neuronal oxidative stress through an N-methyl-D-aspartate receptor-dependent mechanism that is blocked by the Alzheimer drug memantine. *J. Biol. Chem.* **2007**, *282*, 11590–11601. [CrossRef] [PubMed]
32. Pourbadie, H.G.; Naderi, N.; Mehranfard, N.; Janahmadi, M.; Khodagholi, F.; Motamedi, F. Preventing Effect of L-Type Calcium Channel Blockade on Electrophysiological Alterations in Dentate Gyrus Granule Cells Induced by Entorhinal Amyloid Pathology. *PLoS ONE* **2015**, *10*, e0117555. [CrossRef] [PubMed]
33. Stutzmann, G.E.; Smith, I.; Caccamo, A.; Oddo, S.; LaFerla, F.M.; Parker, I. Enhanced Ryanodine Receptor Recruitment Contributes to Ca^{2+} Disruptions in Young, Adult, and Aged Alzheimer's Disease Mice. *J. Neurosci.* **2006**, *26*, 5180–5189. [CrossRef] [PubMed]
34. Lawlor, B.A.; Kennelly, S.; O'Dwyer, S.; Cregg, F.; Walsh, C.D.; Cohen, R.; Kenny, R.A.; Howard, R.; Murphy, C.; Adams, J.; et al. NILVAD protocol: A European multicentre double-blind placebo-controlled trial of nilvadipine in mild-to-moderate Alzheimer's disease. *BMJ Open* **2014**, *4*, e006364. [CrossRef] [PubMed]
35. Lawlor, B.A.; Segurado, R.; Kennelly, S.; Rikkert, M.G.M.O.; Howard, R.; Pasquier, F.; Börjesson-Hanson, A.; Tsolaki, M.; Lucca, U.; Molloy, D.W.; et al. Nilvadipine in mild to moderate Alzheimer disease: A randomised controlled trial. *PLoS Med.* **2018**, *15*, e1002660. [CrossRef] [PubMed]
36. Esteras, N.; Kundel, F.; Amodeo, G.F.; Pavlov, E.V.; Klenerman, D.; Abramov, A.Y. Insoluble tau aggregates induce neuronal death through modification of membrane ion conductance, activation of voltage-gated calcium channels and NADPH oxidase. *FEBS J.* **2020**. [CrossRef] [PubMed]
37. Dreses-Werringloer, U. A polymorphism in CALHM1 influences Ca^{2+} homeostasis, Abeta levels, and Alzheimer's disease risk. *Cell* **2008**, *133*, 1149–1161. [CrossRef] [PubMed]
38. Minster, R.L.; Demirci, F.Y.; DeKosky, S.T.; Kamboh, M.I. No association betweenCALHM1variation and risk of Alzheimer disease. *Hum. Mutat.* **2009**, *30*, E566–E569. [CrossRef]
39. Tan, E.; Ho, P.; Cheng, S.; Yih, Y.; Li, H.; Fook-Chong, S.; Lee, W.; Zhao, Y. CALHM1 variant is not associated with Alzheimer's disease among Asians. *Neurobiol. Aging* **2011**, *32*, 546.e11–546.e12. [CrossRef]
40. Vingtdeux, V.; Chang, E.H.; Frattini, S.A.; Zhao, H.; Chandakkar, P.; Adrien, L.; Strohl, J.J.; Gibson, E.L.; Ohmoto, M.; Matsumoto, I.; et al. CALHM1 deficiency impairs cerebral neuron activity and memory flexibility in mice. *Sci. Rep.* **2016**, *6*, 24250. [CrossRef]
41. Liu, J.; Chang, L.; Song, Y.; Li, H.; Wu, Y. The Role of NMDA Receptors in Alzheimer's Disease. *Front. Neurosci.* **2019**, *13*, 43. [CrossRef] [PubMed]

42. Hardingham, G.E.; Bading, H. Synaptic versus extrasynaptic NMDA receptor signalling: Implications for neurodegenerative disorders. *Nat. Rev. Neurosci.* **2010**, *11*, 682–696. [CrossRef] [PubMed]
43. Ferreira, I.; Bajouco, L.; Mota, S.; Auberson, Y.; Oliveira, C.; Rego, A.C. Amyloid beta peptide 1–42 disturbs intracellular calcium homeostasis through activation of GluN2B-containing N-methyl-d-aspartate receptors in cortical cultures. *Cell Calcium* **2012**, *51*, 95–106. [CrossRef] [PubMed]
44. Ronicke, R.; Mikhaylova, M.; Ronicke, S.; Meinhardt, J.; Schroder, U.H.; Fandrich, M.; Reiser, G.; Kreutz, M.R.; Reymann, K.G. Early neuronal dysfunction by amyloid beta oligomers depends on activation of NR2B-containing NMDA receptors. *Neurobiol. Aging* **2011**, *32*, 2219–2228. [CrossRef]
45. Zhang, Y.; Li, P.; Feng, J.; Wu, M. Dysfunction of NMDA receptors in Alzheimer's disease. *Neurol. Sci.* **2016**, *37*, 1039–1047. [CrossRef]
46. Roberson, E.D.; Scearce-Levie, K.; Palop, J.J.; Yan, F.; Cheng, I.H.; Wu, T.; Gerstein, H.; Yu, G.Q.; Mucke, L. Reducing endogenous tau ameliorates amyloid beta-induced deficits in an Alzheimer's disease mouse model. *Science* **2007**, *316*, 750–754. [CrossRef]
47. Decker, J.M.; Krüger, L.; Sydow, A.; Dennissen, F.J.; Siskova, Z.; Mandelkow, E.; Mandelkow, E. The Tau/A152T mutation, a risk factor for frontotemporal-spectrum disorders, leads to NR2B receptor-mediated excitotoxicity. *EMBO Rep.* **2016**, *17*, 552–569. [CrossRef]
48. Maeda, S.; Djukic, B.; Taneja, P.; Yu, G.; Lo, I.; Davis, A.; Craft, R.; Guo, W.; Wang, X.; Kim, D.; et al. Expression of A152T human tau causes age-dependent neuronal dysfunction and loss in transgenic mice. *EMBO Rep.* **2016**, *17*, 530–551. [CrossRef]
49. Esclaire, F.; Lesort, M.; Blanchard, C.; Hugon, J. Glutamate toxicity enhances tau gene expression in neuronal cultures. *J. Neurosci. Res.* **1997**, *49*, 309–318. [CrossRef]
50. Sindou, P.; Lesort, M.; Couratier, P.; Yardin, C.; Esclaire, F.; Hugon, J. Glutamate increases tau phosphorylation in primary neuronal cultures from fetal rat cerebral cortex. *Brain Res.* **1994**, *646*, 124–128. [CrossRef]
51. Sun, X.-Y.; Tuo, Q.-Z.; Liuyang, Z.-Y.; Xie, A.-J.; Feng, X.-L.; Yan, X.; Qiu, M.; Li, S.; Wang, X.-L.; Cao, F.-Y.; et al. Extrasynaptic NMDA receptor-induced tau overexpression mediates neuronal death through suppressing survival signaling ERK phosphorylation. *Cell Death Dis.* **2016**, *7*, e2449. [CrossRef] [PubMed]
52. Lipton, S.A. Failures and successes of NMDA receptor antagonists: Molecular basis for the use of open-channel blockers like memantine in the treatment of acute and chronic neurologic insults. *NeuroRx* **2004**, *1*, 101–110. [CrossRef] [PubMed]
53. Kishi, T.; Matsunaga, S.; Oya, K.; Nomura, I.; Ikuta, T.; Iwata, N. Memantine for Alzheimer's Disease: An Updated Systematic Review and Meta-analysis. *J. Alzheimer's Dis.* **2017**, *60*, 401–425. [CrossRef] [PubMed]
54. Figueiredo, C.P.; Clarke, J.R.; Ledo, J.H.; Ribeiro, F.C.; Costa, C.V.; Melo, H.M.; Mota-Sales, A.P.; Saraiva, L.M.; Klein, W.L.; Sebollela, A. et al.; et al. Memantine rescues transient cognitive impairment caused by high-molecular-weight abeta oligomers but not the persistent impairment induced by low-molecular-weight oligomers. *J. Neurosci.* **2013**, *33*, 9626–9634. [CrossRef]
55. Song, M.S.; Rauw, G.; Baker, G.B.; Kar, S. Memantine protects rat cortical cultured neurons against β-amyloid-induced toxicity by attenuating tau phosphorylation. *Eur. J. Neurosci.* **2008**, *28*, 1989–2002. [CrossRef]
56. Wevers, A.; Monteggia, L.; Nowacki, S.; Bloch, W.; Schutz, U.; Lindstrom, J.; Pereira, E.F.R.; Eisenberg, H.; Giacobini, E.; De Vos, R.A.I.; et al. Expression of nicotinic acetylcholine receptor subunits in the cerebral cortex in Alzheimer's disease: Histotopographical correlation with amyloid plaques and hyperphosphorylated-tau protein. *Eur. J. Neurosci.* **1999**, *11*, 2551–2565. [CrossRef]
57. Ferreira-Vieira, T.H.; Guimaraes, I.M.; Silva, F.R.; Ribeiro, F.M. Alzheimer's disease: Targeting the Cholinergic System. *Curr. Neuropharmacol.* **2016**, *14*, 101–115. [CrossRef]
58. Lahiri, D.K.; Rogers, J.T.; Greig, N.H.; Sambamurti, K. Rationale for the Development of Cholinesterase Inhibitors as Anti- Alzheimer Agents. *Curr. Pharm. Des.* **2004**, *10*, 3111–3119. [CrossRef]
59. Herholz, K. Acetylcholine esterase activity in mild cognitive impairment and Alzheimer's disease. *Eur. J. Nucl. Med. Mol. Imaging* **2008**, *35*, 25–29. [CrossRef]
60. Patel, L.; Grossberg, G.T. Combination Therapy for Alzheimer's Disease. *Drugs Aging* **2011**, *28*, 539–546. [CrossRef]
61. Gómez-Ramos, A.; Díaz-Hernández, M.; Rubio, A.; Miras-Portugal, M.T.; Avila, J. Extracellular tau promotes intracellular calcium increase through M1 and M3 muscarinic receptors in neuronal cells. *Mol. Cell. Neurosci.* **2008**, *37*, 673–681. [CrossRef] [PubMed]

62. Hartigan, J.A.; Johnson, G.V.W. Transient Increases in Intracellular Calcium Result in Prolonged Site-selective Increases in Tau Phosphorylation through a Glycogen Synthase Kinase 3β-dependent Pathway. *J. Biol. Chem.* **1999**, *274*, 21395–21401. [CrossRef] [PubMed]

63. Giacobini, E.; Gold, G. Alzheimer disease therapy—Moving from amyloid-beta to tau. *Nat. Rev. Neurol.* **2013**, *9*, 677–686. [CrossRef]

64. Kang, S.; Son, S.M.; Baik, S.H.; Yang, J.; Mook-Jung, I. Autophagy-Mediated Secretory Pathway is Responsible for Both Normal and Pathological Tau in Neurons. *J. Alzheimer's Dis.* **2019**, *70*, 667–680. [CrossRef] [PubMed]

65. Clavaguera, F.; Bolmont, T.; Crowther, R.A.; Abramowski, D.; Frank, S.; Probst, A.; Fraser, G.; Stalder, A.K.; Beibel, M.; Staufenbiel, M.; et al. Transmission and spreading of tauopathy in transgenic mouse brain. *Nat. Cell Biol.* **2009**, *11*, 909–913. [CrossRef] [PubMed]

66. Devos, S.L.; Corjuc, B.T.; Oakley, D.H.; Nobuhara, C.K.; Bannon, R.N.; Chase, A.; Commins, C.; Gonzalez, J.A.; Dooley, P.M.; Frosch, M.P.; et al. Synaptic Tau Seeding Precedes Tau Pathology in Human Alzheimer's Disease Brain. *Front. Neurosci.* **2018**, *12*, 267. [CrossRef]

67. Friedhoff, P.; Von Bergen, M.; Mandelkow, E.-M.; Davies, P. A nucleated assembly mechanism of Alzheimer paired helical filaments. *Proc. Nat. Acad. Sci. USA* **1998**, *95*, 15712–15717. [CrossRef]

68. Colin, M.; Dujardin, S.; Schraen-Maschke, S.; Meno-Tetang, G.; Duyckaerts, C.; Courade, J.-P.; Buée, L. From the prion-like propagation hypothesis to therapeutic strategies of anti-tau immunotherapy. *Acta Neuropathol.* **2020**, *139*, 3–25. [CrossRef]

69. Congdon, E.E.; Sigurdsson, E.M. Tau-targeting therapies for Alzheimer disease. *Nat. Rev. Neurol.* **2018**, *14*, 399–415. [CrossRef]

70. Jadhav, S.; Avila, J.; Schöll, M.; Kovacs, G.G.; Kövari, E.; Skrabana, R.; Evans, L.D.; Kontsekova, E.; Malawska, B.; De Silva, R.; et al. A walk through tau therapeutic strategies. *Acta Neuropathol. Commun.* **2019**, *7*, 1–31. [CrossRef]

71. Cummings, J.; Lee, G.; Ritter, A.; Sabbagh, M.; Zhong, K. Alzheimer's disease drug development pipeline. *Alzheimer's Dement. Transl. Res. Clin. Interv.* **2019**, *5*, 272–293. [CrossRef]

72. Cirrito, J.R.; Yamada, K.A.; Finn, M.B.; Sloviter, R.S.; Bales, K.R.; May, P.C.; Schoepp, D.D.; Paul, S.M.; Mennerick, S.; Holtzman, D.M. Synaptic activity regulates interstitial fluid amyloid-beta levels in vivo. *Neuron* **2005**, *48*, 913–922. [CrossRef] [PubMed]

73. Kamenetz, F.; Tomita, T.; Hsieh, H.; Seabrook, G.; Borchelt, D.; Iwatsubo, T.; Sisodia, S.; Malinow, R. APP Processing and Synaptic Function. *Neuron* **2003**, *37*, 925–937. [CrossRef]

74. Palop, J.J. Epilepsy and Cognitive Impairments in Alzheimer Disease. *Arch. Neurol.* **2009**, *66*, 435–440. [CrossRef] [PubMed]

75. Sanchez, P.E.; Zhu, L.; Verret, L.; Vossel, K.A.; Orr, A.G.; Cirrito, J.R.; Devidze, N.; Ho, K.; Yu, G.-Q.; Palop, J.J.; et al. Levetiracetam suppresses neuronal network dysfunction and reverses synaptic and cognitive deficits in an Alzheimer's disease model. *Proc. Nat. Acad. Sci. USA* **2012**, *109*, E2895–E2903. [CrossRef]

76. Kastanenka, K.V.; Bussiere, T.; Shakerdge, N.; Qian, F.; Weinreb, P.H.; Rhodes, K.; Bacskai, B.J. Immunotherapy with Aducanumab Restores Calcium Homeostasis in Tg2576 Mice. *J. Neurosci.* **2016**, *36*, 12549–12558. [CrossRef]

77. Schneider, L. A resurrection of aducanumab for Alzheimer's disease. *Lancet Neurol.* **2020**, *19*, 111–112. [CrossRef]

78. Mekahli, D.; Bultynck, G.; Parys, J.B.; De Smedt, H.; Missiaen, L. Endoplasmic-Reticulum Calcium Depletion and Disease. *Cold Spring Harb. Perspect. Biol.* **2011**, *3*, a004317. [CrossRef]

79. Ito, E.; Oka, K.; Etcheberrigaray, R.; Nelson, T.J.; McPHIE, D.L.; Tofel-Grehl, B.; Gibson, G.E.; Alkon, D.L. Internal Ca^{2+} mobilization is altered in fibroblasts from patients with Alzheimer disease. *Proc. Nat. Acad. Sci. USA* **1994**, *91*, 534–538. [CrossRef]

80. Etcheberrigaray, R.; Hirashima, N.; Nee, L.; Prince, J.; Govoni, S.; Racchi, M.; Tanzi, R.E.; Alkon, D.L. Calcium responses in fibroblasts from asymptomatic members of Alzheimer's disease families. *Neurobiol. Dis.* **1998**, *5*, 37–45. [CrossRef]

81. Stutzmann, G.E.; Caccamo, A.; LaFerla, F.M.; Parker, I. Dysregulated IP3 Signaling in Cortical Neurons of Knock-In Mice Expressing an Alzheimer's-Linked Mutation in Presenilin1 Results in Exaggerated Ca^{2+} Signals and Altered Membrane Excitability. *J. Neurosci.* **2004**, *24*, 508–513. [CrossRef] [PubMed]

82. Shilling, D.; Müller, M.; Takano, H.; Mak, D.-O.D.; Abel, T.; Coulter, D.A.; Foskett, J.K. Suppression of InsP3 Receptor-Mediated Ca^{2+} Signaling Alleviates Mutant Presenilin-Linked Familial Alzheimer's Disease Pathogenesis. *J. Neurosci.* **2014**, *34*, 6910–6923. [CrossRef]

83. De Strooper, B.; Annaert, W.; Cupers, P.; Saftig, P.; Craessaerts, K.; Mumm, J.S.; Schroeter, E.H.; Schrijvers, V.; Wolfe, M.S.; Ray, W.J.; et al. A presenilin-1-dependent gamma-secretase-like protease mediates release of Notch intracellular domain. *Nature* **1999**, *398*, 518–522. [CrossRef]

84. Struhl, G.; Greenwald, I. Presenilin is required for activity and nuclear access of Notch in Drosophila. *Nat. Cell Biol.* **1999**, *398*, 522–525. [CrossRef]

85. A Saura, C.; Choi, S.-Y.; Beglopoulos, V.; Malkani, S.; Zhang, D.; Rao, B.S.S.; Chattarji, S.; Kelleher, R.J.; Kandel, E.R.; Duff, K.; et al. Loss of Presenilin Function Causes Impairments of Memory and Synaptic Plasticity Followed by Age-Dependent Neurodegeneration. *Neuron* **2004**, *42*, 23–36. [CrossRef]

86. Watanabe, H.; Xia, D.; Kanekiyo, T.; Kelleher, R.J.; Shen, J. Familial Frontotemporal Dementia-Associated Presenilin-1 c.548G>T Mutation Causes Decreased mRNA Expression and Reduced Presenilin Function in Knock-In Mice. *J. Neurosci.* **2012**, *32*, 5085–5096. [CrossRef] [PubMed]

87. Xia, W.; Zhang, J.; Kholodenko, D.; Citron, M.; Podlisny, M.B.; Teplow, D.B.; Haass, C.; Seubert, P.; Koo, E.H.; Selkoe, D.J. Enhanced Production and Oligomerization of the 42-residue Amyloid β-Protein by Chinese Hamster Ovary Cells Stably Expressing Mutant Presenilins. *J. Biol. Chem.* **1997**, *272*, 7977–7982. [CrossRef] [PubMed]

88. Duff, K.; Eckman, C.; Zehr, C.; Yu, X.; Prada, C.M.; Perez-tur, J.; Hutton, M.; Buee, L.; Harigaya, Y.; Yager, D.; et al. Increased amyloid-beta42(43) in brains of mice expressing mutant presenilin. *Nature* **1996**, *383*, 710–713. [CrossRef]

89. Scheuner, D.; Eckman, C.; Jensen, M.; Song, X.; Citron, M.; Suzuki, N.; Bird, T.D.; Hardy, J.; Hutton, M.; Kukull, W.; et al. Secreted amyloid beta-protein similar to that in the senile plaques of Alzheimer's disease is increased in vivo by the presenilin 1 and 2 and APP mutations linked to familial Alzheimer's disease. *Nat. Med.* **1996**, *2*, 864–870. [CrossRef]

90. Borchelt, D.R.; Thinakaran, G.; Eckman, C.B.; Lee, M.K.; Davenport, F.; Ratovitsky, T.; Prada, C.M.; Kim, G.; Seekins, S.; Yager, D.; et al. Familial Alzheimer's disease-linked presenilin 1 variants elevate Abeta1-42/1-40 ratio in vitro and in vivo. *Neuron* **1996**, *17*, 1005–1013. [CrossRef]

91. Shen, J.; Iii, R.J.K. The presenilin hypothesis of Alzheimer's disease: Evidence for a loss-of-function pathogenic mechanism. *Proc. Nat. Acad. Sci. USA* **2007**, *104*, 403–409. [CrossRef] [PubMed]

92. Tu, H.; Nelson, O.; Bezprozvanny, A.; Wang, Z.; Lee, S.-F.; Hao, Y.-H.; Serneels, L.; De Strooper, B.; Yu, G.; Bezprozvanny, I. Presenilins Form ER Ca^{2+} Leak Channels, a Function Disrupted by Familial Alzheimer's Disease-Linked Mutations. *Cell* **2006**, *126*, 981–993. [CrossRef] [PubMed]

93. McCombs, J.E.; Gibson, E.A.; Palmer, A.E. Using a genetically targeted sensor to investigate the role of presenilin-1 in ER Ca^{2+} levels and dynamics. *Mol. Biosyst.* **2010**, *6*, 1640–1649. [CrossRef] [PubMed]

94. Shilling, D.; Mak, D.-O.D.; Kang, D.E.; Foskett, J.K. Lack of Evidence for Presenilins as Endoplasmic Reticulum Ca^{2+} Leak Channels. *J. Biol. Chem.* **2012**, *287*, 10933–10944. [CrossRef]

95. Wang, Q.C.; Zheng, Q.; Tan, H.; Zhang, B.; Li, X.; Yang, Y.; Yu, J.; Liu, Y.; Chai, H.; Wang, X.; et al. TMCO1 Is an ER Ca^{2+} Load-Activated Ca^{2+} Channel. *Cell* **2016**, *165*, 1454–1466. [CrossRef]

96. Bruno, A.M.; Huang, J.Y.; Bennett, D.A.; Marr, R.A.; Hastings, M.L.; Stutzmann, G.E. Altered ryanodine receptor expression in mild cognitive impairment and Alzheimer's disease. *Neurobiol. Aging* **2012**, *33*, 1001.e1–1001.e6. [CrossRef]

97. Chan, S.L.; Mayne, M.; Holden, C.P.; Geiger, J.D.; Mattson, M.P. Presenilin-1 Mutations Increase Levels of Ryanodine Receptors and Calcium Release in PC12 Cells and Cortical Neurons. *J. Biol. Chem.* **2000**, *275*, 18195–18200. [CrossRef]

98. Briggs, C.A.; Chakroborty, S.; Stutzmann, G.E. Emerging pathways driving early synaptic pathology in Alzheimer's disease. *Biochem. Biophys. Res. Commun.* **2017**, *483*, 988–997. [CrossRef]

99. Rybalchenko, V.; Hwang, S.-Y.; Rybalchenko, N.; Koulen, P. The cytosolic N-terminus of presenilin-1 potentiates mouse ryanodine receptor single channel activity. *Int. J. Biochem. Cell Biol.* **2008**, *40*, 84–97. [CrossRef]

100. Wu, B.; Yamaguchi, H.; Lai, F.A.; Shen, J. Presenilins regulate calcium homeostasis and presynaptic function via ryanodine receptors in hippocampal neurons. *Proc. Nat. Acad. Sci. USA* **2013**, *110*, 15091–15096. [CrossRef]

101. Oulès, B.; Del Prete, D.; Greco, B.; Zhang, X.; Lauritzen, I.; Sevalle, J.; Moreno, S.; Paterlini-Bréchot, P.; Trebak, M.; Checler, F.; et al. Ryanodine receptor blockade reduces amyloid-beta load and memory impairments in Tg2576 mouse model of Alzheimer disease. *J. Neurosci.* **2012**, *32*, 11820–11834. [CrossRef] [PubMed]

102. Zahradník, I.; Györke, S.; Zahradníková, A. Calcium Activation of Ryanodine Receptor Channels—Reconciling RyR Gating Models with Tetrameric Channel Structure. *J. Gen. Physiol.* **2005**, *126*, 515–527. [CrossRef] [PubMed]

103. Ferreiro, E.; Oliveira, C.R.; Pereira, C. Involvement of endoplasmic reticulum Ca^{2+} release through ryanodine and inositol 1,4,5-triphosphate receptors in the neurotoxic effects induced by the amyloid-beta peptide. *J. Neurosci. Res.* **2004**, *76*, 872–880. [CrossRef] [PubMed]

104. Supnet, C.; Grant, J.; Kong, H.; Westaway, D.; Mayne, M. Amyloid-β-(1-42) Increases Ryanodine Receptor-3 Expression and Function in Neurons of TgCRND8 Mice. *J. Biol. Chem.* **2006**, *281*, 38440–38447. [CrossRef]

105. Bussiere, R.; Lacampagne, A.; Reiken, S.; Liu, X.; Scheuerman, V.; Zalk, R.; Martin, C.; Checler, F.; Marks, A.R.; Chami, M. Amyloid beta production is regulated by beta2-adrenergic signaling-mediated post-translational modifications of the ryanodine receptor. *J. Biol. Chem.* **2017**, *292*, 10153–10168. [CrossRef]

106. Lacampagne, A.; Liu, X.; Reiken, S.; Bussiere, R.; Meli, A.C.; Lauritzen, I.; Teich, A.F.; Zalk, R.; Saint, N.; Arancio, O.; et al. Post-translational remodeling of ryanodine receptor induces calcium leak leading to Alzheimer's disease-like pathologies and cognitive deficits. *Acta Neuropathol.* **2017**, *134*, 749–767. [CrossRef]

107. Brunello, L.; Zampese, E.; Florean, C.; Pozzan, T.; Pizzo, P.; Fasolato, C. Presenilin-2 dampens intracellular Ca^{2+} stores by increasing Ca^{2+} leakage and reducing Ca^{2+} uptake. *J. Cell. Mol. Med.* **2009**, *13*, 3358–3369. [CrossRef]

108. Green, K.N.; Demuro, A.; Akbari, Y.; Hitt, B.D.; Smith, I.F.; Parker, I.; LaFerla, F.M. SERCA pump activity is physiologically regulated by presenilin and regulates amyloid beta production. *J. Cell Biol.* **2008**, *181*, 1107–1116. [CrossRef]

109. Peng, J.; Liang, G.; Inan, S.; Wu, Z.; Joseph, D.J.; Meng, Q.; Peng, Y.; Eckenhoff, M.F.; Wei, H. Dantrolene ameliorates cognitive decline and neuropathology in Alzheimer triple transgenic mice. *Neurosci. Lett.* **2012**, *516*, 274–279. [CrossRef]

110. Wu, Z.; Yang, B.; Liu, C.; Liang, G.; Liu, W.; Pickup, S.; Meng, Q.; Tian, Y.; Li, S.; Eckenhoff, M.F.; et al. Long-term Dantrolene Treatment Reduced Intraneuronal Amyloid in Aged Alzheimer Triple Transgenic Mice. *Alzheimer Dis. Assoc. Disord.* **2015**, *29*, 184–191. [CrossRef]

111. Bollimuntha, S.; Pani, B.; Singh, B.B. Neurological and Motor Disorders: Neuronal Store-Operated Ca^{2+} Signaling: An Overview and Its Function. *Atherosclerosis* **2017**, *993*, 535–556. [CrossRef]

112. Kraft, R. STIM and ORAI proteins in the nervous system. *Channels* **2015**, *9*, 245–252. [CrossRef] [PubMed]

113. Putney, J.W. The Physiological Function of Store-operated Calcium Entry. *Neurochem. Res.* **2011**, *36*, 1157–1165. [CrossRef] [PubMed]

114. Leissring, M.A.; Akbari, Y.; Fanger, C.M.; Cahalan, M.D.; Mattson, M.P.; LaFerla, F.M. Capacitative Calcium Entry Deficits and Elevated Luminal Calcium Content in Mutant Presenilin-1 Knockin Mice. *J. Cell Biol.* **2000**, *149*, 793–798. [CrossRef]

115. Smith, I.; Boyle, J.; Vaughan, P.; Pearson, H.; Cowburn, R.; Peers, C. Ca^{2+} Stores and capacitative Ca^{2+} entry in human neuroblastoma (SH-SY5Y) cells expressing a familial Alzheimer's disease presenilin-1 mutation. *Brain Res.* **2002**, *949*, 105–111. [CrossRef]

116. Calvo-Rodriguez, M.; Hernando-Perez, E.; Nunez, L.; Villalobos, C. Amyloid beta Oligomers Increase ER-Mitochondria Ca^{2+} Cross Talk in Young Hippocampal Neurons and Exacerbate Aging-Induced Intracellular Ca^{2+} Remodeling. *Front. Cell. Neurosci.* **2019**, *13*, 22. [CrossRef] [PubMed]

117. Sun, S.; Zhang, H.; Liu, J.; Popugaeva, E.; Xu, N.-J.; Feske, S.; White, C.L.; Bezprozvanny, I. Reduced Synaptic STIM2 Expression and Impaired Store-Operated Calcium Entry Cause Destabilization of Mature Spines in Mutant Presenilin Mice. *Neuron* **2014**, *82*, 79–93. [CrossRef]

118. Yoo, A.S.; Cheng, I.; Chung, S.; Grenfell, T.Z.; Lee, H.; Pack-Chung, E.; Handler, M.; Shen, J.; Xia, W.; Tesco, G.; et al. Presenilin-Mediated Modulation of Capacitative Calcium Entry. *Neuron* **2000**, *27*, 561–572. [CrossRef]

119. Zhang, H.; Wu, L.; Pchitskaya, E.; Zakharova, O.D.; Saito, T.; Saido, T.C.; Bezprozvanny, I.B. Neuronal Store-Operated Calcium Entry and Mushroom Spine Loss in Amyloid Precursor Protein Knock-In Mouse Model of Alzheimer's Disease. *J. Neurosci.* **2015**, *35*, 13275–13286. [CrossRef]

120. Tsukamoto, A.; Kaneko, Y. Thapsigargin, a Ca^{2+}-ATPase inhibitor, depletes the intracellular Ca^{2+} pool and induces apoptosis in human hepatoma cells. *Cell Biol. Int.* **1993**, *17*, 969–970. [CrossRef]

121. Soboloff, J.; Berger, S.A. Sustained ER Ca^{2+} Depletion Suppresses Protein Synthesis and Induces Activation-enhanced Cell Death in Mast Cells. *J. Biol. Chem.* **2002**, *277*, 13812–13820. [CrossRef] [PubMed]

122. Popugaeva, E.; Pchitskaya, E.; Speshilova, A.B.; Alexandrov, S.; Zhang, H.; Vlasova, O.; Bezprozvanny, I.B. STIM2 protects hippocampal mushroom spines from amyloid synaptotoxicity. *Mol. Neurodegener.* **2015**, *10*, 37. [CrossRef] [PubMed]

123. Li, M.; Liu, E.; Zhou, Q.; Li, S.; Wang, X.; Liu, Y.; Wang, L.; Sun, D.; Ye, J.; Gao, Y.; et al. TRPC1 Null Exacerbates Memory Deficit and Apoptosis Induced by Amyloid-beta. *J. Alzheimer's Dis. JAD* **2018**, *63*, 761–772. [CrossRef] [PubMed]

124. Mishra, P.; Chan, D.C. Metabolic regulation of mitochondrial dynamics. *J. Cell Biol.* **2016**, *212*, 379–387. [CrossRef]

125. Werth, J.L.; A Thayer, S. Mitochondria buffer physiological calcium loads in cultured rat dorsal root ganglion neurons. *J. Neurosci.* **1994**, *14*, 348–356. [CrossRef]

126. Billups, B.; Forsythe, I.D. Presynaptic Mitochondrial Calcium Sequestration Influences Transmission at Mammalian Central Synapses. *J. Neurosci.* **2002**, *22*, 5840–5847. [CrossRef]

127. Wescott, A.P.; Kao, J.P.Y.; Lederer, W.J.; Boyman, L. Voltage-energized calcium-sensitive ATP production by mitochondria. *Nat. Metab.* **2019**, *1*, 975–984. [CrossRef] [PubMed]

128. Baughman, J.M.; Perocchi, F.; Girgis, H.S.; Plovanich, M.; Belcher-Timme, C.A.; Sancak, Y.; Bao, X.R.; Strittmatter, L.; A Goldberger, O.; Bogorad, R.L.; et al. Integrative genomics identifies MCU as an essential component of the mitochondrial calcium uniporter. *Nat. Cell Biol.* **2011**, *476*, 341–345. [CrossRef] [PubMed]

129. De Stefani, D.; Raffaello, A.; Teardo, E.; Szabò, I.; Rizzuto, R. A forty-kilodalton protein of the inner membrane is the mitochondrial calcium uniporter. *Nat. Cell Biol.* **2011**, *476*, 336–340. [CrossRef] [PubMed]

130. Raffaello, A.; De Stefani, D.; Sabbadin, D.; Teardo, E.; Merli, G.; Picard, A.; Checchetto, V.; Moro, S.; Szabò, I.; Rizzuto, R. The mitochondrial calcium uniporter is a multimer that can include a dominant-negative pore-forming subunit. *EMBO J.* **2013**, *32*, 2362–2376. [CrossRef] [PubMed]

131. Lambert, J.P.; Luongo, T.S.; Tomar, D.; Jadiya, P.; Gao, E.; Zhang, X.; Lucchese, A.M.; Kolmetzky, D.W.; Shah, N.S.; Elrod, J.W. MCUB Regulates the Molecular Composition of the Mitochondrial Calcium Uniporter Channel to Limit Mitochondrial Calcium Overload During Stress. *Circulation* **2019**, *140*, 1720–1733. [CrossRef] [PubMed]

132. Sancak, Y.; Markhard, A.L.; Kitami, T.; Kovács-Bogdán, E.; Kamer, K.J.; Udeshi, N.D.; Carr, S.A.; Chaudhuri, D.; Clapham, D.E.; Li, A.A.; et al. EMRE Is an Essential Component of the Mitochondrial Calcium Uniporter Complex. *Science* **2013**, *342*, 1379–1382. [CrossRef] [PubMed]

133. Perocchi, F.; Gohil, V.M.; Girgis, H.S.; Bao, X.R.; McCombs, J.E.; Palmer, A.E.; Mootha, V.K. MICU1 encodes a mitochondrial EF hand protein required for Ca^{2+} uptake. *Nat. Cell Biol.* **2010**, *467*, 291–296. [CrossRef] [PubMed]

134. Mallilankaraman, K.; Doonan, P.J.; Cárdenas, C.; Chandramoorthy, H.C.; Müller, M.; Miller, R.; Hoffman, N.E.; Gandhirajan, R.K.; Molgó, J.; Birnbaum, M.J.; et al. MICU1 Is an Essential Gatekeeper for MCU-Mediated Mitochondrial Ca^{2+} Uptake that Regulates Cell Survival. *Cell* **2012**, *151*, 630–644. [CrossRef] [PubMed]

135. Patron, M.; Checchetto, V.; Raffaello, A.; Teardo, E.; Reane, D.V.; Mantoan, M.; Granatiero, V.; Szabò, I.; De Stefani, D.; Rizzuto, R. MICU1 and MICU2 Finely Tune the Mitochondrial Ca^{2+} Uniporter by Exerting Opposite Effects on MCU Activity. *Mol. Cell* **2014**, *53*, 726–737. [CrossRef] [PubMed]

136. Csordás, G.; Golenár, T.; Seifert, E.L.; Kamer, K.J.; Sancak, Y.; Perocchi, F.; Moffat, C.; Weaver, D.; Perez, S.D.L.F.; Bogorad, R.; et al. MICU1 Controls Both the Threshold and Cooperative Activation of the Mitochondrial Ca^{2+} Uniporter. *Cell Metab.* **2013**, *17*, 976–987. [CrossRef]

137. Plovanich, M.; Bogorad, R.L.; Sancak, Y.; Kamer, K.J.; Strittmatter, L.; Li, A.A.; Girgis, H.S.; Kuchimanchi, S.; De Groot, J.; Speciner, L.; et al. MICU2, a paralog of MICU1, resides within the mitochondrial uniporter complex to regulate calcium handling. *PLoS ONE* **2013**, *8*, e55785. [CrossRef]

138. Patron, M.; Granatiero, V.; Espino, J.; Rizzuto, R.; De Stefani, D. MICU3 is a tissue-specific enhancer of mitochondrial calcium uptake. *Cell Death Differ.* **2018**, *26*, 179–195. [CrossRef]

139. Mallilankaraman, K.; Cárdenas, C.; Doonan, P.J.; Chandramoorthy, H.C.; Irrinki, K.M.; Golenár, T.; Csordás, G.; Madireddi, P.; Yang, J.; Müller, M.; et al. MCUR1 is an essential component of mitochondrial Ca^{2+} uptake that regulates cellular metabolism. *Nat. Cell Biol.* **2012**, *14*, 1336–1343. [CrossRef]

140. Hoffman, N.E.; Chandramoorthy, H.C.; Shanmughapriya, S.; Zhang, X.Q.; Vallem, S.; Doonan, P.J.; Malliankaraman, K.; Guo, S.; Rajan, S.; Elrod, J.W.; et al. SLC25A23 augments mitochondrial Ca^{2+} uptake, interacts with MCU, and induces oxidative stress–mediated cell death. *Mol. Biol. Cell* **2014**, *25*, 936–947. [CrossRef]

141. Palty, R.; Silverman, W.F.; Hershfinkel, M.; Caporale, T.; Sensi, S.L.; Parnis, J.; Nolte, C.; Fishman, D.; Shoshan-Barmatz, V.; Herrmann, S.; et al. NCLX is an essential component of mitochondrial Na^+/Ca^{2+} exchange. *Proc. Nat. Acad. Sci. USA* **2010**, *107*, 436–441. [CrossRef] [PubMed]

142. Jiang, D.; Zhao, L.; Clapham, D.E. Genome-wide RNAi screen identifies Letm1 as a mitochondrial Ca^{2+}/H^+ antiporter. *Science* **2009**, *326*, 144–147. [CrossRef] [PubMed]

143. Baumgartner, H.K.; Gerasimenko, J.V.; Thorne, C.; Ferdek, P.; Pozzan, T.; Tepikin, A.V.; Petersen, O.H.; Sutton, R.; Watson, A.J.; Gerasimenko, O.V.; et al. Calcium elevation in mitochondria is the main Ca^{2+} requirement for mitochondrial permeability transition pore (mPTP) opening. *J. Biol. Chem.* **2009**, *284*, 20796–20803. [CrossRef] [PubMed]

144. Hengartner, M.O. The biochemistry of apoptosis. *Nat. Cell Biol.* **2000**, *407*, 770–776. [CrossRef]

145. Swerdlow, R.H.; Khan, S.M. A "mitochondrial cascade hypothesis" for sporadic Alzheimer's disease. *Med. Hypotheses* **2004**, *63*, 8–20. [CrossRef]

146. Hirai, K.; Aliev, G.; Nunomura, A.; Fujioka, H.; Russell, R.L.; Atwood, C.S.; Johnson, A.B.; Kress, Y.; Vinters, H.V.; Tabaton, M.; et al. Mitochondrial Abnormalities in Alzheimer's Disease. *J. Neurosci.* **2001**, *21*, 3017–3023. [CrossRef]

147. Rice, A.C.; Keeney, P.M.; Algarzae, N.K.; Ladd, A.C.; Thomas, R.R.; Bennett, J.P., Jr. Mitochondrial DNA Copy Numbers in Pyramidal Neurons are Decreased and Mitochondrial Biogenesis Transcriptome Signaling is Disrupted in Alzheimer's Disease Hippocampi. *J. Alzheimer's Dis.* **2014**, *40*, 319–330. [CrossRef]

148. Butterfield, D.A.; Halliwell, B. Oxidative stress, dysfunctional glucose metabolism and Alzheimer disease. *Nat. Rev. Neurosci.* **2019**, *20*, 148–160. [CrossRef]

149. Wang, J.; Markesbery, W.R.; Lovell, M.A. Increased oxidative damage in nuclear and mitochondrial DNA in mild cognitive impairment. *J. Neurochem.* **2006**, *96*, 825–832. [CrossRef]

150. Area-Gomez, E.; De Groof, A.; Bonilla, E.; Montesinos, J.; Tanji, K.; Boldogh, I.; Pon, L.; Schon, E.A. A key role for MAM in mediating mitochondrial dysfunction in Alzheimer disease. *Cell Death Dis.* **2018**, *9*, 1–10. [CrossRef]

151. Calkins, M.J.; Reddy, P.H. Amyloid beta impairs mitochondrial anterograde transport and degenerates synapses in Alzheimer's disease neurons. *Biochim. Biophys. Acta (BBA) Mol. Basis Dis.* **2011**, *1812*, 507–513. [CrossRef] [PubMed]

152. Hansson Petersen, C.A.; Alikhani, N.; Behbahani, H.; Wiehager, B.; Pavlov, P.F.; Alafuzoff, I.; Leinonen, V.; Ito, A.; Winblad, B.; Glaser, E.; et al. The amyloid beta-peptide is imported into mitochondria via the TOM import machinery and localized to mitochondrial cristae. *Proc. Natl. Acad. Sci. USA* **2008**, *105*, 13145–13150. [CrossRef] [PubMed]

153. Cieri, D.; Vicario, M.; Vallese, F.; D'Orsi, B.; Berto, P.; Grinzato, A.; Catoni, C.; De Stefani, D.; Rizzuto, R.; Brini, M.; et al. Tau localises within mitochondrial sub-compartments and its caspase cleavage affects ER-mitochondria interactions and cellular Ca^{2+} handling. *Biochim. Biophys. Acta (BBA) Mol. Basis Dis.* **2018**, *1864*, 3247–3256. [CrossRef] [PubMed]

154. Amadoro, G.; Corsetti, V.; Stringaro, A.; Colone, M.; D'Aguanno, S.; Meli, G.; Ciotti, M.; Sancesario, G.; Cattaneo, A.; Bussani, R.; et al. A NH2 Tau Fragment Targets Neuronal Mitochondria at AD Synapses: Possible Implications for Neurodegeneration. *J. Alzheimer's Dis.* **2010**, *21*, 445–470. [CrossRef] [PubMed]

155. Calvo-Rodríguez, M.; García-Durillo, M.; Villalobos, C.; Núñez, L. Aging Enables Ca^{2+} Overload and Apoptosis Induced by Amyloid-beta Oligomers in Rat Hippocampal Neurons: Neuroprotection by Non-Steroidal Anti-Inflammatory Drugs and R-Flurbiprofen in Aging Neurons. *J. Alzheimer's Dis. JAD* **2016**, *54*, 207–221. [CrossRef]

156. Sanz-Blasco, S.; Valero, R.A.; Rodriguez-Crespo, I.; Villalobos, C.; Nunez, L. Mitochondrial Ca^{2+} overload underlies Abeta oligomers neurotoxicity providing an unexpected mechanism of neuroprotection by NSAIDs. *PLoS ONE* **2008**, *3*, e2718. [CrossRef]

157. Calvo-Rodriguez, M.; Hou, S.S.; Snyder, A.C.; Kharitonova, E.K.; Russ, A.N.; Das, S.; Fan, Z.; Muzikansky, A.; Garcia-Alloza, M.; Serrano-Pozo, A.; et al. Increased mitochondrial calcium levels associated with neuronal death in a mouse model of Alzheimer's disease. *Nat. Commun.* **2020**, *11*, 1–17. [CrossRef]

158. Yan, Y.; Gong, K.; Ma, T.; Zhang, L.; Zhao, N.; Zhang, X.; Tang, P.; Gong, Y. Protective effect of edaravone against Alzheimer's disease-relevant insults in neuroblastoma N2a cells. *Neurosci. Lett.* **2012**, *531*, 160–165. [CrossRef]

159. Moreira, P.I.; Santos, M.S.; Moreno, A.; Oliveira, C. Amyloid β-Peptide Promotes Permeability Transition Pore in Brain Mitochondria. *Biosci. Rep.* **2001**, *21*, 789–800. [CrossRef]

160. Du, H.; Guo, L.; Fang, F.; Chen, D.; A Sosunov, A.; McKhann, G.M.; Yan, Y.; Wang, C.; Zhang, H.; Molkentin, J.D.; et al. Cyclophilin D deficiency attenuates mitochondrial and neuronal perturbation and ameliorates learning and memory in Alzheimer's disease. *Nat. Med.* **2008**, *14*, 1097–1105. [CrossRef]

161. Jadiya, P.; Kolmetzky, D.W.; Tomar, D.; Di Meco, A.; Lombardi, A.A.; Lambert, J.P.; Luongo, T.S.; Ludtmann, M.H.; Praticò, D.; Elrod, J.W. Impaired mitochondrial calcium efflux contributes to disease progression in models of Alzheimer's disease. *Nat. Commun.* **2019**, *10*, 1–14. [CrossRef] [PubMed]

162. Kern, S.; Skoog, I.; Östling, S.; Kern, J.; Börjesson-Hanson, A. Does low-dose acetylsalicylic acid prevent cognitive decline in women with high cardiovascular risk? A 5-year follow-up of a non-demented population-based cohort of Swedish elderly women. *BMJ Open* **2012**, *2*, e001288. [CrossRef] [PubMed]

163. Zhang, C.; Wang, Y.; Wang, D.; Zhang, J.; Zhang, F. NSAID Exposure and Risk of Alzheimer's Disease: An Updated Meta-Analysis From Cohort Studies. *Front. Aging Neurosci.* **2018**, *10*, 83. [CrossRef] [PubMed]

164. Breitner, J.; Meyer, P.-F. Author response: INTREPAD: A randomized trial of naproxen to slow progress of presymptomatic Alzheimer disease. *Neurology* **2020**, *94*, 594. [CrossRef] [PubMed]

165. Sanz-Blasco, S.; Calvo-Rodriguez, M.; Caballero, E.; Garcia-Durillo, M.; Nunez, L.; Villalobos, C. Is it All Said for NSAIDs in Alzheimer's Disease? Role of Mitochondrial Calcium Uptake. *Curr. Alzheimer Res.* **2018**, *15*, 504–510. [CrossRef] [PubMed]

166. Gutknecht, J. Salicylates and proton transport through lipid bilayer membranes: A model for salicylate-induced uncoupling and swelling in mitochondria. *J. Membr. Biol.* **1990**, *115*, 253–260. [CrossRef]

167. Núñez, L.; Valero, R.A.; Senovilla, L.; Sanz-Blasco, S.; García-Sancho, J.; Villalobos, C. Cell proliferation depends on mitochondrial Ca^{2+} uptake: Inhibition by salicylate. *J. Physiol.* **2006**, *571*, 57–73. [CrossRef]

168. Calvo-Rodríguez, M.; Sanz-Blasco, S.; Caballero, E.; Villalobos, C.; Núñez, L. Susceptibility to excitotoxicity in aged hippocampal cultures and neuroprotection by non-steroidal anti-inflammatory drugs: Role of mitochondrial calcium. *J. Neurochem.* **2015**, *132*, 403–417. [CrossRef]

169. Angelova, P.R.; Vinogradova, D.; Neganova, M.E.; Serkova, T.P.; Sokolov, V.V.; Bachurin, S.O.; Shevtsova, E.F.; Abramov, A.Y. Pharmacological Sequestration of Mitochondrial Calcium Uptake Protects Neurons Against Glutamate Excitotoxicity. *Mol. Neurobiol.* **2019**, *56*, 2244–2255. [CrossRef]

170. Dubey, M.; Chaudhury, P.; Kabiru, H.; Shea, T.B. Tau inhibits anterograde axonal transport and perturbs stability in growing axonal neurites in part by displacing kinesin cargo: Neurofilaments attenuate tau-mediated neurite instability. *Cell Motil. Cytoskelet.* **2008**, *65*, 89–99. [CrossRef]

171. DuBoff, B.; Götz, J.; Feany, M.B. Tau Promotes Neurodegeneration via DRP1 Mislocalization In Vivo. *Neuron* **2012**, *75*, 618–632. [CrossRef] [PubMed]

172. Quintanilla, R.A.; Matthews-Roberson, T.A.; Dolan, P.J.; Johnson, G.V. Caspase-cleaved Tau Expression Induces Mitochondrial Dysfunction in Immortalized Cortical Neurons. *J. Biol. Chem.* **2009**, *284*, 18754–18766. [CrossRef] [PubMed]

173. Britti, E.; Ros, J.; Esteras, N.; Abramov, A.Y. Tau inhibits mitochondrial calcium efflux and makes neurons vulnerable to calcium-induced cell death. *Cell Calcium* **2020**, *86*, 102150. [CrossRef] [PubMed]

174. Manczak, M.; Reddy, P.H. Abnormal interaction between the mitochondrial fission protein Drp1 and hyperphosphorylated tau in Alzheimer's disease neurons: Implications for mitochondrial dysfunction and neuronal damage. *Hum. Mol. Genet.* **2012**, *21*, 2538–2547. [CrossRef]

175. García-Pérez, C.; Hajnóczky, G.; Csordás, G. Physical Coupling Supports the Local Ca^{2+} Transfer between Sarcoplasmic Reticulum Subdomains and the Mitochondria in Heart Muscle. *J. Biol. Chem.* **2008**, *283*, 32771–32780. [CrossRef]

176. Hayashi, T.; Su, T.-P. Sigma-1 Receptor Chaperones at the ER- Mitochondrion Interface Regulate Ca^{2+} Signaling and Cell Survival. *Cell* **2007**, *131*, 596–610. [CrossRef]

177. Csordas, G.; Varnai, P.; Golenar, T.; Roy, S.; Purkins, G.; Schneider, T.G.; Balla, T.; Hajnoczky, G. Imaging interorganelle contacts and local calcium dynamics at the ER-mitochondrial interface. *Mol. Cell* **2010**, *39*, 121–132. [CrossRef]

178. Area-Gomez, E.; Schon, E.A. On the Pathogenesis of Alzheimer's Disease: The MAM Hypothesis. *FASEB J.* **2017**, *31*, 864–867. [CrossRef]

179. Hedskog, L.; Pinho, C.M.; Filadi, R.; Rönnbäck, A.; Hertwig, L.; Wiehager, B.; Larssen, P.; Gellhaar, S.; Sandebring, A.; Westerlund, M.; et al. Modulation of the endoplasmic reticulum-mitochondria interface in Alzheimer's disease and related models. *Proc. Nat. Acad. Sci. USA* **2013**, *110*, 7916–7921. [CrossRef]

180. Perreault, S.; Bousquet, O.; Lauzon, M.; Paiement, J.; Leclerc, N. Increased Association Between Rough Endoplasmic Reticulum Membranes and Mitochondria in Transgenic Mice That Express P301L Tau. *J. Neuropathol. Exp. Neurol.* **2009**, *68*, 503–514. [CrossRef]

181. Area-Gomez, E.; Castillo, M.D.C.L.; Tambini, M.D.; Guardia-Laguarta, C.; De Groof, A.J.C.; Madra, M.; Ikenouchi, J.; Umeda, M.; Bird, T.D.; Sturley, S.L.; et al. Upregulated function of mitochondria-associated ER membranes in Alzheimer disease. *EMBO J.* **2012**, *31*, 4106–4123. [CrossRef] [PubMed]

182. Ferreiro, E.; Oliveira, C.R.; Pereira, C.F. The release of calcium from the endoplasmic reticulum induced by amyloid-beta and prion peptides activates the mitochondrial apoptotic pathway. *Neurobiol. Dis.* **2008**, *30*, 331–342. [CrossRef] [PubMed]

183. Zampese, E.; Fasolato, C.; Kipanyula, M.J.; Bortolozzi, M.; Pozzan, T.; Pizzo, P. Presenilin 2 modulates endoplasmic reticulum (ER)-mitochondria interactions and Ca^{2+} cross-talk. *Proc. Nat. Acad. Sci. USA* **2011**, *108*, 2777–2782. [CrossRef] [PubMed]

184. Zhu, X.; Smith, M.A.; Perry, G.H.; Aliev, G. Mitochondrial failures in Alzheimer's disease. *Am. J. Alzheimer's Dis. Other Dement.* **2004**, *19*, 345–352. [CrossRef] [PubMed]

185. Cenini, G.; Voos, W. Mitochondria as Potential Targets in Alzheimer Disease Therapy: An Update. *Front. Pharmacol.* **2019**, *10*, 902. [CrossRef] [PubMed]

186. Kelso, G.F.; Porteous, C.M.; Coulter, C.V.; Hughes, G.; Porteous, W.K.; Ledgerwood, E.C.; Smith, R.A.J.; Murphy, M.P. Selective targeting of a redox-active ubiquinone to mitochondria within cells: Antioxidant and antiapoptotic properties. *J. Biol. Chem.* **2001**, *276*, 4588–4596. [CrossRef]

187. Ortiz, J.M.P.; Swerdlow, R.H. Mitochondrial dysfunction in Alzheimer's disease: Role in pathogenesis and novel therapeutic opportunities. *Br. J. Pharmacol.* **2019**, *176*, 3489–3507. [CrossRef]

188. Szeto, H.H. Cell-permeable, mitochondrial-targeted, peptide antioxidants. *AAPS J.* **2006**, *8*, E277–E283. [CrossRef]

189. Birk, A.V.; Chao, W.M.; Bracken, C.; Warren, J.D.; Szeto, H.H. Targeting mitochondrial cardiolipin and the cytochromec/cardiolipin complex to promote electron transport and optimize mitochondrial ATP synthesis. *Br. J. Pharmacol.* **2014**, *171*, 2017–2028. [CrossRef]

190. Reddy, P.H.; Manczak, M.; Kandimalla, R. Mitochondria-targeted small molecule SS31: A potential candidate for the treatment of Alzheimer's disease. *Hum. Mol. Genet.* **2017**, *26*, 1597. [CrossRef]

191. Reddy, P.H.; Manczak, M.; Yin, X.; Reddy, A.P. Synergistic Protective Effects of Mitochondrial Division Inhibitor 1 and Mitochondria-Targeted Small Peptide SS31 in Alzheimer's Disease. *J. Alzheimer's Dis.* **2018**, *62*, 1549–1565. [CrossRef] [PubMed]

192. Wilkins, H.M.; Morris, J.K. New Therapeutics to Modulate Mitochondrial Function in Neurodegenerative Disorders. *Curr. Pharm. Des.* **2017**, *23*, 731–752. [CrossRef] [PubMed]

193. Lloyd-Evans, E.; Platt, F.M. Lysosomal Ca^{2+} homeostasis: Role in pathogenesis of lysosomal storage diseases. *Cell Calcium* **2011**, *50*, 200–205. [CrossRef] [PubMed]

194. Tian, X.; Gala, U.; Zhang, Y.; Shang, W.; Jaiswal, S.N.; Di Ronza, A.; Jaiswal, M.; Yamamoto, S.; Sandoval, H.; DuRaine, L.; et al. A voltage-gated calcium channel regulates lysosomal fusion with endosomes and autophagosomes and is required for neuronal homeostasis. *PLoS Biol.* **2015**, *13*, e1002103. [CrossRef] [PubMed]

195. Patel, S.; Docampo, R. Acidic calcium stores open for business: Expanding the potential for intracellular Ca^{2+} signaling. *Trends Cell Biol.* **2010**, *20*, 277–286. [CrossRef] [PubMed]

196. Garrity, A.G.; Wang, W.; Collier, C.M.; A Levey, S.; Gao, Q.; Xu, H. The endoplasmic reticulum, not the pH gradient, drives calcium refilling of lysosomes. *eLife* **2016**, *5*, e15887. [CrossRef] [PubMed]

197. Mindell, J.A. Lysosomal Acidification Mechanisms. *Annu. Rev. Physiol.* **2012**, *74*, 69–86. [CrossRef]

198. Bach, G.; Chen, C.-S.; E Pagano, R. Elevated lysosomal pH in Mucolipidosis type IV cells. *Clin. Chim. Acta* **1999**, *280*, 173–179. [CrossRef]

199. Soyombo, A.A.; Tjon-Kon-Sang, S.; Rbaibi, Y.; Bashllari, E.; Bisceglia, J.; Muallem, S.; Kiselyov, K. TRP-ML1 Regulates Lysosomal pH and Acidic Lysosomal Lipid Hydrolytic Activity. *J. Biol. Chem.* **2006**, *281*, 7294–7301. [CrossRef]

200. Medina, D.L.; Di Paola, S.; Peluso, I.; Armani, A.; De Stefani, D.; Venditti, R.; Montefusco, S.; Scotto-Rosato, A.; Prezioso, C.; Forrester, A.; et al. Lysosomal calcium signalling regulates autophagy through calcineurin and TFEB. *Nat. Cell Biol.* **2015**, *17*, 288–299. [CrossRef]

201. Lee, J.-H.; McBrayer, M.K.; Wolfe, D.M.; Haslett, L.J.; Kumar, A.; Sato, Y.; Lie, P.P.Y.; Mohan, P.S.; Coffey, E.E.; Kompella, U.B.; et al. Presenilin 1 Maintains Lysosomal Ca^{2+} Homeostasis via TRPML1 by Regulating vATPase-Mediated Lysosome Acidification. *Cell Rep.* **2015**, *12*, 1430–1444. [CrossRef] [PubMed]

202. Neely, K.M.; Green, K.N.; LaFerla, F.M. Presenilin is necessary for efficient proteolysis through the autophagy-lysosome system in a gamma-secretase-independent manner. *J. Neurosci.* **2011**, *31*, 2781–2791. [CrossRef] [PubMed]

203. Greotti, E.; Capitanio, P.; Wong, A.; Pozzan, T.; Pizzo, P.; Pendin, D. Familial Alzheimer's disease-linked presenilin mutants and intracellular Ca^{2+} handling: A single-organelle, FRET-based analysis. *Cell Calcium* **2019**, *79*, 44–56. [CrossRef] [PubMed]

204. Glick, D.; Barth, S.; MacLeod, K.F. Autophagy: Cellular and molecular mechanisms. *J. Pathol.* **2010**, *221*, 3–12. [CrossRef]

205. Nixon, R.A.; Wegiel, J.; Kumar, A.; Yu, W.H.; Peterhoff, C.; Cataldo, A.; Cuervo, A.M. Extensive Involvement of Autophagy in Alzheimer Disease: An Immuno-Electron Microscopy Study. *J. Neuropathol. Exp. Neurol.* **2005**, *64*, 113–122. [CrossRef]

206. Pickford, F.; Masliah, E.; Britschgi, M.; Lucin, K.; Narasimhan, R.; Jaeger, P.A.; Small, S.; Spencer, B.; Rockenstein, E.; Levine, B.; et al. The autophagy-related protein beclin 1 shows reduced expression in early Alzheimer disease and regulates amyloid β accumulation in mice. *J. Clin. Investig.* **2008**, *118*, 2190–2199. [CrossRef]

207. Wolfe, D.M.; Lee, J.-H.; Kumar, A.; Lee, S.; Orenstein, S.J.; Nixon, R.A. Autophagy failure in Alzheimer's disease and the role of defective lysosomal acidification. *Eur. J. Neurosci.* **2013**, *37*, 1949–1961. [CrossRef]

208. Medeiros, R.; LaFerla, F.M. Astrocytes: Conductors of the Alzheimer disease neuroinflammatory symphony. *Exp. Neurol.* **2013**, *239*, 133–138. [CrossRef]

209. Verkhratsky, A.; Parpura, V. Recent advances in (patho)physiology of astroglia. *Acta Pharmacol. Sin.* **2010**, *31*, 1044–1054. [CrossRef]

210. Abbott, N.J.; Rönnbäck, L.; Hansson, E. Astrocyte–endothelial interactions at the blood–brain barrier. *Nat. Rev. Neurosci.* **2006**, *7*, 41–53. [CrossRef]

211. Guerra-Gomes, S.; Sousa, N.; Pinto, L.; Oliveira, J.F. Functional Roles of Astrocyte Calcium Elevations: From Synapses to Behavior. *Front. Cell. Neurosci.* **2018**, *11*, 427. [CrossRef] [PubMed]

212. Verkhratsky, A.; Nedergaard, M. Physiology of Astroglia. *Physiol. Rev.* **2018**, *98*, 239–389. [CrossRef] [PubMed]

213. Perez-Nievas, B.G.; Serrano-Pozo, A. Deciphering the Astrocyte Reaction in Alzheimer's Disease. *Front. Aging Neurosci.* **2018**, *10*, 114. [CrossRef] [PubMed]

214. Lim, D.; Iyer, A.; Ronco, V.; A Grolla, A.; Canonico, P.L.; Aronica, E.; Genazzani, A.A. Amyloid beta deregulates astroglial mGluR5-mediated calcium signaling via calcineurin and Nf-kB. *Glia* **2013**, *61*, 1134–1145. [CrossRef]

215. Grolla, A.A.; Sim, J.A.; Lim, D.; Rodriguez, J.J.; Genazzani, A.A.; Verkhratsky, A. Amyloid-beta and Alzheimer's disease type pathology differentially affects the calcium signalling toolkit in astrocytes from different brain regions. *Cell Death Dis.* **2013**, *4*, e623. [CrossRef]

216. Alberdi, E.; Wyssenbach, A.; Alberdi, M.; Sanchez-Gomez, M.V.; Cavaliere, F.; Rodriguez, J.J.; Verkhratsky, A.; Matute, C. Ca^{2+}-dependent endoplasmic reticulum stress correlates with astrogliosis in oligomeric amyloid beta-treated astrocytes and in a model of Alzheimer's disease. *Aging Cell* **2013**, *12*, 292–302. [CrossRef]

217. Abramov, A.Y.; Canevari, L.; Duchen, M.R. Changes in Intracellular Calcium and Glutathione in Astrocytes as the Primary Mechanism of Amyloid Neurotoxicity. *J. Neurosci.* **2003**, *23*, 5088–5095. [CrossRef]

218. Toivari, E.; Manninen, T.; Nahata, A.K.; Jalonen, T.O.; Linne, M.-L. Effects of Transmitters and Amyloid-Beta Peptide on Calcium Signals in Rat Cortical Astrocytes: Fura-2AM Measurements and Stochastic Model Simulations. *PLoS ONE* **2011**, *6*, e17914. [CrossRef]

219. Larramona-Arcas, R.; González-Arias, C.; Perea, G.; Gutiérrez, A.; Vitorica, J.; García-Barrera, T.; Gómez-Ariza, J.L.; Pascua-Maestro, R.; Ganfornina, M.D.; Kara, E.; et al. Sex-dependent calcium hyperactivity due to lysosomal-related dysfunction in astrocytes from APOE4 versus APOE3 gene targeted replacement mice. *Mol. Neurodegener.* **2020**, *15*, 1–23. [CrossRef]

220. Hirase, H.; Qian, L.; Barthó, P.; Buzsáki, G. Calcium Dynamics of Cortical Astrocytic Networks In Vivo. *PLoS Biol.* **2004**, *2*, e96. [CrossRef]

221. Torres, A.; Wang, F.; Xu, Q.; Fujita, T.; Dobrowolski, R.; Willecke, K.; Takano, T.; Nedergaard, M. Extracellular Ca^{2+} Acts as a Mediator of Communication from Neurons to Glia. *Sci. Signal.* **2012**, *5*, ra8. [CrossRef] [PubMed]

222. Paolicelli, R.C.; Bolasco, G.; Pagani, F.; Maggi, L.; Scianni, M.; Panzanelli, P.; Giustetto, M.; Ferreira, T.A.; Guiducci, E.; Dumas, L.; et al. Synaptic Pruning by Microglia Is Necessary for Normal Brain Development. *Science* **2011**, *333*, 1456–1458. [CrossRef] [PubMed]

223. Von Bernhardi, R.; Tichauer, J.E.; Eugenín, J. Aging-dependent changes of microglial cells and their relevance for neurodegenerative disorders. *J. Neurochem.* **2010**, *112*, 1099–1114. [CrossRef] [PubMed]

224. Hoffmann, A.; Kann, O.; Ohlemeyer, C.; Hanisch, U.-K.; Kettenmann, H. Elevation of Basal Intracellular Calcium as a Central Element in the Activation of Brain Macrophages (Microglia): Suppression of Receptor-Evoked Calcium Signaling and Control of Release Function. *J. Neurosci.* **2003**, *23*, 4410–4419. [CrossRef] [PubMed]

225. Franciosi, S.; Choi, H.B.; Kim, S.U.; McLarnon, J.G. Interferon-? acutely induces calcium influx in human microglia. *J. Neurosci. Res.* **2002**, *69*, 607–613. [CrossRef]

226. Goghari, V.; Franciosi, S.; Kim, S.U.; Lee, Y.B.; McLarnon, J.G. Acute application of interleukin-1beta induces Ca^{2+} responses in human microglia. *Neurosci. Lett.* **2000**, *281*, 83–86. [CrossRef]

227. McLarnon, J.G.; Franciosi, S.; Wang, X.; Bae, J.; Choi, H.; Kim, S. Acute actions of tumor necrosis factor-α on intracellular Ca^{2+} and K$^+$ currents in human microglia. *Neuroscience* **2001**, *104*, 1175–1184. [CrossRef]

228. Turovskaya, M.V.; Turovsky, E.A.; Zinchenko, V.P.; Levin, S.G.; Godukhin, O.V. Interleukin-10 modulates [Ca^{2+}]i response induced by repeated NMDA receptor activation with brief hypoxia through inhibition of InsP3-sensitive internal stores in hippocampal neurons. *Neurosci. Lett.* **2012**, *516*, 151–155. [CrossRef]

229. Itagaki, S.; McGeer, P.; Akiyama, H.; Zhu, S.; Selkoe, D. Relationship of microglia and astrocytes to amyloid deposits of Alzheimer disease. *J. Neuroimmunol.* **1989**, *24*, 173–182. [CrossRef]

230. Meyer-Luehmann, M.; Spires-Jones, T.L.; Prada, C.; Garcia-Alloza, M.; de Calignon, A.; Rozkalne, A.; Koenigsknecht-Talboo, J.; Holtzman, D.M.; Bacskai, B.J.; Hyman, B.T. Rapid appearance and local toxicity of amyloid-beta plaques in a mouse model of Alzheimer's disease. *Nature* **2008**, *451*, 720–724. [CrossRef]

231. Perez-Nievas, B.G.; Stein, T.D.; Tai, H.-C.; Dols-Icardo, O.; Scotton, T.C.; Barroeta-Espar, I.; Fernandez-Carballo, L.; De Munain, E.L.; Perez, J.; Marquie, M.; et al. Dissecting phenotypic traits linked to human resilience to Alzheimer's pathology. *Brain* **2013**, *136*, 2510–2526. [CrossRef] [PubMed]

232. Bezzi, P.; Domercq, M.; Brambilla, L.; Galli, R.; Schols, D.; De Clercq, E.; Vescovi, A.; Bagetta, G.; Kollias, G.; Meldolesi, J.; et al. CXCR4-activated astrocyte glutamate release via TNFalpha: Amplification by microglia triggers neurotoxicity. *Nat. Neurosci.* **2001**, *4*, 702–710. [CrossRef] [PubMed]

233. Agulhon, C.; Sun, M.-Y.; Murphy, T.; Myers, T.; Lauderdale, K.; Fiacco, T.A. Calcium Signaling and Gliotransmission in Normal vs. Reactive Astrocytes. *Front. Pharmacol.* **2012**, *3*, 139. [CrossRef] [PubMed]

234. Brawek, B.; Garaschuk, O. Network-wide dysregulation of calcium homeostasis in Alzheimer's disease. *Cell Tissue Res.* **2014**, *357*, 427–438. [CrossRef] [PubMed]

235. Silei, V.; Fabrizi, C.; Venturini, G.; Salmona, M.; Bugiani, O.; Tagliavini, F.; Lauro, G.M. Activation of microglial cells by PrP and beta-amyloid fragments raises intracellular calcium through L-type voltage sensitive calcium channels. *Brain Res.* **1999**, *818*, 168–170. [CrossRef]

236. McLarnon, J.G.; Choi, H.B.; Lue, L.-F.; Walker, D.G.; Kim, S.U. Perturbations in calcium-mediated signal transduction in microglia from Alzheimer's disease patients. *J. Neurosci. Res.* **2005**, *81*, 426–435. [CrossRef] [PubMed]

237. Eichhoff, G.; Brawek, B.; Garaschuk, O. Microglial calcium signal acts as a rapid sensor of single neuron damage in vivo. *Biochim. Biophys. Acta (BBA) Bioenerg.* **2011**, *1813*, 1014–1024. [CrossRef]

238. Heo, D.K.; Lim, H.M.; Nam, J.H.; Lee, M.G.; Kim, J.Y. Regulation of phagocytosis and cytokine secretion by store-operated calcium entry in primary isolated murine microglia. *Cell. Signal.* **2015**, *27*, 177–186. [CrossRef]

239. Michaelis, M.; Nieswandt, B.; Stegner, D.; Eilers, J.; Kraft, R. STIM1, STIM2, and Orai1 regulate store-operated calcium entry and purinergic activation of microglia. *Glia* **2015**, *63*, 652–663. [CrossRef]

240. Klegeris, A.; Choi, H.B.; McLarnon, J.G.; McGeer, P.L. Functional ryanodine receptors are expressed by human microglia and THP-1 cells: Their possible involvement in modulation of neurotoxicity. *J. Neurosci. Res.* **2007**, *85*, 2207–2215. [CrossRef]

241. Lim, H.M.; Woon, H.; Han, J.W.; Baba, Y.; Kurosaki, T.; Lee, M.G.; Kim, J.Y. UDP-Induced Phagocytosis and ATP-Stimulated Chemotactic Migration Are Impaired in STIM1(-/-) Microglia In Vitro and In Vivo. *Mediat. Inflamm.* **2017**, *2017*, 8158514. [CrossRef] [PubMed]

242. Burnstock, G. Purinergic Signalling: Therapeutic Developments. *Front. Pharmacol.* **2017**, *8*, 661. [CrossRef] [PubMed]

243. Harada, H.; Chan, C.M.; Loesch, A.; Unwin, R.; Burnstock, G. Induction of proliferation and apoptotic cell death via P2Y and P2X receptors, respectively, in rat glomerular mesangial cells. *Kidney Int.* **2000**, *57*, 949–958. [CrossRef] [PubMed]

244. McLarnon, J.G.; Ryu, J.K.; Walker, D.G.; Choi, H.B. Upregulated expression of purinergic P2X(7) receptor in Alzheimer disease and amyloid-beta peptide-treated microglia and in peptide-injected rat hippocampus. *Exp. Neurol.* **2006**, *65*, 1090–1097.

245. Parvathenani, L.K.; Tertyshnikova, S.; Greco, C.R.; Roberts, S.B.; Robertson, B.; Posmantur, R. P2 × 7 mediates superoxide production in primary microglia and is up-regulated in a transgenic mouse model of Alzheimer's disease. *J. Biol. Chem.* **2003**, *278*, 13309–13317. [CrossRef]

246. Sanz, J.M.; Chiozzi, P.; Ferrari, D.; Colaianna, M.; Idzko, M.; Falzoni, S.; Fellin, R.; Trabace, L.; Di Virgilio, F. Activation of microglia by amyloid {beta} requires P2 × 7 receptor expression. *J. Immunol. Res.* **2009**, *182*, 4378–4385.

247. Illes, P.; Rubini, P.; Huang, L.; Tang, Y. The P2 × 7 receptor: A new therapeutic target in Alzheimer's disease. *Expert Opin.* **2019**, *23*, 165–176.

248. Jana, M.K.; Cappai, R.; Ciccotosto, G.D. Oligomeric Amyloid-beta Toxicity Can Be Inhibited by Blocking Its Cellular Binding in Cortical Neuronal Cultures with Addition of the Triphenylmethane Dye Brilliant Blue, G. *ACS Chem. Biol.* **2016**, *7*, 1141–1147.

249. Diaz-Hernandez, J.I.; Gomez-Villafuertes, R.; Leon-Otegui, M.; Hontecillas-Prieto, L.; Del Puerto, A.; Trejo, J.L.; Lucas, J.J.; Garrido, J.J.; Gualix, J.; Miras-Portugal, M.T.; et al. In vivo P2 × 7 inhibition reduces amyloid plaques in Alzheimer's disease through GSK3beta and secretases. *Neurobiol. Aging* **2012**, *33*, 1816–1828. [CrossRef]

250. Martin, E.; Amar, M.; Dalle, C.; Youssef, I.; Boucher, C.; Le Duigou, C.; Bruckner, M.; Prigent, A.; Sazdovitch, V.; Halle, A.; et al. New role of P2 × 7 receptor in an Alzheimer's disease mouse model. *Mol. Psychiatry* **2019**, *24*, 108–125. [CrossRef]

251. Lee, G.-S.; Subramanian, N.; Kim, A.I.; Aksentijevich, I.; Goldbach-Mansky, R.; Sacks, D.B.; Germain, R.N.; Kastner, D.L.; Chae, J.J. The calcium-sensing receptor regulates the NLRP3 inflammasome through Ca^{2+} and cAMP. *Nat. Cell Biol.* **2012**, *492*, 123–127. [CrossRef] [PubMed]

252. Murakami, T.; Ockinger, J.; Yu, J.; Byles, V.; McColl, A.; Hofer, A.M.; Horng, T. Critical role for calcium mobilization in activation of the NLRP3 inflammasome. *Proc. Nat. Acad. Sci. USA* **2012**, *109*, 11282–11287. [CrossRef] [PubMed]

253. Halle, A.; Hornung, V.; Petzold, G.C.; Stewart, C.R.; Monks, B.G.; Reinheckel, T.; Fitzgerald, K.A.; Latz, E.; Moore, K.J.; Golenbock, D.T. The NALP3 inflammasome is involved in the innate immune response to amyloid-beta. *Nat. Immunol.* **2008**, *9*, 857–865. [CrossRef] [PubMed]

254. Heneka, M.T.; Kummer, M.P.; Stutz, A.; Delekate, A.; Schwartz, S.; Vieira-Saecker, A.; Griep, A.; Axt, D.; Remus, A.; Tzeng, T.-C.; et al. NLRP3 is activated in Alzheimer's disease and contributes to pathology in APP/PS1 mice. *Nat. Cell Biol.* **2013**, *493*, 674–678. [CrossRef]

255. Jonsson, T.; Stefansson, H.; Steinberg, S.; Jonsdottir, I.; Jonsson, P.V.; Snaedal, J.; Bjornsson, S.; Huttenlocher-Moser, J.; Levey, A.I.; Lah, J.J.; et al. Variant of TREM2 Associated with the Risk of Alzheimer's Disease. *N. Engl. J. Med.* **2013**, *368*, 107–116. [CrossRef]

256. Jiang, T.; Tan, L.; Zhu, X.-C.; Zhang, Q.-Q.; Cao, L.; Tan, M.-S.; Gu, L.-Z.; Wang, H.-F.; Ding, Z.-Z.; Zhang, Y.-D.; et al. Upregulation of TREM2 Ameliorates Neuropathology and Rescues Spatial Cognitive Impairment in a Transgenic Mouse Model of Alzheimer's Disease. *Neuropsychopharmacology* **2014**, *39*, 2949–2962. [CrossRef] [PubMed]

257. Bouchon, A.; Hernández-Munain, C.; Cella, M.; Colonna, M. A Dap12-Mediated Pathway Regulates Expression of Cc Chemokine Receptor 7 and Maturation of Human Dendritic Cells. *J. Exp. Med.* **2001**, *194*, 1111–1122. [CrossRef] [PubMed]
258. Miyanohara, A.; Kamizato, K.; Juhas, S.; Juhasova, J.; Navarro, M.; Marsala, S.; Lukacova, N.; Hruska-Plochan, M.; Curtis, E.; Gabel, B.; et al. Potent spinal parenchymal AAV9-mediated gene delivery by subpial injection in adult rats and pigs. *Mol. Ther. Methods Clin. Dev.* **2016**, *3*, 16046. [CrossRef]
259. Challis, R.C.; Kumar, S.R.; Chan, K.Y.; Challis, C.; Beadle, K.; Jang, M.J.; Kim, H.M.; Rajendran, P.S.; Tompkins, J.D.; Shivkumar, K.; et al. Systemic AAV vectors for widespread and targeted gene delivery in rodents. *Nat. Protoc.* **2019**, *14*, 379–414. [CrossRef] [PubMed]
260. Alves, S.; Fol, R.; Cartier, N. Gene Therapy Strategies for Alzheimer's Disease: An Overview. *Hum. Gene Ther.* **2016**, *27*, 100–107. [CrossRef]
261. Zhao, L.; Gottesdiener, A.J.; Parmar, M.; Li, M.; Kaminsky, S.M.; Chiuchiolo, M.J.; Sondhi, D.; Sullivan, P.M.; Holtzman, D.M.; Crystal, R.G.; et al. Intracerebral adeno-associated virus gene delivery of apolipoprotein E2 markedly reduces brain amyloid pathology in Alzheimer's disease mouse models. *Neurobiol. Aging* **2016**, *44*, 159–172. [CrossRef] [PubMed]
262. Rosenberg, J.B.; Kaplitt, M.G.; De, B.P.; Chen, A.; Flagiello, T.; Salami, C.; Pey, E.; Zhao, L.; Ricart Arbona, R.J.; Monette, S.; et al. AAVrh.10-Mediated APOE2 Central Nervous System Gene Therapy for APOE4-Associated Alzheimer's Disease. *Hum. Gen. Ther. Clin. Dev.* **2018**, *29*, 24–47. [CrossRef] [PubMed]

Publisher's Note: MDPI stays neutral with regard to jurisdictional claims in published maps and institutional affiliations.

MDPI

St. Alban-Anlage 66

4052 Basel

Switzerland

Tel. +41 61 683 77 34

Fax +41 61 302 89 18

www.mdpi.com

Cells Editorial Office

E-mail: cells@mdpi.com

www.mdpi.com/journal/cells